Perovskites

Fundamentals, properties, preparation and applications

Online at: https://doi.org/10.1088/978-0-7503-5427-1

Perovskites

Fundamentals, properties, preparation and applications

N M Ravindra
New Jersey Institute of Technology, Newark, NJ, USA

Priyanka Singh
National Physical Laboratory, New Delhi, India

Leqi Lin
Shanghai Jiao Tong University, Shanghai, China

Pankaj Kumar
CSIR-NPL, New Delhi, India

IOP Publishing, Bristol, UK

ISBN 978-0-7503-5427-1 (ebook)
ISBN 978-0-7503-5425-7 (print)
ISBN 978-0-7503-5428-8 (myPrint)
ISBN 978-0-7503-5426-4 (mobi)

DOI 10.1088/978-0-7503-5427-1

Version: 20251201

IOP ebooks

British Library Cataloguing-in-Publication Data: A catalogue record for this book is available from the British Library.

Published by IOP Publishing, wholly owned by The Institute of Physics, London

IOP Publishing, No.2 The Distillery, Glassfields, Avon Street, Bristol, BS2 0GR, UK

US Office: IOP Publishing, Inc., 190 North Independence Mall West, Suite 601, Philadelphia, PA 19106, USA

Dedicated to Mother Nature who inspires creativity, collaboration and innovation, and the passion to share and acquire knowledge

Contents

Preface

Perovskite semiconductors have gained special attention for their versatile properties and wide-ranging applications in electronic devices such as sensors, light-emitting diodes (LEDs), lasers, solar cells, and thin-film transistors. Among them, organometal halide perovskites have shown exceptional promise for photovoltaic applications, with solar cell power conversion efficiencies surpassing 27%, exceeding even those of conventional silicon solar cells. Their unique crystal structure, tunable properties, and low-cost processing methods continue to open new opportunities for both fundamental research and practical applications. An important advantage of perovskites lies in their simple solution processability. They can be deposited on both rigid and flexible substrates, and large-area thin films can be fabricated using scalable printing and coating methods, making them highly cost-effective for device manufacturing. However, despite their outstanding performance, perovskite materials face serious challenges with long-term stability. Under real operating conditions, they tend to degrade rapidly, leading to significant losses in device efficiency and reliability. This degradation can occur throughout the device stack, from the top to the bottom electrodes, and arises from a combination of intrinsic material instabilities and extrinsic environmental factors. Overcoming these limitations remains one of the most critical hurdles for achieving successful commercialization of perovskite technologies.

The book explores the fundamental properties of perovskite semiconductors, including their crystallographic structure, growth mechanisms, processability, physical properties, degradation mechanisms, and their use in various electronic devices, especially perovskite solar cells (PSCs). It also reviews processing approaches ranging from laboratory-scale fabrication to industrial-scale production methods, highlighting strategies that are aimed at enhancing the stability of PSCs and advancing their prospects for commercial viability.

The implementation of artificial intelligence/machine learning-assisted tools into accelerating the discovery of perovskite materials [1] and predicting the characteristics of high-performance perovskite solar cells [2] has just begun. Already, teams from across the world have begun to collaborate on this excellent endeavor. An example of such an effort is illustrated by the ongoing collaboration between researchers at EPFL, Shanghai University and Université Catholique de Louvain [3]. On another note, scientists from the National Renewable Energy Laboratory are collaborating with CubicPV on the development of perovskite minimodules [4].

From a broader and futuristic perspective, the convergence of enhanced and reproducible conversion efficiency of perovskite solar cells, combined with the US Defense Advanced Research Project Agency's (DARPA's) Program on wireless transmission 'Persistent Optical Wireless Energy Relay' (POWER) [5] represents the future of energy transmission and renewable energy—beyond the use of the grid.

We sincerely hope that this book will be useful both for beginners as an easy guide and for researchers as a helpful reference, supporting the continued advancement of perovskite science and technology.

References

[1] Wang H *et al* 2024 High-quality data enabling universality of band gap descriptor and discovery of photovoltaic perovskites *J. Am. Chem. Soc.* **146** 17636–45

[2] Peleg R 2025 Machine learning optimizes hybrid perovskite solar cells/thermoelectric systems design *Perovskite-Info* https://perovskite-info.com/machine-learning-optimizes-hybrid-perovskite-solar-cells-thermoelectric-systems

[3] Papageorgiou N 2024 From The Swiss Federal Institute of Technology in Lausanne [EPFL-École Polytechnique Fédérale de Lausanne] (CH): 'Machine learning accelerates discovery of solar-cell perovskites' *sciencesprings* https://sciencesprings.wordpress.com/2024/05/20/from-the-swiss-federal-institute-of-technology-in-lausanne-epfl-ecole-polytechnique-federale-de-lausanne-ch-machine-learning-accelerates-discovery-of-solar-cell-perovskites/ (Accessed: 20 May 2024)

[4] Hicks W 2025 Collaboration between NREL and CubicPV pushes perovskite minimodule performance to new heights *NREL* https://nrel.gov/news/detail/program/2025/collaboration-between-nrel-and-cubicpv-pushes-perovskite-minimodule-performance-to-new-heights (Accessed: 9 July 2025)

[5] DARPA 2025 POWER: Persistent Optical Wireless Energy Relay *DAPRA* https://darpa.mil/research/programs/power (Accessed: 7 September 2025)

Acknowledgements

Ravi (N M Ravindra) acknowledges with thanks the participation of hundreds of graduate students over the years at the New Jersey Institute of Technology (NJIT). In particular, he is very appreciative of his long-term collaborations with Dr Priyanka Singh, Ms Leqi Lin and Dr Aneer Lamichhane.

Priyanka Singh expresses her sincere gratitude to all those who have contributed, directly or indirectly, to the completion of this book. Priyanka is deeply grateful to her mentors, whose guidance, encouragement, and advice have strongly influenced both her doctoral studies and her later research at NJIT. She is also grateful to the National Physical Laboratory (NPL), her current institution, and to the Department of Science and Technology (DST) for financial support. She extends heartfelt thanks to her family for their patience, encouragement, and constant support, which have been her greatest source of strength throughout this journey.

Leqi Lin wishes to convey her sincere gratitude to all those whose efforts and support have contributed to the successful completion of this book. She is profoundly appreciative of Dr Ravi, whose guidance, encouragement, and support have shaped her research journey at NJIT from 2018 and over the years since. She also extends her heartfelt appreciation to her family for their patience, encouragement, and unwavering support, which have remained her greatest source of strength throughout this endeavor.

Pankaj Kumar would like to thank everyone who helped him in contributing to this book. He would like to thank his research students and collaborators from CSIR-National Physical Laboratory (CSIR-NPL), New Delhi, and other institutes (especially Professor Lokendra Kumar from Allahabad University, Prayagraj, and Dr Aparna Tripathi from NIT, New Delhi) for their fruitful discussions and new ideas towards improving the performance of perovskite electronic devices. He is grateful to Professor Paul Dastoor, Director, Centre for Organic Electronics (COE), University of Newcastle, Australia for giving him the opportunity to work at COE for the development of flexible printable electronic devices. He is also grateful to Dr Chhinder Bilen and other colleagues from COE for their wonderful company and fruitful scientific discussions.

Pankaj Kumar would also like to thank all his friends, relatives and family members especially his mother Kusum Lata, brothers Prashant and Abhishek, sister Abhilasha Anand and brother-in-law Vikas Anand for providing him tremendous love, support and blessings during preparation of this book. He wishes to extend his sincere thanks to uncle Chhinder Bilen and aunty Resham Bilen from Newcastle, Australia, for their unconditional care and support. He wants to give a special thanks to his nephew Mr Avik Anand and nieces Ms Amaira Anand and Ms Drishana for their funny and naughty activities, which made every moment he spent with them special for him.

The authors thank Ms Mia Foulkes and the IOP team for their continuous help, guidance, and support.

Author biographies

N M Ravindra

N M Ravindra (Ravi) is Professor of Physics at the New Jersey Institute of Technology (NJIT). He was the Chair of the Physics Department (2009–13) and Director, Interdisciplinary Program in Materials Science and Engineering at NJIT (2009–16). Before joining NJIT in 1987, Ravi had been associated with Vanderbilt University, the Microelectronics Center of North Carolina (MCNC), North Carolina State University, The Abdus Salam International Center for Theoretical Physics (ICTP–Trieste), Politecnico di Torino, CNRS associated labs in Paris and Montpellier. Ravi holds a PhD in Physics from Indian Institute of Technology (Roorkee, India), and MS and BS degrees in Physics from Bangalore University, India. Ravi is an author or coauthor of more than 400 papers and 9 books, in addition to several book chapters. He is the founding editor of *Emerging Materials Research* and serves on the editorial board of several international journals.

Priyanka Singh

Dr Priyanka Singh is a Scientist at the National Physical Laboratory, New Delhi, India. She previously worked as a Research Scientist with Professor N M Ravindra in the Department of Physics at NJIT, USA. She has coauthored several research papers and two books in the field of solar cells and semiconductors.

Leqi Lin

Ms Leqi Lin is a senior PhD student at Shanghai Jiao Tong University. Previously, Leqi obtained her Master's degree in civil engineering at NJIT and mechanical engineering at State University of New York–Buffalo. Leqi is a coauthor of more than 15 papers in the areas of AI4S for materials, energy and sensors.

Pankaj Kumar

Dr Pankaj Kumar is Senior Principal Scientist at CSIR-NPL, New Delhi and Associate Professor at the Academy of Scientific and Innovative Research (AcSIR), India. He has published over 100 research articles, five book chapters and one single-authored book. He is the recipient of several national/international awards including the Indo-Australia early career research fellowship from the Indian National Science Academy (INSA), New Delhi, in 2013, CSIR-Young Scientist Award by CSIR, India, in 2016 and Faculty of Science Staff Excellence Award by University of Newcastle, Australia, in 2020.

Chapter 1

Introduction

In this chapter, we present an overview of the contents of this book as well as the importance of perovskites for materials engineers, scientists, students and industry practitioners. From a materials perspective, perovskites are important worldwide not only for solar cell applications but also in several areas of industry including energy, ferroelectrics, healthcare, optoelectronics, piezoelectrics, sensors, thermoelectrics, and a variety of devices, circuits and systems. Perovskite materials are evolving rapidly, with ongoing research pushing their boundaries into new and innovative applications, making them one of the most promising classes of materials in modern science and technology. Thus, this book presents a detailed development of perovskite materials in a range of topics that include their fundamentals, properties, preparation and applications in the form of case studies, examples, patents and products. This book also includes an in-depth exploration of the physical simulation of perovskite materials, a critical area of research that leverages a wide range of computational methodologies, tools and techniques. Researchers worldwide employ diverse analytical and numerical simulation methods to understand, predict, and optimize the properties and behaviors of perovskite materials across various scales, from the atomic to macroscopic levels.

1.1 Background

The renaissance of perovskite materials represents a convergence of serendipity and rational design. At the atomic scale, the corner-sharing BX_6 octahedral framework creates a delicate balance between ionic and covalent bonding. This structural duality enables properties that are generally unattainable in conventional semiconductors: for example, methylammonium lead iodide ($CH_3NH_3PbI_3$) exhibits ambipolar charge transport with electron/hole mobilities exceeding 10 $cm^2\ V^{-1}\ s^{-1}$, while maintaining solution processability at 100 °C. Such paradoxical combinations have fueled the explosive growth across various disciplines—the Web of Science records over 25 000 perovskite-related publications in 2023 alone. Perovskites, with

doi:10.1088/978-0-7503-5427-1ch1

their intriguing and special physical properties, have become a cornerstone of modern material sciences, captivating the attention of researchers across diverse fields. These materials, named after the common name for the mineral calcium titanium oxide (or calcium titanate, $CaTiO_3$), are renowned for their remarkable versatility and exceptional performance in various applications. The unique combination of their structural, optical, electronic, and mechanical properties has driven intense research efforts to understand, optimize, and deploy perovskites in cutting-edge technologies. This book provides a thorough exploration of these fascinating materials, bridging the gap between foundational science and practical applications.

The book begins by introducing the foundational aspects of perovskites, focusing on their crystal structures, bandgap and electronic structure. At the heart of their unique properties lies the perovskite crystal structure, characterized by the general formula ABX_3, where A and B are cations of different sizes, and X is an anion. This structural framework provides exceptional flexibility, enabling the synthesis of a wide range of compositions and tailored properties. By examining their historical development, we gain insights into how these materials have evolved from mineral curiosities to essential components of modern science and technology. The historical trajectory of perovskites highlights their transformative potential. First discovered in the early nineteenth century, natural perovskite minerals were initially of only geological interest. However, with the advent of synthetic perovskites, particularly in the middle of the twentieth century, their role in ferroelectric materials, catalysis, and other applications began to unfold. Over the past two decades, perovskites have experienced a renaissance, driven by their exceptional optoelectronic properties, which have positioned them as game-changers in areas such as photovoltaics and light-emitting devices. Their rapid ascent to prominence underscores their significance in addressing the global challenges such as renewable energy and sustainable technologies.

One of the most compelling aspects of perovskites is their remarkable properties ranging from appealing optical, electronic, thermal and mechanical properties, which stem from their ability to exhibit a wide range of physical and chemical properties. Organic–inorganic hybrid perovskites, for example, combine the best of both worlds—the tunability of organic materials and the stability of inorganic frameworks. This has made them particularly attractive for applications in optoelectronics including photovoltaics and light-emitting diodes (LEDs). Similarly, all-inorganic perovskites, such as cesium lead halides, offer enhanced thermal stability, making them suitable for high-temperature applications. These classifications highlight the adaptability of perovskites, enabling researchers to engineer materials for specific functionalities. The electronic properties of perovskites further amplify their appeal. Their direct bandgap, high absorption coefficients, and long carrier diffusion lengths contribute to their outstanding performance in optoelectronic devices. For instance, perovskite solar cells (PSCs) have achieved record-breaking efficiencies, surpassed traditional silicon-based solar cells while contributing to their advantages in terms of cost and processing. These properties are not only pivotal for energy-related applications but also extend to areas such as sensors and photodetectors, where sensitivity and precision are critical.

However, the promise of perovskites is not without challenges. Stability remains a significant hurdle, as many perovskite materials are prone to degradation when exposed to moisture, heat, and ultraviolet light. Addressing these stability concerns is essential for the commercialization of perovskite-based technologies. Researchers are exploring innovative strategies, such as compositional engineering, surface passivation, and encapsulation, to enhance the durability of these materials. By overcoming these obstacles, perovskites can transition from the laboratory to real-world applications, unlocking their full potential. The book further progresses to an in-depth discussion of the remarkable properties of perovskites, encompassing the optical, electronic, thermal, mechanical, and environmental aspects. Their optical properties, such as high photoluminescence quantum yields and tunable emission spectra, make them ideal candidates for applications in displays and lighting. Meanwhile, their mechanical properties, including flexibility and robustness, open up possibilities for use in flexible electronics and wearable devices. Environmental sensitivity, while posing challenges, also provides opportunities for innovation in applications such as environmental sensing and remediation.

A detailed overview of the preparation methods equips readers with practical insights into the synthesis of perovskites. Solution-based techniques, such as spin coating and inkjet printing, are widely used for their simplicity and scalability. These methods enable the fabrication of high-quality thin films, which are essential for device performance. Advanced techniques such as vapor deposition offer precise control over film morphology and composition, catering to the demands of high-performance applications. The chapter also explores crystallization and growth mechanisms, providing a comprehensive understanding of how perovskite structures form and evolve. Characterization techniques are integral to the study of perovskites, offering a toolkit for investigating their structural, optical, and electronic properties. Techniques such as x-ray diffraction (XRD), transmission electron microscopy (TEM), and scanning electron microscopy (SEM) provide valuable insights into the crystal structure and morphology of perovskites. Optical measurements, including UV–vis spectroscopy and photoluminescence studies, shed light on their electronic behavior. Advanced spectroscopic techniques further complement these analyses, enabling a deeper understanding of perovskite materials.

Simulation methods represent another frontier in perovskite research, enabling researchers to model complex phenomena and predict material behavior. Techniques such as density functional theory (DFT) and device simulations provide insights into band structures, optical properties, and device performance characteristics. These computational tools are invaluable for guiding experimental efforts and accelerating the development of perovskite-based technologies. By integrating simulations with experimental data, researchers can achieve a more holistic understanding of perovskites and their applications. The transformative potential of perovskites is exemplified by their diverse applications across various fields. In photovoltaics, perovskite solar cells have achieved power conversion efficiencies exceeding 25%, making them strong contenders for next-generation energy solutions. Flexible and tandem solar cells offer new opportunities for integration into wearable electronics and building integrated photovoltaics. Beyond energy, perovskites are making strides in optoelectronics, catalysis, and energy storage, showcasing their versatility and impact.

Recent breakthroughs in perovskites underscore the material's versatility. In photonics, perovskite nanocrystals have achieved 99% photoluminescence quantum yields through surface ligand engineering, outperforming commercial quantum dots. Energy storage applications leverage the oxygen evolution activity of $LaNiO_3$, achieving water-splitting efficiencies rivaling noble metal catalysts. Even in quantum technologies, Majorana fermions (conceptual fermions that are both particles and their own antiparticles) have been observed in superconducting Sr_2RuO_4 thin films. However, perovskites face a critical juncture: translating laboratory success into prototyping and subsequently, commercial products requires solving the stability–toxicity–scalability trilemma. Encapsulation strategies utilizing Al_2O_3 films, deposited by atomic layer deposition (ALD), and lead-sequestering polymers have extended the perovskite solar cell lifetimes beyond IEC 61215 standards. Scaled slot-die coatings have achieved 30 cm s^{-1} production speeds. This chapter deconstructs the perovskite phenomenon through three lenses: historical evolution, structure–property relationships, and technological translation.

1.2 Overview of perovskite materials

Perovskite mineral was first discovered by the Russian mineralogist Gustav Rose in 1839 and later named after the Russian mineralogist Lev Perovski (1792–1856). In general, perovskites represent a specific crystal structure; however, from a historic point of view, perovskite referred to particular mineral, calcium titanium oxide (or calcium titanate, $CaTiO_3$) from the very beginning and later expanded to structures with the overall formula of ABC_3 and similar crystallographic arrangements $A^{2+}B^{4+}(X^{2-})_3$ [1, 2]. As shown in figure 1.1, a perovskite structure is a cubic unit cell having the general formula ABO_3, where titanium atoms are found at the junctions, atoms of oxygen at the centers of the boundaries, and atom of calcium in the center. The defining feature of perovskite materials lies in their crystallographic adaptability. The ideal cubic structure (space group Pm3m), with a Goldschmidt tolerance factor $t \approx 1$, undergoes systematic distortions as atomic radii deviate. For instance, replacing Ca^{2+} (1.34 Å) with smaller Sr^{2+} (1.44 Å) in $SrTiO_3$ induces tetragonal distortion ($c/a = 1.0006$), while substituting with larger Ba^{2+} (1.61 Å) creates cubic symmetry. Such geometric flexibility enables isovalent/alloying substitutions across the periodic table—over 90% of metallic elements have been incorporated into perovskite frameworks.

The Goldschmidt tolerance factor (t), is a critical predictor of perovskite stability. When $t \approx 1$, the cubic structure is favored, but deviations lead to lower-symmetry phases. For example, substituting A-site cations with smaller ions (e.g. La^{3+} in $LaAlO_3$) reduces t, triggering rhombohedral distortions, while larger ions (e.g. K^+ in $KTaO_3$) increase t, stabilizing the cubic phase. This tunability allows perovskites to host exotic phenomena, such as ferroelectricity in $BaTiO_3$ ($t = 1.06$) and colossal magnetoresistance in $La_{1-x}Sr_xMnO_3$ ($t = 0.96$) [3]. $MgSiO_3$ and $FeSiO_3$ are the most abundant perovskite compounds found in the Earth's crust [3]. With the vigorous

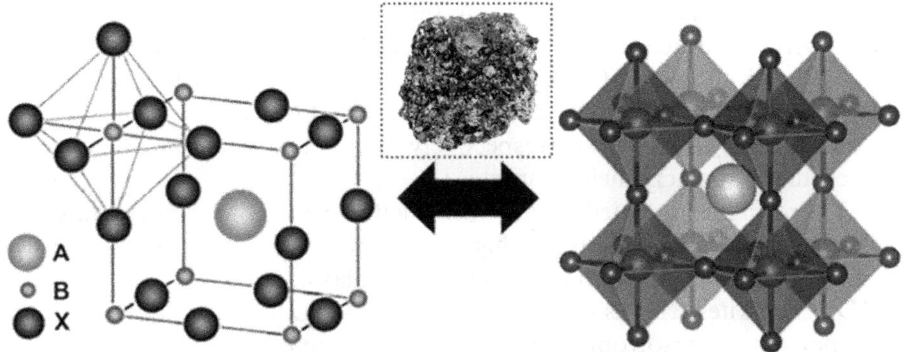

Figure 1.1. General structure of perovskites; the inset shows the naturally occurring mineral $CaTiO_3$. (Reproduced with permission from [15]. Copyright 2021 Elsevier.)

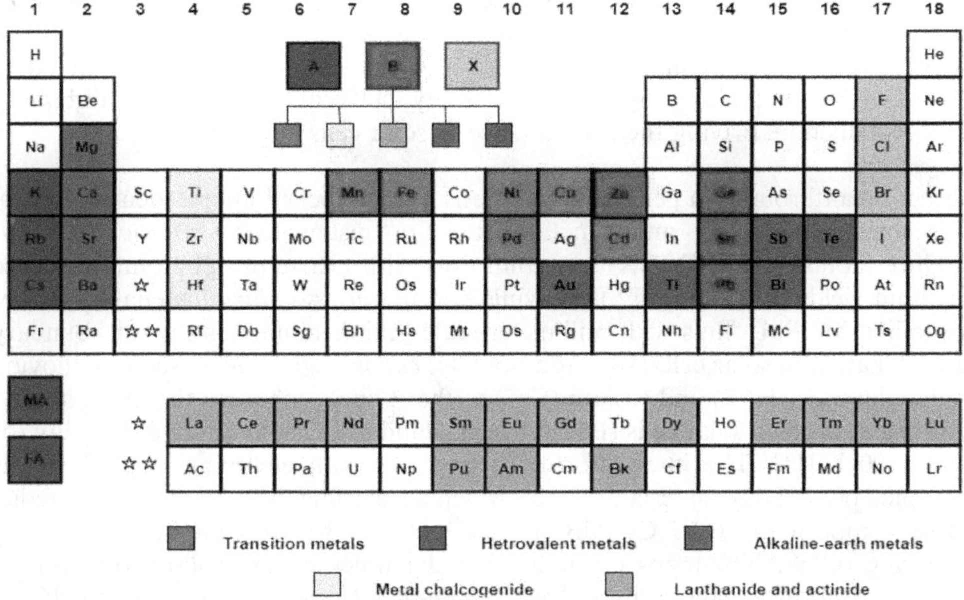

Figure 1.2. Classifications of perovskites materials with ABX_3 structure. (Reproduced with permission from [16]. Copyright 2024 Elsevier.)

development of materials technology in recent years, the perovskite materials have been expanded to various categories as shown in figure 1.2—with detailed classification of perovskites [4].

The cubic structure of perovskites undergoes distortion with the substitution or the partial replacement of cations/anion in the perovskites structure, leading to various crystallographic arrangements. The characteristics of perovskite materials depend mainly on the atomic configuration of substitutional elements. Any alteration in those atomic configuration can lead to many phase transitions, and further result in impressive electrical, optical and chemical properties [5, 6]. Perovskites exhibit four dimensionalities from zero- to three-dimensional, such as the following: 0D ABX3 perovskites (e.g. $CaTiO_3$, $CH_3NH_3PbI_3$); 1D A_2BX_4 layered perovskites (e.g. Cs_2PbI_4, K_2NiF_4); 2D $A_2BB'X_6$ double perovskites (e.g. Sr_2FeMoO_6, Ba_2TiRuO_6, $MAPb_1{-}xSnxI_3$); and 3D $A_2A'B_2B'X_9$ triple perovskites (e.g. $Ba_2KNaTe_2O_9$, $La_2SrCo_2FeO_9$), respectively [7, 8]. The dimensionality of perovskites profoundly impacts their properties:

- 0D perovskites (isolated octahedra), such as Cs_4PbBr_6 nanocrystals, exhibit quantum confinement effects; this enables size-tunable photoluminescence for single-photon emitters in quantum dot displays.
- 1D perovskites (chains of edge-sharing octahedra), such as $(C_4N_2H_{14})PbBr_4$, demonstrate anisotropic charge transport, making them candidates for polarized LEDs.
- 2D perovskites (layered sheets separated by organic spacers), exemplified by $(PEA)_2PbI_4$ (PEA = phenethylammonium), combine high moisture resistance with strong exciton binding energies (<200 meV); these are ideal candidates for stable blue LEDs.
- 3D perovskites (continuous network of corner-sharing octahedra), including $CH_3NH_3PbI_3$, offer high carrier mobility ($>10\ cm^2\ V^{-1}\ s^{-1}$) and ambipolar conduction, driving their dominance in solar cells and x-ray detectors.

Recent breakthroughs in perovskite synthesis have unlocked unprecedented control over their morphology and functionality. For instance, *hot-injection methods* produce monodisperse $CsPbX_3$ quantum dots with near-unity photoluminescence quantum yields (PLQY < 95%) [9], while *antisolvent crystallization* enables single-crystalline $MAPbI_3$ films with millimeter-scale grains, minimizing grain boundary recombination in solar cells [10]. Additionally, *cation engineering*—such as alloying formamidinium (FA^+) and cesium (Cs^+) at the A-site—enhances thermal stability, enabling perovskite solar cells (PSCs) to withstand 85 °C/85% relative humidity for over 1000 h [11]. The integration of perovskites with emerging technologies is reshaping present day materials science. Machine learning (ML) models now predict stable compositions (e.g. Cs_2PdBr_6) [12], while *in situ* grazing-incidence x-ray scattering (GIWAXS) maps crystallization dynamics during roll-to-roll printing [13]. As the field advances, hybrid perovskite-metal organic frameworks (MOFs) and chiral perovskites are emerging for applications in spin-polarized optoelectronics and asymmetric catalysis [14]. The classifications of perovskites materials with ABX_3 structure is summarized in figure 1.2.

1.3 Historical background

The first sample of the mineral, $CaTiO_3$, was transferred from Saint Petersburg to Berlin in 1839 by the Russian mineralogist Alexander Kämmerer, who gave the sample for further investigation to the German mineralogist and crystallographer Gustav Rose. Rose determined its properties and chemical composition and named the mineral after the Russian politician and mineralogist Lev Perovski [17, 18]. This naming not only honored Perovski's contributions to mineralogy but also reflected the nineteenth-century European thinking and tradition of commemorating patrons in scientific nomenclature; this was similar to Berzelius's naming of thorium after the Norse god Thor. For the chemical properties of these special minerals, Gustav Rose collaborated with his older brother Heinrich Rose, a professor of chemistry at the University of Berlin. Heinrich Rose rediscovered niobium in 1844 and coined the name of this element by using the isomorphism concept (identical morphology found in two or more compounds; it has been well known that niobium can substitute for other elements such as iron, silicon and titanium in several compounds and minerals). His work on isomorphism—where different elements occupy the same crystallographic site—laid the groundwork for understanding perovskite's structural versatility, a principle later formalized by Vegard's law in 1921. For the first time in the literature, the Rose system offered a way to organize minerals based on crystalline form and composition. Later on, two classical books on the topic—*Elemente der Krystallographie: Nebst Einer Tabellarischen Uebersicht der Mineralien Nach den Krystallformen* (Elements of Crystallography: a Tabular Overview of the Minerals According to the Crystal Forms) and *Das Krystallo-chemische Mineralsystem* (The Crystal Chemical Mineral System)—were published in 1833 [19] and 1852 [20], respectively. The modern understanding of the perovskite structure, composed of various ions, serves as an excellent illustration of the practical application of the concept of isomorphism. Perovskite's notable crystal structure was first described by Victor Goldschmidt in 1926 in his work on tolerance factors [21]. The crystal structure was later published in 1945 from x-ray diffraction data on barium titanate by Helen Dick Megaw [22]. Portraits of Gustav Rose and Lev Perovski are presented in figure 1.3.

The post-World War II materials revolution (1945–1970) witnessed the transition of perovskites from geological curiosities to functional materials. The functional materials era (1950–2000) saw explosive growth. The discovery of the ferroelectric properties of $BaTiO_3$ by Merz and Anderson in 1950, at Bell Labs, sparked interest in perovskites for capacitors and transducers. The discovery of PZT ceramics ($PbZr_xTi_{1-x}O_3$) in 1954, also at Bell Labs, revolutionized piezoelectric devices,

Figure 1.3. Photographs of (a) Gustav Rose and (b) Lev Perovski. (These Gustav Rose and Lev Perovski images have been obtained by the author(s) from the Wikimedia website [25, 26], where they are stated to have been released into the public domain. They are included within this book on that basis.)

enabling their applications from ultrasonic imaging (1950s) to fuel injection systems (1980s) [23]. Bednorz and Müller's 1986 discovery of LaBaCuO (LBCO) superconductors ($T_c = 35$ K) shattered previous temperature records, triggering a global race for high-T_c materials [24]. The subsequent synthesis of $YBa_2Cu_3O_7$ (YBCO; $T_c = 93$ K) in 1987 marked the first superconductors operating above liquid nitrogen temperatures, revolutionizing the design of MRI magnets. Parallel developments in catalysis witnessed $LaCoO_3$ emerge as a three-way catalyst for automotive exhaust purification, reducing NO_x emissions by 95%. By the 1990s, perovskite-based catalysts dominated catalytic converters, with Ce-doped $LaMnO_3$ enhancing oxygen storage capacity for lean-burn engines.

The investigation of organic–inorganic halide perovskites was initiated in 1978 by D Weber at the University of Stuttgart, where he analysed the structural characteristics of $MAZX_3$ (MA: methyl ammonium; Z: Pb/Sn; X: Cl/Br/I) [27]. Later on in 1994, researchers at the IBM T J Watson Research Center successfully developed light-emitting devices based on luminescent organic–inorganic halide perovskites [28]. Subsequently, M Chikao *et al* from the National Institute of Advanced Industrial Science and Technology AIST, Japan, published findings on the application of R_2CuO_4 (R for rare earth elements) perovskite materials and an oxide having a perovskite crystal structure represented by the formula $XZ_2Cu_3O_6$ (wherein X is yttrium, lanthanum or praseodymium, and Z is an alkali earth metal) as an optical absorption layer in solar cells in 1999 [29]. At the end of the twentieth century, Mitzi conducted extensive studies on various compositions incorporating both small and large cations, focusing on their unique properties [16, 30]. The nanotechnology era (2009–present) began with Miyasaka's 2009 report of 3.8% efficient perovskite-sensitized solar cells [31, 32]. Subsequent breakthroughs in sequential deposition (2013) and solvent engineering (2015) pushed solar cell efficiencies past 20%, while recent tandem configurations with silicon achieved 33.9% certification [33]. Concurrently, Kovalenko's 2015 synthesis of $CsPbX_3$ quantum dots established perovskites as premier optoelectronic nanomaterials, with narrow emission linewidths (< 20 nm) surpassing II–VI semiconductors [9]. To date, the perovskite materials can be categorized into a variety of compositions in ABX_3 systems as shown in figure 1.4 [34]. Emerging frontiers include lead-free double perovskites (e.g. $Cs_2AgBiBr_6$), chiral perovskites for circularly polarized light emission, and metal–halide frameworks for photocatalytic CO_2 reduction—each addressing critical challenges in sustainability and multifunctionality.

1.4 Significance in modern science and technology

Perovskite materials have emerged as a transformative platform in modern science and technology, bridging fundamental research and industrial applications through their unparalleled structural and compositional adaptability. By tailoring the compositions of the perovskite compounds, they can be utilized as a bridge between fundamental and applied science due to the tunable properties and unique crystallographic perovskite structure. These revolutionized materials are fascinating with a variety of interesting physical properties, such as ferroelectric, dielectric,

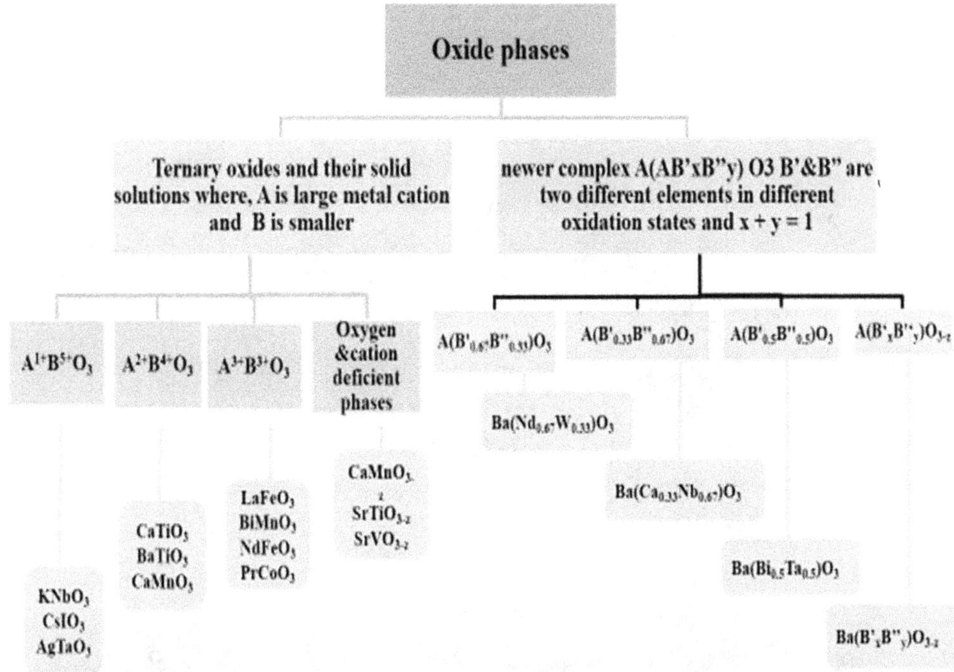

Figure 1.4. A diagram of a variety of compositions in ABX_3 systems. (Reproduced with permission from [35]. Copyright 2019 Springer Nature.)

pyroelectric, piezoelectric, magnetic, catalytic, photovoltaic, electronic-conducting, ionic-conducting, mixed conducting and superconducting properties, by enabling breakthroughs in bandgap engineering, ion migration studies, and defect passivation strategies [36–39]. This versatility arises from the perovskite crystal structure's inherent flexibility, where substitutions at the A-, B-, or X-sites modulate bond angles, lattice distortions, and charge-carrier dynamics. For instance, replacing organic A-site cations (e.g. methylammonium) with inorganic ions (e.g. cesium) enhances thermal stability, while alloying lead (Pb^{2+}) with tin (Sn^{2+}) at the B-site provides tunable bandgaps for optimal light absorption. Such precise control over material behavior has unlocked applications spanning energy harvesting, quantum computing, and environmental catalysis.

A wide range of potential applications of perovskites, including those in energy and devices, based on these properties, are being pursued. This is illustrated in figure 1.5. Both oxide perovskites and halide perovskites can be used in sensor applications [40]. For example, oxide perovskites such as $LaFeO_3$ serve as redox--active catalysts in automotive exhaust systems, leveraging their oxygen vacancy-rich surfaces to convert harmful gases such as NO_x. In contrast, halide perovskites such as $CsPbBr_3$ exhibit near-unity photoluminescence quantum yields, enabling ultra-bright LEDs for displays and lighting. Hybrid organic–inorganic perovskites (e.g. $CH_3NH_3PbI_3$) further blur traditional material boundaries, combining the solution processability of polymers with the high carrier mobility of inorganic

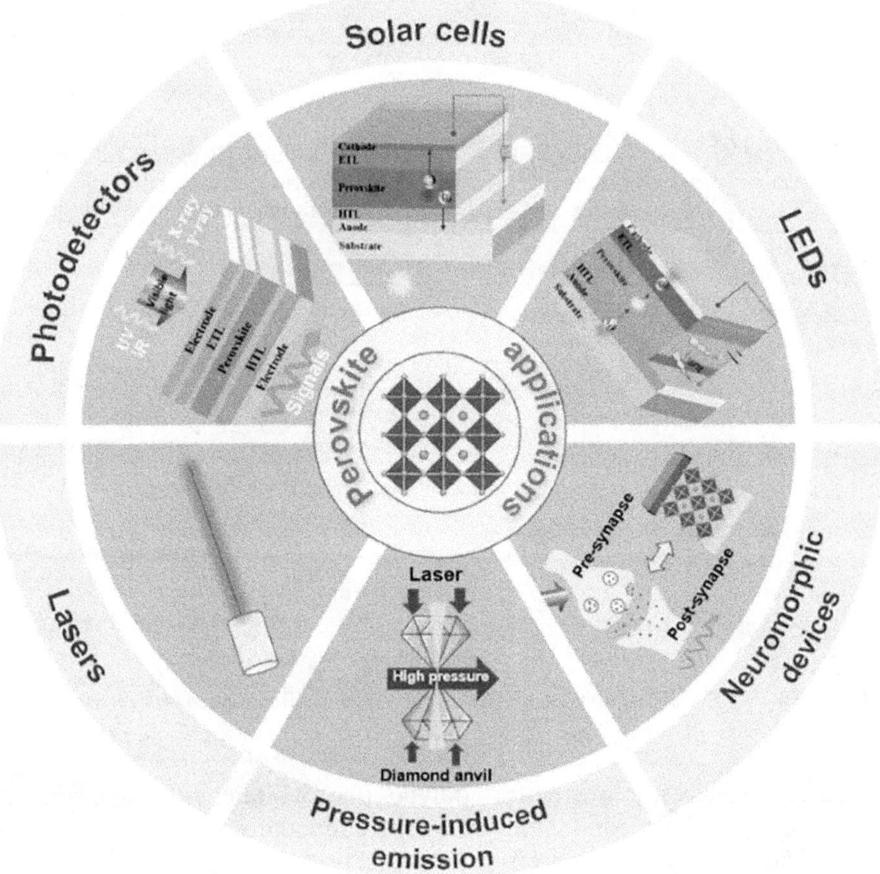

Figure 1.5. Advances in the application of perovskite materials. (Reproduced with permission from [43]. CC BY 4.0.)

semiconductors. This dual functionality has propelled their use in flexible electronics, where lightweight, bendable perovskite films are integrated into wearable health monitors and foldable solar panels.

More recently, a type of organic–inorganic lead halide perovskites has presented highly promising applications in solar energy conversion with the development of new types of solar cells called perovskite solar cells (PSCs), which have become the hottest research area in solar energy utilization [15, 41, 42]. The meteoric rise of PSCs is attributed to their exceptional optoelectronic properties. The extraordinary optoelectronic properties of perovskites have positioned them as strong contenders in renewable energy technologies, particularly in solar cells, where they have achieved efficiency levels surpassing 26%, rivaling traditional silicon-based photovoltaics [42]. Additionally, they are making significant contributions to modern electronics and photonics, such as high-performance LEDs, photodetectors, and lasers, due to their tunable emission spectra and high carrier mobility.

Research on perovskites integrates multiple scientific disciplines, with chemistry contributing to material synthesis and stability improvements, physics playing a crucial role in understanding their electronic structure and optical properties, and engineering focusing on device fabrication and large-scale integration for commercial applications. Their versatility allows for cross-disciplinary innovation, leading to novel applications in sensing, catalysis, and quantum computing. With the development of science and technology in various fields, interdisciplinary integration can achieve results beyond the traditional single-discipline research. Artificial intelligence (AI) is a topic of enormous significance today, and the combination of various fields and computer technology has many successful applications [44, 45].

Compared to conventional semiconductor materials, perovskites offer a cost-effective and sustainable solution, as they can be fabricated using low-temperature, scalable processes [46, 47]. Their lightweight and flexible nature make them ideal for wearable electronics and the next-generation energy solutions. However, challenges related to stability, scalability, and environmental safety (especially lead toxicity) remain, driving ongoing research into lead-free perovskites [48, 49] and green manufacturing approaches for sustainable deployment [50, 51]. Environmental concerns are being systematically addressed. Life cycle analyses show perovskite PV modules achieve energy payback in 0.3 years versus silicon's 2.1 years. Lead sequestration technologies using Eu^{3+}-doped $CsPbI_3$ reduce Pb leakage below 50 ppb, while tin-based perovskites ($CH_3NH_3SnI_3$) demonstrate 9% efficiencies with 100h stability. These developments position perovskites as sustainable, high-performance materials for the twenty-first century. By addressing these scientific, technological, and economic aspects, perovskite materials continue to shape the future of advanced materials research and global energy solutions.

The evolution of perovskite in solar cell applications is illustrated in figure 1.6 [52]. As can be seen in this figure, the 20 years of research in perovskite photovoltaics has revolutionized the solar cell industry. Longi has reported the perovskite/silicon tandem solar cell efficiency of 34.85% in April 2025 [53] while Liu *et al* [54] have reported efficiencies >29% for all-perovskite tandem solar cells. Several industry players are involved in the development of perovskite solar cells. Some of the highlights of these ongoing activities include the following: Oxford PV and Trina have entered into a licensing agreement for the manufacture and sale of perovskite–

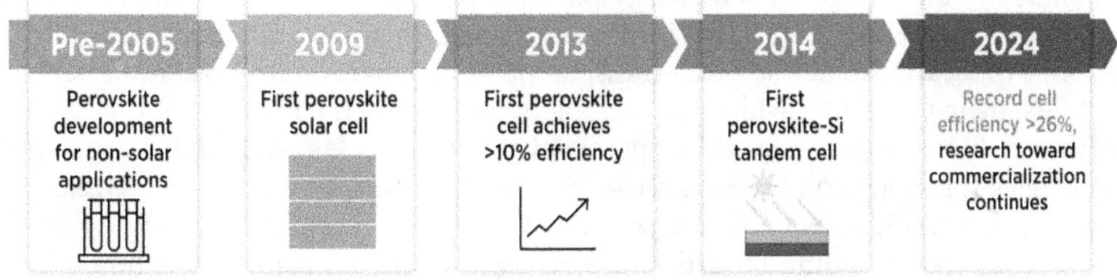

Figure 1.6. Evolution of perovskites in solar cell applications. (Reproduced with permission from [52]. Image credit: US Department of Energy.)

silicon tandem PVs [55]. Qcells has reported successful stress test validation for its tandem modules, consistent with IEC and UL certification standards [56].

1.5 Conclusion

Perovskite materials, characterized by their adaptable ABX_3 structure, stand at the forefront of modern materials science due to their unparalleled ability to host diverse cations and anions, enabling precise tuning of electronic, optical, and ionic properties. From their 1839 discovery as a rare mineral to their current status as engineered quantum materials, perovskites have undergone an extensive transformative journey. Early oxide perovskites like $BaTiO_3$ revolutionized capacitors with their ferroelectric behavior, while recent organic–inorganic hybrids (e.g. $CH_3NH_3PbI_3$) blend solution processability with semiconductor performance, driving breakthroughs in photovoltaics and optoelectronics. This evolution reflects a paradigm shift from accidental discovery to rational design, guided by computational modeling and machine learning.

Key milestones highlight their expanding role: oxide perovskites laid the groundwork for high-temperature superconductivity in the 1950s, while the 2009 integration of lead halide perovskites into solar cells ignited a photovoltaics revolution. By 2023, perovskite solar cells (PSCs) achieved efficiencies surpassing 26%, rivaling silicon through bandgap tunability and defect tolerance. Concurrently, halide perovskites such as $CsPbBr_3$ emerged as high-purity light emitters for LEDs, outperforming conventional semiconductors in color purity. These advancements stem from interdisciplinary synergy—chemists developed passivation strategies to stabilize metastable phases, physicists decoded charge-carrier dynamics via ultrafast spectroscopy, and engineers scaled production using roll-to-roll printing.

Perovskites redefine multifunctionality: $CsPbBr_3$ serves as both a gamma-ray detector and green-light emitter, while two-dimensional variants enable ultrathin photonic circuits. Their compatibility with AI-driven design accelerates innovation, such as predicting lead-free compositions (e.g. $Cs_2AgBiBr_6$) or optimizing nanocrystal synthesis with robotic platforms. However, challenges persist. Stability under heat, light, and humidity demands advanced encapsulation, while lead toxicity concerns drive research into alternatives such as tin-based perovskites ($CH_3NH_3SnI_3$). Scalability hurdles require novel deposition techniques to ensure uniformity in large-area films. Despite these challenges, perovskites offer compelling sustainability advantages. Their energy payback period of 0.3 years and recyclability protocols—recovering >95% of lead—underscore their environmental edge over silicon. Applications such as semi-transparent solar windows and wearable sensors highlight their versatility, while lifecycle analyses validate their low-carbon potential.

Future directions are expansive: perovskite/silicon tandem solar cells target 40% efficiencies for space and terrestrial use, and chiral perovskites promise advances in spintronics. Quantum technologies leverage $CsPbX_3$ quantum dots as single-photon emitters for secure communications. Real-time characterization, combined with machine learning, will enable closed-loop optimization, transitioning perovskites from lab curiosities to global technologies.

This chapter has illustrated how perovskites epitomize the convergence of structural ingenuity and interdisciplinary innovation. Their journey mirrors material science's evolution toward atomic-scale tailoring for global challenges. Subsequent chapters will explore their fundamental principles, stability mechanisms, and transformative applications, cementing their role in shaping the twenty-first century energy and technology.

References

[1] Wang J *et al* 2021 Redirecting dynamic surface restructuring of a layered transition metal oxide catalyst for superior water oxidation *Nat. Catal.* **4** 212–22

[2] María Luisa M 1997 Structural, magnetic and electronic properties of $RNiO_3$ perovskites (R = rare earth) *J. Phys. Condens. Matter* **9** 1679

[3] Wentzcovitch R M, Karki B B, Karato S and Da Silva C R S 1998 High pressure elastic anisotropy of $MgSiO_3$ perovskite and geophysical implications *Earth Planet. Sci. Lett.* **164** 371–8

[4] Monama G R, Ramohlola K E, Iwuoha E I and Modibane K D 2022 Progress on perovskite materials for energy application *Results Chem.* **4** 100321

[5] Whitfield P *et al* 2016 Structures, phase transitions and tricritical behavior of the hybrid perovskite methyl ammonium lead iodide *Sci. Rep.* **6** 35685

[6] Wei W-J *et al* 2019 Phase transition, optical and dielectric properties regulated by anion-substitution in a homologous series of 2D hybrid organic–inorganic perovskites *J. Mater. Chem.* C **7** 11964–71

[7] Gonzalez-Carrero S, Galian R E and Pérez-Prieto J 2016 Organic–inorganic and all-inorganic lead halide nanoparticles *Opt. Express* **24** A285–301

[8] Yin W-J *et al* 2019 Oxide perovskites, double perovskites and derivatives for electrocatalysis, photocatalysis, and photovoltaics *Energy Environ. Sci.* **12** 442–62

[9] Protesescu L *et al* 2015 Nanocrystals of cesium lead halide perovskites ($CsPbX_3$, X = Cl, Br, and I): novel optoelectronic materials showing bright emission with wide color gamut *Nano Lett.* **15** 3692–6

[10] Nie W *et al* 2015 High-efficiency solution-processed perovskite solar cells with millimeter-scale grains *Science* **347** 522–5

[11] Saliba M *et al* 2016 Incorporation of rubidium cations into perovskite solar cells improves photovoltaic performance *Science* **354** 206–9

[12] Sendek A D *et al* 2017 Holistic computational structure screening of more than 12000 candidates for solid lithium-ion conductor materials *Energy Environ. Sci.* **10** 306–20

[13] Hoffman J M *et al* 2020 *In situ* grazing-incidence wide-angle scattering reveals mechanisms for phase distribution and disorientation in 2D halide perovskite films *Adv. Mater.* **32** 2002812

[14] Long G *et al* 2020 Chiral-perovskite optoelectronics *Nat. Rev. Mater.* **5** 423–39

[15] Goel P *et al* 2021 Perovskite materials as superior and powerful platforms for energy conversion and storage applications *Nano Energy* **80** 105552

[16] Elangovan N K *et al* 2024 Recent developments in perovskite materials, fabrication techniques, band gap engineering, and the stability of perovskite solar cells *Energy Rep.* **11** 1171–90

[17] Katz E A 2020 Perovskite: name puzzle and German–Russian odyssey of discovery *Helv. Chim. Acta* **103** e2000061

[18] Rose G 1839 Beschreibung einiger neuen Mineralien des Urals *Ann. Phys.* **124** 551–73
[19] Rose G 1833 *Elemente der Krystallographie: Nebst Einer Tabellarischen Uebersicht der Mineralien Nach den Krystallformen* **vol 1** (ES Mittler)
[20] Rose G 1852 *Das Krystallo-chemische Mineralsystem* (W Engelmann)
[21] Goldschmidt V M 1926 Die Gesetze der Krystallochemie *Naturwissenschaften* **14** 477–85
[22] Forrester W and Hinde R 1945 Crystal structure of barium titanate *Nature* **156** 177–7
[23] Jaffe B, Roth R S and Marzullo S 1954 Piezoelectric properties of lead zirconate-lead titanate solid-solution ceramics *J. Appl. Phys.* **25** 809–10
[24] Chu C W *et al* 1987 Evidence for superconductivity above 40 K in the La–Ba–Cu–O compound system *Phys. Rev. Lett.* **58** 405–7
[25] Wikimedia Commons 2008 File:Gustav Rose.jpg https://commons.wikimedia.org/wiki/File:Gustav_Rose.jpg
[26] Wikimedia Commons 2017 File:Lev Perovsky.jpg https://commons.wikimedia.org/wiki/File:Lev_Perovsky.jpg
[27] Toshniwal A and Kheraj V 2017 Development of organic–inorganic tin halide perovskites: a review *Sol. Energy* **149** 54–9
[28] Mahapatra A, Kumar P and Pradhan B 2024 *Perovskite Optoelectronic Devices* ed P Basudev (Springer International Publishing) pp 1–17
[29] 花村 C M and 村 A 安 Y 矢 H 1999 Material for forming optical absorption layer of solar cell *Japan Patent*
[30] Mitzi D B 1996 Synthesis, crystal structure, and optical and thermal properties of $(C_4H_9NH_3)_2MI_4$ (M = Ge, Sn, Pb) *Chem. Mater.* **8** 791–800
[31] Nair S, Patel S B and Gohel J V 2020 Recent trends in efficiency-stability improvement in perovskite solar cells *Mater. Today Energy* **17** 100449
[32] Kojima A, Ikegami M, Teshima K and Miyasaka T 2012 Highly luminescent lead bromide perovskite nanoparticles synthesized with porous alumina media *Chem. Lett.* **41** 397–9
[33] Liu J *et al* 2024 Perovskite/silicon tandem solar cells with bilayer interface passivation *Nature* **635** 596–603
[34] Guesmia Kaouthar S A 2021 Elaboration et caractérisation des oxydes de type pérovskite $XFeO_3$ dopés *MS Thesis* Université Mohamed Khider de Biskra
[35] Assirey E A R 2019 Perovskite synthesis, properties and their related biochemical and industrial application *Saudi Pharm. J.* **27** 817–29
[36] Zheng T *et al* 2016 Bandgap modulation and magnetic switching in $PbTiO_3$ ferroelectrics by transition elements doping *Ceram. Int.* **42** 6033–8
[37] Lv X-H *et al* 2016 Dielectric and photoluminescence properties of a layered perovskite-type organic–inorganic hybrid phase transition compound: $NH_3(CH_2)_5NH_3MnCl_4$ *J. Mater. Chem.* C **4** 1881–5
[38] Shi C, Meidong L, Churong L, Yike Z and Da Costa J 2000 Investigation of crystallographic and pyroelectric properties of lead-based perovskite-type structure ferroelectric thin films *Thin Solid Films* **375** 288–91
[39] Tsiakaras P *et al* 1998 Methane activation on a $La_{0.6}Sr_{0.4}Co_{0.8}Fe_{0.2}O_3$ perovskite: catalytic and electrocatalytic results *Appl. Catal.* A **169** 249–61
[40] Wang Z *et al* 2022 Applications of machine learning in perovskite materials *Adv. Compos. Hybrid Mater.* **5** 2700–20
[41] Ali N *et al* 2021 A review on perovskite materials with solar cell prospective *Int. J. Energy Res.* **45** 19729–45

[42] Yu X, Sun X, Zhu Z and Li Z 2025 Stabilization strategies of buried interface for efficient SAM-based inverted perovskite solar cells *Angew. Chem. Int. Ed.* **64** e202419608

[43] Zhang L *et al* 2023 Advances in the application of perovskite materials *Nano-Micro Lett.* **15** 177

[44] Zhang H *et al* 2019 Machine learning for novel thermal-materials discovery: early successes, opportunities, and challenges arXiv:1901.05801

[45] Kumbhar A, Dhawale P G, Kumbhar S, Patil U and Magdum P 2021 A comprehensive review: machine learning and its application in integrated power system *Energy Rep.* **7** 5467–74

[46] Lin L and Ravindra N M 2020 CIGS and perovskite solar cells—an overview *Emerg. Mater. Res.* **9** 812–24

[47] Lin L and Ravindra N M 2020 Temperature dependence of CIGS and perovskite solar cell performance: an overview *SN Appl. Sci.* **2** 1361

[48] Wang X, Zhang T, Lou Y and Zhao Y 2019 All-inorganic lead-free perovskites for optoelectronic applications *Mater. Chem. Front.* **3** 365–75

[49] Xiao Z, Song Z and Yan Y 2019 From lead halide perovskites to lead-free metal halide perovskites and perovskite derivatives *Adv. Mater.* **31** 1803792

[50] Podapangi S K *et al* 2023 Green solvents, materials, and lead-free semiconductors for sustainable fabrication of perovskite solar cells *RSC Adv.* **13** 18165–206

[51] Jain S M, Edvinsson T and Durrant J R 2019 Green fabrication of stable lead-free bismuth based perovskite solar cells using a non-toxic solvent *Commun. Chem.* **2** 91

[52] US Department of Energy 2025 Perovskite solar cells *Solar Energies in Technology Office* https://energy.gov/eere/solar/perovskite-solar-cells (Accessed: 17 May 2025)

[53] US Department of Energy 2025 Latest efficiency records: perovskite–silicon tandem solar cells *Solar Energy Technologies Office* https://fluxim.com/research-blogs/perovskite-silicon-tandem-pv-record-updates (Accessed: 17 May 2025)

[54] Liu Z *et al* 2025 All-perovskite tandem solar cells achieving $>$ 29% efficiency with improved (100) orientation in wide-bandgap perovskites *Nat. Mater.* **24** 252–9

[55] Peleg R 2025 Oxford PV and Trina Solar enter patent licensing agreement for perovskite–silicon tandem PVs *Perovskite-info* https://perovskite-info.com/oxford-pv-and-trina-solar-enter-patent-licensing-agreement-perovskite-silicon (Accessed: 18 May 2025)

[56] Peleg R 2025 Qcells achieves key tandem solar module stability milestone *Perovskite-info* https://perovskite-info.com/qcells-achieves-key-tandem-solar-module-stability-milestone (Accessed: 18 May 2025)

Chapter 2

Fundamentals of perovskites

Perovskite materials, defined by the ABX_3 structural motif, have emerged as promising candidates for a variety of applications due to their remarkable optoelectronic properties and structural flexibility. By varying the cations and halides, these materials exhibit tunable bandgaps, high charge carrier mobility, and efficient light absorption. Hybrid organic–inorganic perovskites, such as $MAPbI_3$ and $FAPbI_3$, demonstrate phase transitions that influence their functional performance. Additionally, all-inorganic perovskites such as $CsPbI_3$ display enhanced thermal stability but face challenges with phase instability. Understanding the relationships between structure, composition, and phase transitions is essential for optimizing perovskites for applications in solar cells, LEDs, and other optoelectronic devices. Strategies such as compositional engineering, surface passivation, and advanced structural variants are being continuously developed to improve efficiency, stability, and environmental resilience, making perovskites a versatile platform for next-generation technologies.

2.1 Crystal structure

2.1.1 Basic structure and composition

The term 'perovskite' originates from the mineral calcium titanate ($CaTiO_3$), but the structural motif now encompasses a vast family of materials defined by the general formula ABX_3, where A and B are cations occupying distinct crystallographic sites, and X is an anion (commonly oxygen in oxides or halogens in halide perovskites) [1]. The A-site cation, typically a monovalent species, must balance the charge of the B-site divalent metal cation and X-site anions. For example, in $MAPbI_3$ ($CH_3NH_3PbI_3$), the methylammonium cation (MA^+) provides a +1 charge, compensating for the Pb^{2+} and three I^- ions (each −1 charge), resulting in an overall neutral formula unit. As shown in figure 2.1, the A-site cation, such as cesium (Cs^+), methylammonium ($CH_3NH_3^+$, MA^+), or formamidinium ($HC(NH_2)_2^+$, FA^+), is typically larger in ionic radius (~1.6–2.5 Å) and occupies the cuboctahedral

doi:10.1088/978-0-7503-5427-1ch2

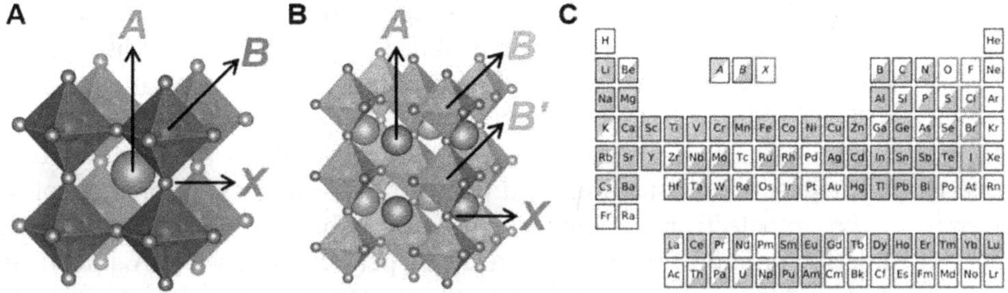

Figure 2.1. Perovskite structure and composition (A) ABX_3, in the cubic single perovskite structure, (B) $A_2BB'X_6$, in the rock salt double perovskite structure and (C) a map of the elements that occupy the A-, B-, and/or X-sites within the 576 compounds experimentally characterized as perovskite or non-perovskite at ambient conditions. (Reproduced from [5]. CC BY 4.0.)

voids formed by a three-dimensional network of corner-sharing $[BX_6]$ octahedra. The smaller B-site cation, often a divalent metal such as lead (Pb^{2+}), tin (Sn^{2+}), or germanium (Ge^{2+}), resides at the center of these octahedra, coordinated by six X anions (e.g. I^-, Br^-, Cl^-). The octahedral coordination geometry around the B-site cation is critical for maintaining structural stability. For instance, Pb^{2+} in $MAPbI_3$ adopts a slightly distorted octahedral environment due to the Jahn–Teller effect, where degenerate d-orbitals split into lower-energy and higher-energy states, causing asymmetric electron distribution. The X-site anions bridge adjacent B cations, forming the rigid inorganic framework that dictates the material's electronic and optical properties. A critical factor governing perovskite stability is the ratio of the ionic size between the A, B, and X ions [2]. The Goldschmidt tolerance factor (t), serves as a predictive tool for structural stability [3]. When $t \approx 1$, the ideal cubic perovskite structure forms, characterized by perfect octahedral symmetry and minimal lattice strain. Deviations from this ideal ratio ($0.8 < t < 1$) lead to distorted structures, such as tetragonal or orthorhombic phases, due to octahedral tilting or cation displacement. For example, $MAPbI_3$ ($t \approx 0.9$) adopts a pseudo-cubic structure at room temperature with slight tetragonal distortion, while $CsPbI_3$ ($t \approx 0.8$) stabilizes in a non-perovskite orthorhombic phase unless heated above 300 °C. Conversely, $t > 1$ often results in hexagonal packing, which lacks the corner-sharing octahedral connectivity that is essential for optoelectronic functionality [4, 5].

The flexibility of the perovskite lattice allows for extensive compositional engineering. For instance, mixing A-site cations (e.g. $Cs_{0.17}FA_{0.83}PbI_3$) can enhance thermal stability, while halide alloying (e.g. $CsPb(I_{0.5}Br_{0.5})_3$) enables precise bandgap tuning. However, mismatched ion sizes or charges can induce strain, leading to phase segregation or defect formation. The B-site also permits substitution: replacing Pb^{2+} with Sn^{2+} reduces toxicity but introduces susceptibility to oxidation, while double perovskites (e.g. $Cs_2AgBiBr_6$) employ alternating B-site cations to mitigate lead usage entirely. The structural adaptability of perovskites is further exemplified in hybrid organic–inorganic variants, where dynamic organic cations (e.g. MA^+, FA^+) introduce rotational degrees of freedom. These organic moieties can stabilize the lattice through hydrogen bonding with the $[BX_6]$

framework but also contribute to anisotropic thermal expansion and phase instability. In contrast, all-inorganic perovskites (e.g. $CsPbBr_3$) exhibit higher symmetry and rigidity due to the spherical, non-polarizable Cs^+ ion, often resulting in enhanced thermal and environmental stability.

The symmetry of perovskite crystals is intrinsically linked to their functional properties, with structural distortions and phase transitions playing pivotal roles in determining the optoelectronic behavior. The ideal cubic perovskite represents the highest symmetry configuration, characterized by perfectly aligned $[BX_6]$ octahedra and a 12-coordinated A-site cation. However, most perovskites deviate from this idealized geometry due to octahedral tilting, A- or B-site cation displacements, or electronic instabilities, resulting in lower-symmetry phases such as tetragonal, orthorhombic, or rhombohedral structures.

2.1.2 Symmetry and phase transitions

Phase transitions in perovskites are governed by thermodynamic and kinetic factors [6]. Thermally induced transitions, such as the cubic-to-tetragonal shift in $MAPbI_3$, result from entropy-driven lattice expansion and the softening of rotational modes. Conversely, pressure can stabilize high-symmetry phases by applying hydrostatic pressure (>0.3 GPa) to $CsPbBr_3$ suppressing the octahedral tilting, forcing a metastable cubic phase even at room temperature [7, 8]. Compositional changes, such as halide alloying (e.g. $CsPb(I_{1-x}Br_x)_3$), modulate transition temperatures by altering lattice parameters and ionic bonding strength.

For instance, the phase transition behavior of perovskites such as $MAPbI_3$ (methylammonium lead iodide) and $FAPbI_3$ (formamidinium lead iodide) illustrates their temperature-dependent structural transformations, which are critical for their optoelectronic properties and device performance. For $MAPbI_3$, the material transitions through three distinct phases are as follows: the low-temperature γ phase with an orthorhombic structure below 160 K, the intermediate β phase with a tetragonal structure from 160 K to 330 K, and the high-temperature α phase with a cubic structure above 330 K. Similarly, $FAPbI_3$ undergoes a series of transitions but with a notable difference. It begins in a β phase with a tetragonal or orthorhombic structure below 130 K, transitions to a non-perovskite δ phase with a hexagonal structure between 130 K and 300 K, and finally achieves a stable α phase with a cubic structure above 300 K. Unlike $MAPbI_3$, the δ phase of $FAPbI_3$ is thermodynamically stable at room temperature but undesirable for photovoltaic applications due to poor charge transport properties. These phase transitions are primarily driven by the temperature-induced tilting and distortion of the PbI_6 octahedra, which affect the material's crystal symmetry and electronic band structure. Understanding and controlling these phase transitions is essential for optimizing perovskite materials for various applications, including solar cells and light-emitting devices, where stability and phase purity are critical for high efficiency and durability [9, 10]. Ma *et al* explore the phase transition dynamics of metal–halide perovskites, specifically focusing on $CsPbI_3$ (cesium lead iodide) nanocrystals. The work employs advanced low-dose imaging techniques to observe in real-time how these nanocrystals transition between different structural phases—namely, the α (cubic), β (tetragonal), and γ (orthorhombic) phases—during *in situ* heating [11].

Structural distortions profoundly influence electronic properties. Tetragonal $MAPbI_3$ exhibits a slightly larger bandgap (~1.6 eV) than its cubic phase (~1.5 eV) due to reduced orbital overlap from octahedral tilting [12]. Similarly, orthorhombic $CsPbBr_3$ shows anisotropic charge transport, with higher hole mobility along the *a*-axis compared to the *b*-axis. Phase transitions can also induce ferroelectricity or piezoelectricity in perovskites such as $BaTiO_3$, where off-center displacements of Ti^4+ ions create spontaneous polarization [13]. In hybrid perovskites, the dynamic motion of organic A-site cations (e.g. MA^+ rotation) introduces transient symmetry-breaking fields, which modulate exciton binding energies and carrier recombination rates. Understanding phase transitions is critical for device optimization. In solar cells, light-induced heating can reversibly alter crystal symmetry, affecting charge extraction efficiency. Phase-stabilization strategies, such as strain engineering via substrate clamping or doping, are employed to lock perovskites in functional phases. For instance, incorporating rubidium (Rb^+) into $CsPbI_3$ suppresses the cubic-to-orthorhombic transition, enabling stable perovskite solar cells under operational conditions. Similarly, layered perovskites (e.g. Ruddlesden–Popper phases) exploit symmetry-breaking interfaces to achieve tunable quantum confinement for light-emitting applications [14–16]. Figure 2.2 illustrates these phase variations.

2.1.3 Defects and disorder

Perovskite semiconductors exhibit a unique defect-tolerant behavior that distinguishes them from conventional semiconductors [17]. Unlike traditional photovoltaic materials where defects typically cause severe performance degradation, lead halide perovskites can maintain good optoelectronic properties even with relatively high defect densities. This remarkable characteristic stems from their hybrid ionic–electronic nature and specific electronic band structure [18]. The most common defects in these materials include ionic vacancies such as iodide vacancies (V_{I^-}), interstitial atoms occupying lattice gaps, and anti-site defects where atoms swap positions (e.g. IPb defects where iodine occupies lead sites) [19].

The impact of these defects varies significantly depending on their energy levels within the bandgap. Shallow defects such as iodide vacancies exhibit low formation energies and minimal effects on carrier lifetimes due to the material's high dielectric constant ($\varepsilon \sim 70$) which provides effective charge screening. In contrast, deep-level defects such as lead vacancies ($V_{Pb^{2-}}$) or IPb anti-site defects act as strong non-radiative recombination centers that can significantly degrade device performance. The defect tolerance mechanism primarily arises from three interrelated factors: dielectric screening that reduces Coulombic attraction between charges and defects, the antibonding character of the conduction band that minimizes trap state formation, and strong spin–orbit coupling in lead-based perovskites that modifies carrier–defect interactions.

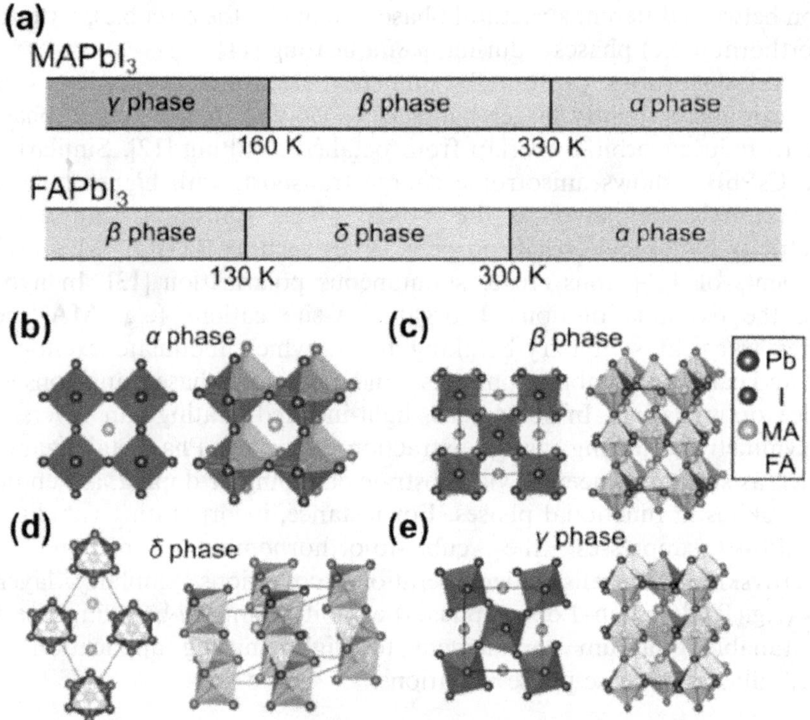

Figure 2.2. (a) The phase variations of $MAPbI_3$ and $FAPbI_3$ at various temperatures. Four different phases: (b) the cubic alpha (α) phase, (c) the tetragonal beta (β) phase, (d) the trigonal delta (δ) phase, and (e) the orthorhombic gamma (γ) phase. (b)–(e) Gray and purple spheres indicate lead and iodine atoms, respectively. The methylammonium (MA) or formamidinium (FA) molecular units are represented by orange spheres. (Reproduced with permission from [9]. Copyright 2020 Springer Nature.)

Recent studies using advanced characterization techniques such as deep-level transient spectroscopy (DLTS) have revealed that this defect tolerance has practical limits [20]. When defect densities exceed 10^{16} cm^{-3} or under prolonged operational stresses, defect clustering can occur, leading to performance degradation. This understanding has driven the development of various defect passivation strategies, particularly for deep-level traps that remain problematic for device stability and efficiency. The unique defect physics of perovskites continues to be an active research area, with ongoing investigations into how different compositions (e.g. mixed-halide or cation systems) and processing conditions affect defect formation and behavior.

2.2 Classification of perovskites

2.2.1 Organic–inorganic hybrid perovskites

Organic–inorganic hybrid perovskites exhibit a unique crystalline structure following the ABX_3 formula, where the A-site contains organic cations such as methylammonium ($CH_3NH_3^+$, MA^+) or formamidinium ($HC(NH_2)_2^+$, FA^+) [21].

The B-site typically accommodates divalent metal ions (Pb^{2+} or Sn^{2+}), while X represents halide anions (I^-, Br^-, Cl^-) [22]. This combination creates a dynamic lattice where organic cations interact with the inorganic [BX_6] octahedral framework through hydrogen bonding, imparting structural flexibility [23]. However, the size disparity between bulky organic cations ($FA^+ \approx 2.8$ Å) and smaller inorganic ions ($Pb^{2+} \approx 1.2$ Å) often induces lattice distortions, leading to non-cubic phases that exhibit phase instability under environmental stress [24].

These materials demonstrate exceptional optoelectronic performance, with tunable direct bandgaps (1.5–2.3 eV) and absorption coefficients exceeding 10^4 cm^{-1}. Their strong spin–orbit coupling and low exciton binding energies (< 50 meV) enable efficient charge separation, resulting in carrier diffusion lengths surpassing 1 μm. These properties have led to remarkable progress in photovoltaic applications, with perovskite solar cells achieving power conversion efficiencies above 25%, while light-emitting devices demonstrate external quantum efficiencies exceeding 20%. The choice of organic cation significantly influences material properties: MA^+-based perovskites tend to form highly crystalline films, whereas FA^+-rich compositions exhibit broader absorption spectra [25].

Solution-processing methods, particularly spin-coating and blade-coating, remain dominant for perovskite synthesis due to their cost-effectiveness and scalability. Researchers employ techniques such as sequential deposition and anti-solvent treatment to control crystallization kinetics, producing dense, pinhole-free films. While vapor deposition enables precise stoichiometric control for high-performance devices, challenges persist regarding solvent toxicity (e.g. dimethylformamide) and moisture sensitivity during processing. Post-deposition treatments including thermal annealing and solvent vapor exposure have proven to be effective for optimizing grain growth and minimizing defect densities [26].

2.2.2 All-inorganic perovskites

All-inorganic perovskites ($CsPbX_3$, X = I, Br, Cl) demonstrate remarkable thermal stability (>400 °C) and reduced ion migration compared to hybrid perovskites, owing to their rigid, rotationally symmetric Cs^+-centered lattice structure. However, their smaller Goldschmidt tolerance factor (~0.8–1.0) presents phase-stabilization challenges, as evidenced by $CsPbI_3$ requiring temperatures above 320 °C to maintain the photoactive α-phase without external stabilization strategies [27]. These materials exhibit exceptional optoelectronic characteristics including narrow emission linewidths (<20 nm FWHM), near-unity photoluminescence quantum yields (>90%), and defect-tolerant charge transport properties with bulk electron mobilities exceeding 100 cm^2 V^{-1}·s. The bandgap can be tuned precisely from 1.7 to 3.0 eV through halide substitution, enabling a wide range of optoelectronic applications. Recent studies have shown that the unique electronic structure of these materials arises from strong spin–orbit coupling and the ionic nature of their chemical bonds [26].

Advanced synthetic approaches have been developed to control the morphology and composition of inorganic perovskites. Colloidal hot-injection methods can

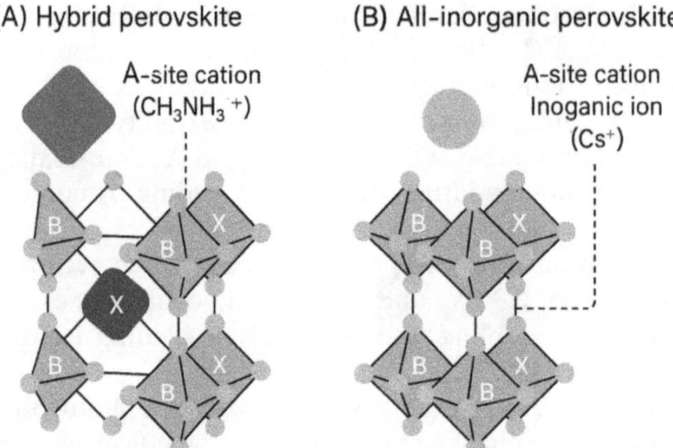

Figure 2.3. The structure of (A) hybrid perovskites that includes an organic molecule as the A-site cation. Specifically, $CH_3NH_3^+$ (methylammonium, MA^+) is used as the A-site cation and (B) all-inorganic perosvkites that uses an inorganic ion (Cs^+) as the A-site cation instead of an organic molecule.

produce monodisperse quantum dots with size dispersions below 5%, while mechanochemical synthesis enables scalable production of bulk single crystals. Post-synthetic halide exchange techniques allow for real-time bandgap tuning across the visible spectrum, significantly expanding their potential applications. Beyond photovoltaics, these materials show great promise in specialized applications: lead-free variants such as $Cs_2AgBiBr_6$ demonstrate excellent performance in gamma-ray detection (mobility-lifetime product $\mu\tau > 10^{-4}$ cm^2 V^{-1}), while Mn^{2+}-doped $CsPbCl_3$ nanocrystals achieve 60% photoluminescence quantum yield for orange emission through efficient energy transfer processes. Heterostructures combining $CsPbBr_3$ with TiO_2 have shown remarkable selectivity (85%) for photocatalytic CO_2 reduction, opening new avenues for sustainable energy applications [28]. These developments highlight the versatility of inorganic perovskites beyond conventional optoelectronic devices. Figure 2.3 illustrates the structure of (A) hybrid perovskites that includes an organic molecule as the A-site cation.

2.3 Bandgap and electronic structure

2.3.1 Fundamental concepts

The electronic structure of perovskites is intrinsically tied to their crystal geometry. In the ideal cubic phase, the valence band (VB) arises primarily from antibonding hybrid orbitals of the B-site metal (e.g. Pb-6s and I-5p in $MAPbI_3$), while the conduction band (CB) is dominated by Pb-6p orbitals. This direct bandgap configuration enables strong light–matter interaction, with optical transitions occurring without phonon assistance, yielding high absorption coefficients ($>10^4$ cm^{-1}). However, structural distortions (e.g. octahedral tilting) split degenerate energy levels, modifying band dispersion. For instance, tetragonal $MAPbI_3$ exhibits a marginally increased bandgap (~1.6 eV) compared to its cubic phase (~1.5 eV) due to reduced orbital overlap. Indirect bandgaps, seen in double perovskites such as

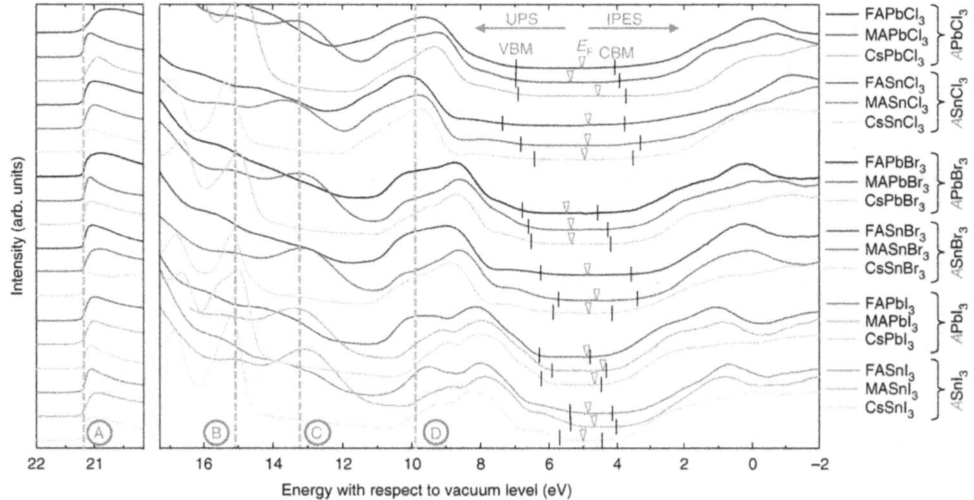

Figure 2.4. Representative UPS and IPES spectra of all 18 metal–halide perovskite systems. The extracted positions of the valence band maximum (VBM) and conduction band minimum (CBM) are given by black vertical markers, and the Fermi level positions are marked by triangles. (Reproduced with permission from [31]. Copyright 2019 Springer Nature.)

$Cs_2AgBiBr_6$, arise from staggered contributions of Ag-4d and Bi-6p orbitals, limiting radiative efficiency but enabling applications in photon upconversion [29, 30].

Bandgap engineering is achieved through compositional modulation. Halide substitution (e.g. replacing I^- with Br^- in $CsPbX_3$) systematically shifts the bandgap from ∼1.7 eV ($CsPbI_3$) to ∼2.3 eV ($CsPbBr_3$) due to changes in electronegativity and orbital hybridization. Cation alloying at the A-site (e.g. $Cs_{0.17}FA_{0.83}PbI_3$) or B-site (e.g. Pb–Sn alloys) introduces strain and alters lattice parameters, enabling sub-1.3 eV bandgaps for tandem solar cells. Quantum confinement in 2D perovskites (e.g. $(PEA)_2PbI_4$) allows precise tuning via layer thickness—monolayers emit blue light (∼3.1 eV), while thicker layers redshift to near-infrared. The representative ultraviolet photoelectron spectroscopy (UPS) and inverse photoemission spectroscopy (IPES) spectra of all 18 metal–halide perovskite systems are shown in figure 2.4.

2.3.2 Electronic properties and charge transport

Perovskites exhibit ambipolar charge transport with balanced electron and hole mobilities (∼10–100 $cm^2\ V^{-1}\cdot s$). Low exciton binding energies (<50 meV) facilitate efficient free-carrier generation at room temperature. However, grain boundaries and defects (e.g. Pb^0 clusters) introduce mid-gap states, accelerating non-radiative recombination. Single-crystal perovskites mitigate this with diffusion lengths exceeding 10 μm, while polycrystalline films require passivation (e.g. PEAI—2-phenylethylammonium iodide treatment) to suppress trap densities. Ion migration under bias further complicates transport, causing hysteresis in solar cell current–voltage curves.

2.4 Stability considerations

2.4.1 Intrinsic material stability

The stability of perovskite materials is governed by both intrinsic material properties and environmental factors. Hybrid perovskites such as $MAPbI_3$ degrade at elevated temperatures (>85 °C) due to organic cation decomposition ($MA^+ \rightarrow CH_3NH_2 + HI$), whereas all-inorganic variants such as $CsPbI_3$ exhibit phase instability under humidity, transitioning from the photoactive cubic phase to an inactive orthorhombic structure. Lead-free alternatives (e.g. $Cs_2AgBiBr_6$) offer improved oxidation resistance but often suffer from hygroscopicity and lower efficiency. Environmental stressors further exacerbate degradation: moisture infiltrates grain boundaries, hydrolyzing Pb–I bonds into PbI_2 and HI, while UV irradiation accelerates halide segregation (e.g. I/Br phase separation), and thermal cycling induces microcracks. Oxygen exposure introduces p-type doping, increasing non-radiative recombination losses.

In order to mitigate these issues, several strategies have been developed. Compositional engineering—such as mixed cations (Cs/FA/MA) and mixed halides (I/Br)—suppresses phase transitions and improves thermal stability [32]. Surface passivation with 2D perovskite capping layers (e.g. PEAI) effectively blocks moisture penetration and reduces surface recombination [33]. Encapsulation techniques, including atomic-layer-deposited Al_2O_3 and glass–glass sealing, have extended device lifetimes beyond 1000 h under IEC 61215 testing conditions [34]. Carbon-electrode architectures demonstrate exceptional stability under damp heat conditions (85 °C/85% RH), retaining >90% initial efficiency after 2000 h without encapsulation. This originates from their intrinsic tolerance to ion migration and elimination of metal electrode corrosion pathways [35]. Accelerated aging studies confirm that moisture—rather than heat—remains the primary failure mode for encapsulated perovskite devices.

2.4.2 Environmental factors

The environmental instability of perovskite materials stems from multiple factors. These include the following:

1. *Moisture* (H_2O): The most dominant degradation pathway, where water infiltrates grain boundaries, hydrolyzing Pb–I bonds into PbI_2 and HI, leading to irreversible decomposition. Even encapsulated devices eventually fail under prolonged humidity exposure [36].
2. *UV light*: Accelerates halide segregation (e.g. I/Br phase separation in mixed-halide perovskites) and generates reactive oxygen species that degrade organic cations [37]. UV filtering or luminescent down-shifting layers can mitigate this effect.
3. *Oxygen* (O_2): Induces p-type doping, increasing non-radiative recombination and trap-assisted degradation. Oxygen reacts with iodide ions (I^-), forming I_2 and leaving behind vacancies [38].
4. *Thermal cycling*: Repeated heating/cooling (e.g. day–night temperature variations) induces mechanical stress, forming microcracks that accelerate moisture and oxygen penetration [39].

5. *Electric bias*: Under operational conditions, ion migration (e.g. I^- vacancies) leads to phase segregation and electrode corrosion [40].
6. *Combined stressors*: Simultaneous exposure to heat, humidity, and light (e.g. outdoor conditions) causes synergistic degradation, often worse than individual factors [41].

2.5 Conclusion

Perovskite materials, characterized by their ABX_3 structural motif, have emerged as a highly versatile class of materials with exceptional optoelectronic properties. The flexibility in composition—ranging from hybrid organic–inorganic perovskites such as $MAPbI_3$ and $FAPbI_3$ to all-inorganic variants such as $CsPbI_3$—enables precise tuning of bandgaps, charge carrier mobility, and light absorption efficiency. However, their functional performance is intrinsically linked to structural stability and phase transitions. For instance, $MAPbI_3$ undergoes temperature-dependent phase transitions (orthorhombic → tetragonal → cubic), while $CsPbI_3$ faces challenges in stabilizing its photoactive cubic phase at room temperature. These phase instabilities, driven by deviations from the ideal Goldschmidt tolerance factor ($t \approx 1$), highlight the delicate balance between ionic radii and lattice strain that governs perovskite crystallinity and functionality.

The unique defect-tolerant nature of perovskites, attributed to their high dielectric screening and antibonding conduction band character, allows them to maintain performance despite moderate defect densities. However, deep-level defects such as lead vacancies or halide interstitials remain critical concerns, necessitating strategies such as surface passivation and compositional engineering. Mixed-cation approaches (e.g. Cs/FA/MA alloys) and halide alloying (e.g. I/Br mixtures) have proven to be effective in enhancing thermal and environmental stability, suppressing phase segregation, and reducing non-radiative recombination. For example, surface treatment with 2D perovskite layers or organic molecules such as PEAI can mitigate moisture ingress and passivate surface traps, extending device lifetimes under operational conditions.

Stability challenges, particularly under moisture, heat, and light exposure, remain significant barriers to commercialization. Hybrid perovskites degrade through organic cation decomposition or hydrolysis of metal–halide bonds, while all-inorganic counterparts suffer from phase transitions under humidity. Encapsulation techniques, such as atomic-layer-deposited oxides or polymer coatings, coupled with carbon-based electrode architectures, have shown promise in enhancing environmental resilience. Lead-free alternatives such as $Cs_2AgBiBr_6$ offer reduced toxicity but require further optimization to address efficiency losses and hygroscopicity.

Looking forward, the structural and compositional adaptability of perovskites positions them as a transformative platform for applications beyond photovoltaics, including LEDs, photodetectors, and photocatalysis. Continued advancements in understanding phase transition dynamics, defect mitigation, and scalable fabrication methods—such as blade-coating or vapor deposition—will be critical for transitioning laboratory-scale innovations to industrial deployment. By addressing stability and toxicity challenges through material design and engineering, perovskites hold the potential to redefine the next-generation optoelectronic technologies, offering a sustainable pathway for high-efficiency, low-cost energy solutions.

References

[1] Noguera Gómez J 2025 Synthesis and characterization of multifunctional nanocomposite-based materials for next-gen energy solutions *PhD thesis* University of Valencia

[2] Yang C *et al* 2024 Achievements, challenges, and future prospects for industrialization of perovskite solar cells *Light Sci. Appl.* **13** 227

[3] Goldschmidt V M 1926 Die Gesetze der Krystallochemie *Naturwissenschaften* **14** 477–85

[4] Han G *et al* 2018 Additive selection strategy for high performance perovskite photovoltaics *J. Phys. Chem.* C **122** 13884–93

[5] Bartel C J *et al* 2019 New tolerance factor to predict the stability of perovskite oxides and halides *Sci. Adv.* **5** eaav0693

[6] Wang S, Li M-H, Jiang Y and Hu J-S 2023 Instability of solution-processed perovskite films: origin and mitigation strategies *Mater. Futures* **2** 012102

[7] Hosen A 2024 Investigating the effects of hydrostatic pressure on the physical properties of cubic Sr_3BCl_3 (B = As, Sb) for improved optoelectronic applications: a DFT study *Heliyon* **10** e35855

[8] Rodrigues A D *et al* 2025 Transformations induced by hydrostatic pressure on lead metasilicate phases arXiv:2502.12274

[9] Kim B, Kim J and Park N 2020 First-principles identification of the charge-shifting mechanism and ferroelectricity in hybrid halide perovskites *Sci. Rep.* **10** 19635

[10] Whitfield P *et al* 2016 Structures, phase transitions and tricritical behavior of the hybrid perovskite methyl ammonium lead iodide *Sci. Rep.* **6** 35685

[11] Ma M *et al* 2023 *In situ* imaging of the atomic phase transition dynamics in metal halide perovskites *Nat. Commun.* **14** 7142

[12] di Vera A 2025 Alternative halide perovskite structures for light-based applications *PhD thesis* University of Trieste

[13] Chung T H 2017 Study of low-temperature sintered lead-free piezoelectric ceramics for multilayer applications *M Phil Thesis*, The Hong Kong University

[14] Xu Z *et al* 2022 $CsPbI_3$-based phase-stable 2D Ruddlesden–Popper perovskites for efficient solar cells *Nano Lett.* **22** 2874–80

[15] Li W *et al* 2019 Band gap evolution in Ruddlesden–Popper phases *Phys. Rev. Mater.* **3** 101601

[16] Yatoo M A and Skinner S J 2022 Ruddlesden–Popper phase materials for solid oxide fuel cell cathodes: a short review *Mater. Today Proc.* **56** 3747–54

[17] Yin W-J, Shi T and Yan Y 2014 Unusual defect physics in $CH_3NH_3PbI_3$ perovskite solar cell absorber *Appl. Phys. Lett.* **104** 063903

[18] Baumann A *et al* 2015 Identification of trap states in perovskite solar cells *J. Phys. Chem. Lett.* **6** 2350–4

[19] Bush K A *et al* 2018 Controlling thin-film stress and wrinkling during perovskite film formation *ACS Energy Lett.* **3** 1225–32

[20] Zhou Y, Poli I, Meggiolaro D, De Angelis F and Petrozza A 2021 Defect activity in metal halide perovskites with wide and narrow bandgap *Nat. Rev. Mater.* **6** 986–1002

[21] Xing G *et al* 2013 Long-range balanced electron-and hole-transport lengths in organic–inorganic $CH_3NH_3PbI_3$ *Science* **342** 344–7

[22] Jeon N J *et al* 2014 Solvent engineering for high-performance inorganic–organic hybrid perovskite solar cells *Nat. Mater.* **13** 897–903

[23] Cao D H, Stoumpos C C, Farha O K, Hupp J T and Kanatzidis M G 2015 2D homologous perovskites as light-absorbing materials for solar cell applications *J. Am. Chem. Soc.* **137** 7843–50

[24] Ke W *et al* 2016 Employing lead thiocyanate additive to reduce the hysteresis and boost the fill factor of planar perovskite solar cells *Adv. Mater.* **28** 5214–21

[25] Li Y, Huang X, Sheriff H K and Forrest S R 2023 Semitransparent organic photovoltaics for building-integrated photovoltaic applications *Nat. Rev. Mater.* **8** 186–201

[26] Sidhik S *et al* 2022 Deterministic fabrication of 3D/2D perovskite bilayer stacks for durable and efficient solar cells *Science* **377** 1425–30

[27] Zhang X, Shen J-X, Turiansky M E and Van de Walle C G 2021 Minimizing hydrogen vacancies to enable highly efficient hybrid perovskites *Nat. Mater.* **20** 971–6

[28] Lee C *et al* 2023 Grooved electrodes for high-power-density fuel cells *Nat. Energy* **8** 685–94

[29] Huang J, Yuan Y, Shao Y and Yan Y 2017 Understanding the physical properties of hybrid perovskites for photovoltaic applications *Nat. Rev. Mater.* **2** 17042

[30] Stranks S D *et al* 2013 Electron–hole diffusion lengths exceeding 1 micrometer in an organometal trihalide perovskite absorber *Science* **342** 341–4

[31] Tao S *et al* 2019 Absolute energy level positions in tin- and lead-based halide perovskites *Nat. Commun.* **10** 2560

[32] McMeekin D P *et al* 2016 A mixed-cation lead mixed-halide perovskite absorber for tandem solar cells *Science* **351** 151–5

[33] Zhao D *et al* 2017 Low-bandgap mixed tin–lead iodide perovskite absorbers with long carrier lifetimes for all-perovskite tandem solar cells *Nat. Energy* **2** 17018

[34] Matlack K H, Serra-Garcia M, Palermo A, Huber S D and Daraio C 2018 Designing perturbative metamaterials from discrete models *Nat. Mater.* **17** 323–8

[35] Rolston N *et al* 2018 Effect of cation composition on the mechanical stability of perovskite solar cells *Adv. Energy Mater.* **8** 1702116

[36] Eperon G E *et al* 2015 The importance of moisture in hybrid lead halide perovskite thin film fabrication *ACS Nano* **9** 9380–93

[37] Mosconi E, Meggiolaro D, Snaith H J, Stranks S D and De Angelis F 2016 Light-induced annihilation of Frenkel defects in organo-lead halide perovskites *Energy Environ. Sci.* **9** 3180–7

[38] Pearson A *et al* 2016 Oxygen degradation in mesoporous $Al_2O_3/CH_3NH_3PbI_{3-x}Cl_x$ perovskite solar cells: kinetics and mechanisms *Adv. Energy Mater.* **6** 1600014

[39] Tan W, Bowring A R, Meng A C, McGehee M D and McIntyre P C 2018 Thermal stability of mixed cation metal halide perovskites in air *ACS Appl. Mater. Interfaces* **10** 5485–91

[40] Yuan Y and Huang J 2016 Ion migration in organometal trihalide perovskite and its impact on photovoltaic efficiency and stability *Acc. Chem. Res.* **49** 286–93

[41] Park N-G, Grätzel M, Miyasaka T, Zhu K and Emery K 2016 Towards stable and commercially available perovskite solar cells *Nat. Energy* **1** 1–8

Chapter 3

Properties of perovskites

Metal–halide perovskites (general formula ABX_3) have emerged as versatile semiconductors for application in photovoltaics, light emission, sensing, and photo detection. In particular, the hybrid organic–inorganic and all-inorganic halide variants are playing a key role in these applications. During the past decade, perovskite solar cell power-conversion efficiency has increased from a few percent to over 25%, thanks to their desirable optoelectronic properties such as strong absorption in the visible component of the solar spectrum, tunable direct bandgaps, long carrier diffusion lengths, and low exciton binding energies. This chapter presents a status report of the optical, electronic, thermal, mechanical, and environmental properties of halide perovskites (including lead-based, hybrid, and lead-free variants), with examples from recent literature. The band-gap engineering via compositional control, luminescence efficiency, defect tolerance, charge transport, and stability issues of perovskites are briefly described. The associated challenges that halide perovskites are facing, such as moisture and thermal instability and degradation, are also discussed. For instance, $MAPbI_3$ easily decomposes to PbI_2 above temperatures of ~100 °C, whereas inorganic $CsPbBr_3$ crystals can remain phase-stable up to ~460 °C. Key concepts, with band-structure diagrams, tunable bandgap plots, and schematics that describe stability, are illustrated with examples. This chapter aims to provide a comprehensive, updated overview of perovskite material properties, supported by recent peer-review findings.

3.1 Introduction

Perovskites are a broad class of materials that share the ABX_3 crystal structure (originally known for calcium titanate, $CaTiO_3$) where A and B are cations and X is an anion. In halide perovskites, the B-site is typically a group-IV metal (e.g. Pb^{2+}, Sn^{2+}, Bi^{3+}) coordinated by halides (Cl^-, Br^-, I^-), and the A-site can be an alkali or organic cation (e.g. Cs^+, $CH_3NH_3^+$, $CH(NH_2)_2^+$). The prototypical hybrid perovskites (e.g. $CH_3NH_3PbI_3$, $CH_3NH_3PbBr_3$) are three-dimensional (3D) lattices of

doi:10.1088/978-0-7503-5427-1ch3

corner-sharing PbX_6 octahedra hosting A-site cations. The variants include all-inorganic perovskites (e.g. $CsPbX_3$) and low-dimensional layered perovskites (e.g. Ruddlesden–Popper phases ($A_2B_{n-1}X_{3n+1}$) in which large organic spacers create quasi-2D layers). More recently, lead-free perovskite-inspired materials (e.g. Sn^{2+}-based [$CH_3NH_3SnI_3$] or double perovskites such as $Cs_2AgBiBr_6$) have been explored to reduce toxicity.

From the beneficial standpoint of organic–inorganic semiconductors, metal–halide perovskites (hybrid organic–inorganic perovskites (HOIPs)) have particularly attracted significant interest for applications in optoelectronics. Among several others, the main advantage of these semiconductors is that they can be solution-processed at low temperatures and still provide the benefits of high optical absorption coefficients ($>10^4$–10^5 cm^{-1}) and long charge-carrier lifetimes. Perovskite solar cells (PSCs) have quickly achieved lab efficiencies surpassing 25% [1]. This efficiency surge (from 3.8% in 2009 to 25.7% in ~2023) can be attributed to their favorable bandgaps (~1.1–1.7 eV, tunable by composition), direct-bandgap nature, ambipolar transport, and defect tolerance. In addition to their application in solar cells, perovskites also exhibit bright photoluminescence for applications in LEDs and lasers with emission colors that are tunable across the visible by halide mixing as well as in photodetectors. For example, $CH_3NH_3PbI_3$ absorbs near-infrared to visible light (bandgap ~1.55 eV), whereas replacing I with Br or Cl shifts the bandgap to ~2.3 eV or ~3.0 eV, respectively [2].

The applications of perovskite materials extend beyond solar cells to the following devices: (a) light-emitting diodes since perovskite materials exhibit high external quantum efficiencies, (b) x-ray and gamma-ray detectors that benefit from heavy atom absorption, and (c) gas/humidity sensors due to their ionic conductivity and surface reactivity. Researchers have also fabricated perovskite photodetectors with high responsivity and fast response, as well as perovskite-based lasers (optical gain in the near-IR/visible). However, many halide perovskites suffer from environmental instability; they are degraded under the influence of moisture, heat, light and many other environmental factors [3, 4]. For instance, $CH_3NH_3PbI_3$ transforms into PbI_2 at modest temperatures ($\approx$ 100 °C–150 °C) [4], whereas all-inorganic $CsPbX_3$ can tolerate much higher temperatures before phase transformation [5]. The softness of the perovskite lattice (low Young's modulus) also makes devices flexible but mechanically fragile (grain-boundary related issues) [6–8]. These stability and durability issues have led to implementation of strategies such as compositional mixing, surface passivation, and encapsulation.

Halide perovskites are crystalline semiconductors with tunable optical bandgaps, high absorption, and good charge transport, enabling the possibility to manufacture high-performance solar cells, LEDs, lasers, and sensors. Their properties are highly sensitive to composition and environment, leading to a rich playground for materials engineering, science and technology. Drawing up on the recent peer-reviewed studies in the literature to illustrate these concepts, case studies and examples, the following sections present an analysis of the optical, electronic, thermal, mechanical, and environmental characteristics of these materials.

3.2 Optical properties

Optical properties of perovskites are remarkable due to the direct bandgaps in the visible/near-IR, large absorption coefficients, and strong photoluminescence. The fundamental bandgap (E_g) can be tuned continuously by modifying the chemical composition (halide substitution, A-site cation, 2D/3D dimensionality). For example, exchanging I^- for Br^- or Cl^- in $APbX_3$ increases the bandgap from ~1.5 eV ($MAPbI_3$) to ~2.3 eV ($MAPbBr_3$) or ~3.0 eV ($MAPbCl_3$) [2]. Bandgap tunability is a key optical feature in perovskites. Mixed-halide perovskites (e.g. $CH_3NH_3Pb(I_{1-x}Br_x)_3$) display continuous tuning, UV–vis absorption spectra red-shift or blue-shift as the halide ratio changes. The electronic and optical properties of mixed-halide perovskites ($MAPbX_3$, X = I, Br, Cl), demonstrating (a) a blue-shift in the UV–vis absorption spectra (from 1.6 to 3.1 eV) and visible color transitions (black → orange → transparent), as halides evolve from I to Cl, (b) a schematic band structure revealing increased ionization potential (6.0–7.0 eV) with I → Br substitution, minimal valence band shifts with Br → Cl, and *n*-type behavior (Fermi level near conduction band), and (c) ultraviolet photoelectron spectroscopy (UPS) data quantifying work function (4.58–5.14 eV) and ionization trends, correlating wider bandgaps to higher open-circuit voltage (V_{OC}) but reduced short-circuit current (J_{SC}) in solar cells, are presented in figure 3.1. Figure 3.1(a) shows that as Br content increases, the color of the film transitions from black (I-rich, narrow-gap) to yellow (Br-rich, wide-gap). Such compositional bandgap engineering characteristics are exploited in the fabrication of tandem solar cells (stacking different perovskites) and in LEDs of various color emissions.

Optical excitations in perovskites are generally associated with weakly bound excitons. In 3D perovskites such as $MAPbI_3$, the exciton binding energy is on the order of a few meV to tens of meV (comparable to k_BT (Boltzmann constant k_B × temperature T in kelvins) at room temperature), meaning photoexcited carriers are often free electrons and holes rather than tightly bound excitons [4, 9]. In contrast, low-dimensional (2D or quasi-2D) perovskites have much larger exciton binding energies (hundreds of meV) due to dielectric confinement. This leads to sharp excitonic peaks in the absorption and luminescence for 2D phases, which is particularly useful for light emission in LEDs and lasers. For example, $CsPbBr_3$ nanocrystals exhibit narrow emission (~10–30 meV linewidth) and near-unity photoluminescence quantum yield (PLQY) when well-passivated. In fact, colloidal $CsPbX_3$ (X = Cl, Br, I) nanocrystals have demonstrated PLQYs approaching 100%. [10]. This high emissive efficiency (especially in green/blue) makes perovskites excellent candidates for LEDs and display applications.

The general static dielectric constants for perovskite materials are relatively large compared to many organics, contributing to screening of charged defects. The high dielectric screening and the antibonding character of the valence-band edge (Pb 6s–I 5p antibonding) mean that defects often introduce shallow states rather than deep traps, which is a phenomenon being called 'defect tolerance'. Figure 3.2 demonstrates that the optical bandgap of lead-halide perovskites (e.g. $MAPbI_3$) can be tuned by adjusting the steric size of the cation. Larger cations (e.g. Cs^+) reduce

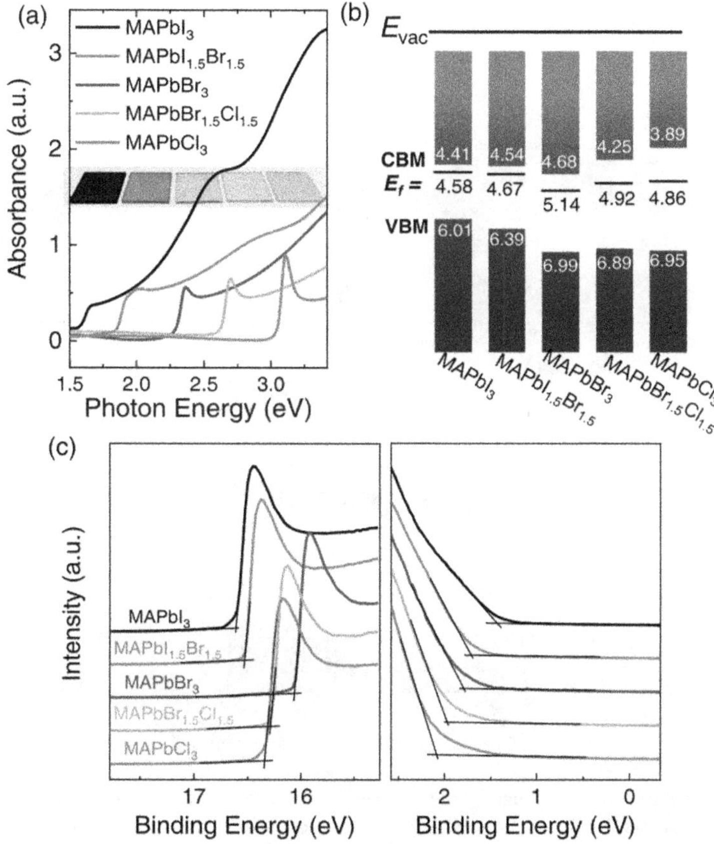

Figure 3.1. (a) UV–vis absorption spectra of mixed-halide perovskite films. Inset: photographs of each perovskite thin film. (b) Schematics of the electronic structure of the mixed-halide perovskite films near the bandgap region. (c) UPS secondary electron edge (SEE) and the low-binding energy region (near the Fermi energy $E_f = 0$ eV) of mixed-halide perovskite films. The onset of ionization of filled states relative to zero binding energy is used to track the shifts of E_f relative to valence-band maxima (VBM). (Reproduced with permission from [2]. Copyright 2021 American Chemical Society.)

octahedral tilting, increasing Pb–I–Pb bond angles (up to ~170 °) and narrowing the bandgap (to ~1.2 eV), while smaller cations (e.g. Li^+) induce greater tilting, reducing bond angles (~130 °) and widening the gap (~1.8 eV). This correlation arises from the influence of the cation on the perovskite lattice geometry, modulating their electronic structure [9]. Experimental validation via mixed Rb/Cs-cation perovskites confirms bandgap blue-shifting with smaller cations, enabling precise control from infrared to visible absorption.

Beyond bandgap tunability, perovskites exhibit strong visible-light absorption and high-efficiency emission, properties that are amplified by their direct bandgaps (1.1–3.0 eV). Halide mixing further broadens the spectral range, offering a wide color gamut for optoelectronic applications. In 2D perovskites, excitonic effects dominate, with sharp absorption and emission spectra highlighting the dimensionality-driven quantum confinement. Remarkably, even polycrystalline perovskite films, despite grain

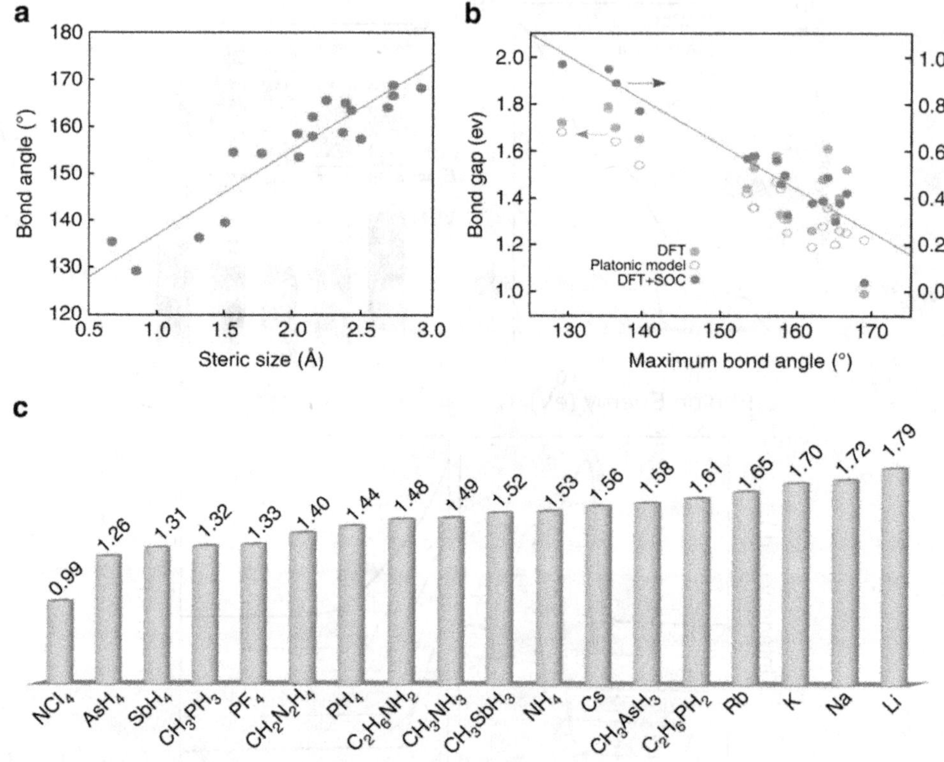

Figure 3.2. Tuning the bandgap of metal–halide perovskites via the steric size of the cation. (a) Correlation between the apical and equatorial bond angles of PbI_3-based perovskites and the steric radius of the cation (discs). (b) Correlation between the density functional theory (DFT) bandgap and the largest metal–halide–metal bond angle in the structure. We show calculations for the Platonic model (circles) and fully optimized structures within scalar relativistic (gray discs) and fully relativistic (blue discs) DFT. The arrows point to the scales corresponding to each set of data. The significant data dispersion at large angles is due to the distortions of the octahedra. (c) Calculated bandgaps (in eV) of all the PbI_3-based perovskites considered in this work. The bandgaps were obtained after full structural optimization within scalar relativistic DFT. (Reproduced with permission from [9]. Copyright 2014 Springer Nature.)

boundaries and defects, retain exceptional optoelectronic performance, including millisecond carrier lifetimes, micron-scale diffusion lengths, and efficient luminescence. This synergy of tunable bandgaps, robust light–matter interaction, and defect tolerance positions perovskites as transformative materials for the fabrication of high-performance solar cells, LEDs, and photonic devices.

3.3 Electronic properties

Perovskites exhibit intriguing electronic behavior as they are direct-gap semiconductors, with conduction band minimum and valence-band maximum at the same k-point. Because of the direct gap and high absorption, perovskites can photo-generate high densities of electron–hole pairs. Carrier mobilities in polycrystalline perovskite films are moderate (typically 1–10 $cm^2\ V^{-1}{\cdot}s^{-1}$) but sufficient for efficient devices. For example, perovskite thin-film transistors have demonstrated ambipolar transport with both electron and hole mobilities exceeding ~10 $cm^2 V^{-1}·s^{-1}$ [11]. In single crystal perovskites, the mobilities can be even larger (tens to hundreds $cm^2 V^{-1}·s^{-1}$) due to fewer grain boundaries. These mobilities enable long carrier diffusion lengths (exceeding 1 μm) and lifetimes (hundreds of ns) in high-quality materials, so that the photogenerated carriers can reach the device contacts before recombination [3].

A remarkable feature of metal–halide perovskites is their unusual tolerance to defects, which manifests in suppressed nonradiative recombination despite relatively high intrinsic defect densities. First-principles studies reveal that native defects such as iodine vacancies or interstitials primarily introduce shallow electronic states near band edges rather than deep traps [12, 13]. Liu *et al* proposed that GeSe exhibits a perovskite-like antibonding valence-band maximum (VBM) via Ge 4s–Se 4p coupling, analogous to Pb-based perovskites, which suppresses deep-level defects and enables defect-tolerant electronic properties. While bulk defects in GeSe are shallow (density ~10^{12} cm^{-3}), surface defects limit device performance. By passivating these with Sb_2Se_3, the authors achieved a certified 5.2% power-conversion efficiency (PCE), a 3.7 × improvement over prior GeSe solar cells. The GeSe devices demonstrated exceptional stability—retaining efficiency for 12 months in ambient conditions and enduring thermal cycling (−40 °C–85 °C)— this has been attributed to the covalent bonding in GeSe and the suppressed oxidation. This work highlights GeSe as a stable, lead-free alternative with tunable electronic structure for application in photovoltaics [14].

Perovskites are direct-bandgap semiconductors with defect-tolerant electronic structures, enabling ambipolar charge transport (balanced electron/hole motion), long carrier diffusion lengths (>1 μm), and suppressed recombination due to shallow trap states. Their intrinsic Fermi level typically lies near mid-gap or slightly *n*-type due to halide vacancy dominance. Doping remains challenging: extrinsic substitutions (e.g. Bi^{3+}/Sb^{3+} for Pb^{2+}) often introduce deep traps, while monovalent cations (e.g. Na^+ on A-sites) minimally perturb band edges. Uncontrolled doping risks instability. Thus, carrier density is often modulated via interfacial engineering or contact optimization. Lead-free variants exhibit distinct properties. Double perovskites such as $Cs_2AgBiBr_6$ have indirect gaps (~2.1 eV) with weaker near-edge absorption compared to lead perovskites, limiting photovoltaic efficiency. Tin-based analogs (e.g. $FASnI_3$) achieve narrower gaps (~1.3 eV) but suffer from $Sn^{2+} \rightarrow Sn^{4+}$ oxidation, inducing p-type self-doping and elevated dark conductivity. Emerging Sb/Bi-based perovskites show improved stability and visible-range emission, although their larger effective masses and indirect gaps hinder their performance [15]. Current research prioritizes balancing stability and optoelectronic metrics in lead-free systems, leveraging defect tolerance and tailored band structures. Carrier mobilities in polycrystalline films typically range from 1–10 $cm^2 V^{-1}·s^{-1}$, with single crystals reaching higher values [11].

3.4 Thermal stability

Thermal stability is a critical concern for application of perovskites. The lattice of perovskites is relatively soft and often relies on volatile components. Thus, heating can induce phase transitions or decomposition. Hybrid organic–inorganic perovskites (e.g. $MAPbI_3$, $FAPbI_3$) generally decompose at much lower temperatures than fully inorganic ones. For instance, $MAPbI_3$ starts to decompose above ~100 °C–150 °C under inert conditions. *In situ* synchrotron studies have shown that heating $MAPbI_3$ to 100 °C for minutes leads to the loss of CH_3NH_2 and HI, leaving PbI_2 on the surface [4]. A thermogravimetric study confirmed that $CH_3NH_3PbI_3$ evolves into CH_3I and NH_3 around 100 °C, consistent with rapid decomposition to PbI_2 [4]. Formamidinium lead iodide ($FAPbI_3$) is somewhat more thermally robust (its α-phase is stable up to ≈150 °C) [16] and has a more favorable bandgap compared to the archetypical methylammonium lead iodide ($MAPbI_3$). However, the photoactive α-phase is not thermodynamically stable at operating temperatures, which is a challenge that must be overcome for the viability of $FAPbI_3$-based photovoltaics [17]. In contrast, inorganic $CsPbX_3$ perovskites exhibit much higher thermal tolerance. For example, $CsPbBr_3$ single crystals remain in the perovskite phase well above 300 °C; one study found no phase degradation up to about 460 °C [5]. where above this temperature, $CsPbBr_3$ converts to related phases such as $CsPb_2Br_5$. Similarly, $CsPbI_3$ (black perovskite phase) has a high decomposition onset (> 400 °C) but tends to revert to a yellow non-perovskite phase at room temperature unless stabilized. The key point is that removing the volatile organic cation greatly improves heat resilience [4]. As shown in figure 3.3, Niu *et al* have demonstrated that Cs-doped perovskite films ($Cs_xMA_{1-x}PbI_3$) exhibit enhanced thermal stability at 120 °C in air. Optimal Cs doping ($x = 0.09$) retains 700 nm absorption better than undoped $MAPbI_3$ ($x = 0$), reducing MA loss and oxidation. Higher Cs ($x = 0.23$) degrades stability due to $CsPbI_3$ phase segregation. Small Cs amounts suppress thermal degradation, improving perovskite resilience [18].

Thermal degradation in devices is often catalyzed by environmental factors. In air, even inert-temperature-stable perovskites can degrade if moisture or oxygen is present during heating. For example, studies of mixed-cation films (FA/MA mixtures), under the influence of humid heat, show synergistic decomposition pathways. Moreover, the mismatch of thermal expansion coefficients between perovskite layers and substrates/electrodes can introduce strain and lead to micro-cracks upon cycling. Phase transitions also play a role: for example, many perovskites undergo structural changes upon heating (e.g. $MAPbI_3$ is tetragonal at room temperature and becomes cubic above ~55 °C). These transitions can affect the bandgap and stability. For example, the desirable perovskite (α) phase of $CsPbI_3$ exists only at high temperature (~300 °C) or under special conditions; at room temperature, it normally converts to a yellow δ-phase unless it is stabilized by strain or mixing. In order to improve the thermal stability of perovskites, numerous strategies have been reported in the literature. These include mixing multiple A-site cations (MA/FA/Cs) to stabilize the lattice, surface passivation with polymers or ceramics, and substituting Pb partially with Sn, Ge or other metals (although Sn introduces other issues). Ji *et al* reported a Mn-doping that significantly enhances the thermal and UV stability of α-$CsPbI_3$ perovskite nanocrystals (NCs) [19]. At an optimal Mn/Pb ratio (5.0%), Mn^{2+}-doped NCs retain 70% photoluminescence (PL) intensity after 22 min of UV irradiation (versus 35% for undoped NCs) and resist phase degradation under ethanol (58% versus 10%). Thermally, doped NCs

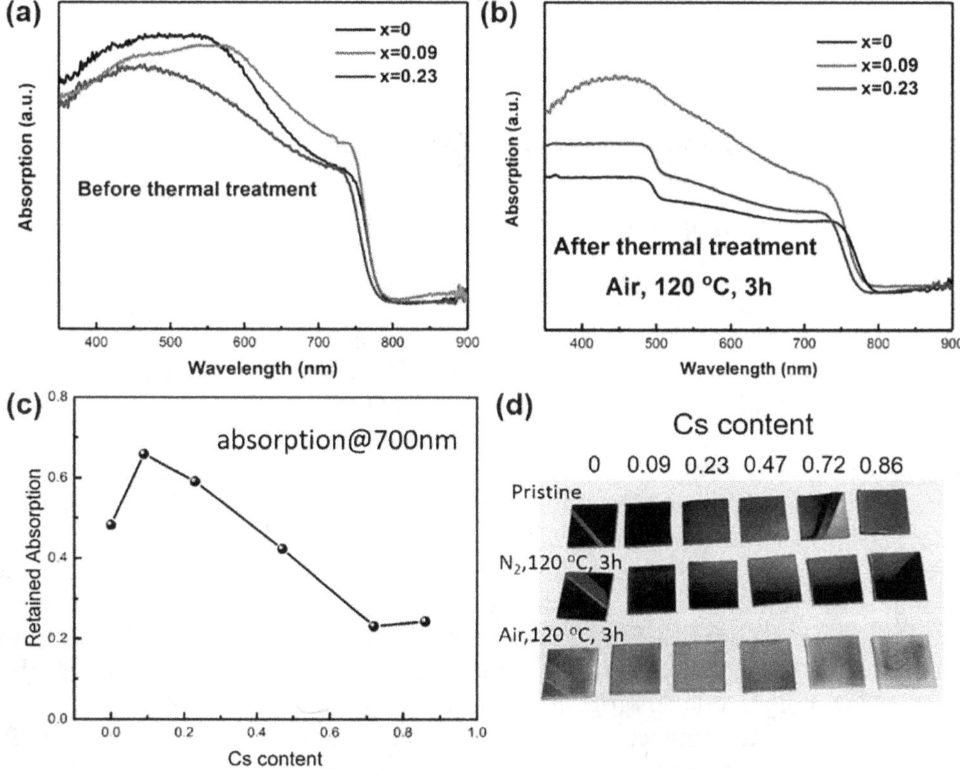

Figure 3.3. Thermal stability study of perovskite films with different Cs content. (a) Perovskite films with compositions of $Cs_xMA_{1-x}PbI_3$ before thermal treatment, where x represents the Cs content. (b) Perovskite films after thermal treatment at 120 °C for 3 h of exposure to air. (c) Retained absorption at 700 nm for perovskite films with x ranging from 0 to 0.86. (d) Photographs of perovskite films exposed to different atmosphere. (Reproduced from [18]. CC BY 3.0.)

maintain cubic-phase stability for ⩾ 7 h at 120 °C, compared to rapid orthorhombic-phase transition in undoped NCs. Doping introduces smaller Mn^{2+} ions (0.067 nm versus 0.119 nm for Pb^{2+}), causing lattice contraction that suppresses ionic migration and halide vacancy formation. This defect reduction also boosts PL quantum yield to 98%. The improved stability under extreme conditions highlights the doping potential of Mn^{2+} for the fabrication of durable optoelectronic devices.

Moreover, as the most efficient α-$FA_{1-x}Cs_xPbI_3$ PSCs require the inclusion of the additive methylammonium chloride, which generates volatile organic residues (methylammonium), it limits device stability at elevated temperatures. Li *et al* have demonstrated enhanced thermal stability in methylammonium-free α-$FA_{0.94}Cs_{0.06}PbI_3$ perovskite solar cells (PSCs) by addressing interfacial Cs^+ accumulation [20]. Conventional devices suffer from Cs^+ segregation at the perovskite/hole transport layer interface, inducing band misalignment and contact losses. By introducing acetate ligand coordination, an intermediate-phase-assisted crystallization pathway is enabled, homogenizing Cs^+ distribution and eliminating defects. The optimized PSCs achieved a certified stabilized efficiency of 25.94% and retained >95% of the initial performance after 2000 h under 85 °C and 60% humidity (ISOS-L-3). This approach suppresses Cs^+ migration and stabilizes the cubic perovskite phase under the influence of heat, offering a route to durable, high-efficiency photovoltaics without the unstable methylammonium additives.

3.5 Mechanical properties

Perovskite materials are generally mechanically soft compared to conventional semiconductors. Their organic components and ionic bonding give them low Young's moduli (~10–20 GPa) and high compressibility. For example, nano-indentation measurements have found that 3D $MAPbI_3$ films have an elastic modulus around 15–20 GPa, whereas purely 2D organic–inorganic perovskites (e.g. $(5\text{-AVA})_2PbI_4$) can be much softer (~6–7 GPa) [21]. This softness aids flexibility (important for rollable/flexible solar cells) but also means perovskites are relatively fragile under stress. Indeed, cracks or grain-boundary defects can form under strain. Interestingly, mixing 2D/3D phases can tailor stiffness: incorporating 2D layers into a 3D matrix can reduce the film modulus (improving flexibility) and simultaneously increase environmental stability (due to the hydrophobic organic spacers). The mechanical behavior is anisotropic in layered perovskites as they are much stiffer in-plane than out-of-plane due to the weaker van der Waals forces that hold the 2D inorganic layers together. Grain size and film morphology also affect their mechanical performance. Larger grains and fewer grain boundaries generally increase their strength. Additives that cross-link grains or heal cracks (e.g. polymers, fullerene derivatives) have been used to improve the mechanical durability of perovskite films and devices. In flexed PSCs, mechanical failure often occurs at the electrodes or interfaces rather than in the bulk perovskite layer; but the perovskite modulus is a key factor in providing matching to the flexible substrates. Brittleness and fracture toughness of perovskites have been less studied quantitatively, but conceptually, perovskites behave more like soft plastics than ceramics. Reports in the literature note that eliminating pinholes and reinforcing grains leads to better mechanical stability of perovskites. In multilayer devices, the relative stiffness of perovskite versus transport layers influences stress distribution; a softer perovskite layer can relieve stress, whereas a rigid perovskite might crack under bending.

The mechanical stability of inorganic perovskite, Ca_3PCl_3, under biaxial strain was investigated via DFT by Islam *et al* [22]. Elastic constants (C_{11}, C_{12}, C_{44}) and Cauchy pressure ($C_{12} - C_{44} > 0$) confirmed ductility, satisfying Born stability criteria ($C_{11} > 0$, $C_{44} > 0$, $C_{11} - C_{12} > 0$). Figure 3.4 shows the strain-induced modulation of the electronic band structure of Ca_3PCl_3. Under compressive strain (−6% to −2%), the direct bandgap at Γ-point narrows as valence (VBM) and conduction (CBM) bands shift toward Fermi level. Tensile strain (+2% to +6%) widens the gap via opposite shifts. Compressive strain (0% to −6%) increases the elastic constants ($C_{11} = 78.87$ GPa at −6%), enhancing stiffness [22]. In addition to the study of the electronic band-structure modulation under strain for Ca_3PCl_3, the mechanical

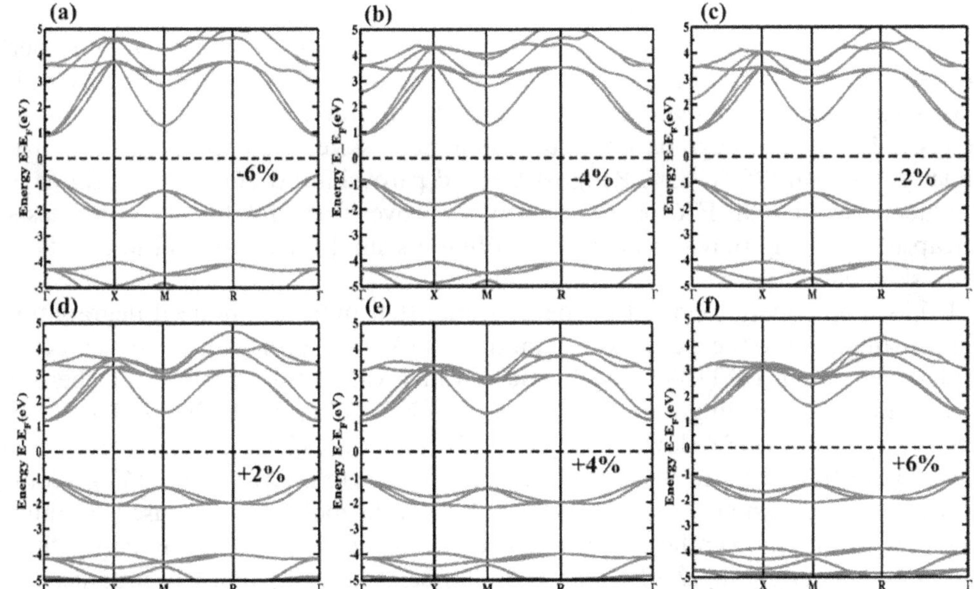

Figure 3.4. The electronic band structure of Ca_3PCl_3 under various applied compressive and tensile strain: (a) −6%, (b) −4%, (c) −2%, (d) +2 %, (e) +4%, and (f) +6 %. (Reproduced with permission from [22]. Copyright 2024 Elsevier.)

stability of Ba_3SbI_3 has been investigated by Rahman *et al* [23] in which the low thermal conductivity (0.247 W $m^{-1}\cdot K^{-1}$) and Debye temperature of 123.78 K highlight the potential of Ba_3SbI_3 for applications in thermal barrier coatings [23]. By utilizing DFT, the authors reveal that the elastic constants (C_{11} = 53.81 GPa, C_{12} = 77.01 GPa, C_{44} = 9.98 GPa) of Ba_3SbI_3 are in compliance with Born's stability criteria ($C_{11} + 2\ C_{12} > 0$, $C_{44} > 0$). Positive Cauchy pressure ($C_{12} - C_{44}$ = 67.03 GPa) indicates ductility. Pugh's ratio (B/G = 1.73) near 1.75 suggests borderline ductile behavior, while Poisson's ratio ($\nu = 0.235$) hints at the brittleness of Ba_3SbI_3. High elastic anisotropy (Universal Anisotropy Index A^U = 1.28) underscores the directional mechanical variability of this perovskite.

3.6 Environmental sensitivity

One of the defining challenges of halide perovskites is their sensitivity to the environment. In particular, moisture and oxygen can rapidly degrade many perovskites. Until now, much of the published literature on the stability of perovskite solar cells has been concerned with the inherent moisture sensitivity demonstrated by the perovskite layer. Oxygen can degrade perovskites especially under light. Photoexcitation in oxygenated air produces a reactive O_2^- (superoxide) that attacks MA^+, abstracting protons and forming water. This feedback loop accelerates the breakdown of perovskites (often synergistic with humidity). Efforts to encapsulate perovskite-based devices often target blocking O_2 diffusion. Strong illumination, particularly exposure to UV, can lead to photo-induced halide segregation in mixed-halide perovskites (leading to bandgap fluctuations) and can drive trap formation. However, under moderate solar illumination and inert atmosphere, well-prepared perovskites are reasonably photostable. Moreover, in real devices, perovskites are exposed to heat, moisture, light, and bias simultaneously. The map of degradation pathways in perovskites is complex. For example, under damp heat (85 °C/85% RH), MA-based perovskites decompose within days. Replacing MA with FA or Cs greatly improves the stability in these tests. Encapsulation (e.g. polymer or glass) is often essential for real-world longevity of perovskite-based devices.

It has been widely reported in the literature that moisture-induced degradation hinders thermochromic perovskite windows (TPWs). Liu *et al* have developed a triple-layer design (MTPW) that balances water vapor transmission for thermochromism while repelling excess moisture. The MTPW retains >20% solar modulation over 45 days (37× slower decay than TPW), reduces lead leakage by 66×, and lowers haze from 90% to 30%, enabling durable, clear smart windows [24]. Some studies have suggested that when sulfonyl chloride molecules are used as intermediates in organic synthesis, their hydrolysis reactions may act as competitive reactions [25, 26]. Liu *et al* propose an ambient-moisture-assisted modification strategy by using a sulfonyl chloride molecule to generate sulfonic acid (PFS) to *in situ* passivate the surface/interface of perovskites through hydrolysis by the moisture in the environment. PFS passivates surface defects and aligns energy levels between perovskite and hole transport layers. This strategy achieves a certified aperture area efficiency of 18.22%, enhancing the stability and performance of PSC under humidity [27].

An example of diffusion of oxygen into a perovskite such as $CH_3NH_3PbI_3$ is shown in figure 3.5. In this case, oxygen diffuses into the perovskite films, forming superoxide (O_2^-) at iodide vacancies under light. This reactive species degrades $CH_3NH_3PbI_3$ into PbI_2, H_2O, CH_3NH_2, and I_2 via deprotonation [28]. Gao *et al* [29] emphasize the critical role of oxygen defect (OD) concentration in $LaBO_3$ perovskites for activating peroxymonosulfate (PMS) to degrade pollutants such as ofloxacin. Their study reveals that higher OD levels—exemplified by the trend $LaNiO_3 > LaMnO_3 > LaFeO_3$—significantly lower the activation energy barrier for PMS, facilitating efficient singlet oxygen ($^1O_2$) generation. These defects promote PMS self-decomposition into $^1O_2$, a key reactive species for pollutant degradation. Among the tested perovskites, $LaNiO_3$ demonstrates superior catalytic performance and stability, attributed to its abundant oxygen defects, which sustain long-term reactivity without structural compromise [29]. In contrast, Bryant *et al* [30] uncover a distinct degradation mechanism in methylammonium lead triiodide ($MAPbI_3$) perovskites, in which the combined exposure to light and oxygen accelerates decomposition. Their work shows that the photoexcited electrons in $MAPbI_3$ interact with ambient oxygen to form superoxide radicals (O_2^-), which destabilize the perovskite structure, ultimately yielding PbI_2. Notably, this degradation pathway is far more rapid under light and oxygen than under moisture alone—for example, perovskite-based devices, exposed to 85% relative humidity in an inert nitrogen atmosphere, degrade at a markedly slower rate. This highlights the critical

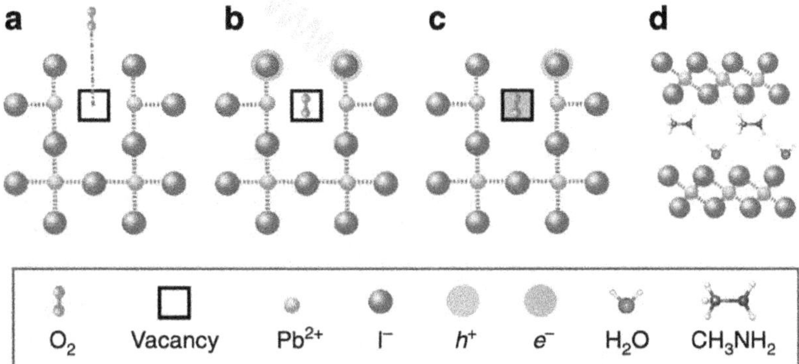

Figure 3.5. Oxygen-induced photo-degradation. Schematic representation of the reaction steps of O_2 with $CH_3NH_3PbI_3$. (a) Oxygen diffusion and incorporation into the lattice, (b) photoexcitation of $CH_3NH_3PbI_3$ to create electrons and holes, (c) superoxide formation from O_2, and (d) reaction and degradation to layered PbI_2, H_2O, I_2 and CH_3NH_2. (Reproduced from [28]. CC BY 4.0.)

role of environmental factors in perovskite stability, with oxygen and light acting synergistically to drive structural breakdown [30].

3.7 Summary of properties

In addition to their abundancy and low-costs, perovskites are easy to process while offering several advantages including their tunable bandgap, light weight, flexible form factors, and, generally, ease of recyclability [31]. From the perspective of light–electricity conversion technologies, the importance and application of metal–halide perovskites, particularly in photovoltaics and optoelectronics, has grown over the years [32]. Some of the fundamental optoelectronic properties of these perovskites, such as energy gap (E_g) and refractive index (n), are summarized in table 3.1 [33].

3.8 Conclusion

Halide perovskites have revolutionized optoelectronics through their exceptional optoelectronic properties, including tunable direct bandgaps, high absorption coefficients, ambipolar charge transport, and defect tolerance. These attributes have propelled perovskite solar cells (PSCs) to efficiencies exceeding 25%, rivaling traditional silicon technologies. The ability to engineer bandgaps via compositional tuning —such as halide substitution (I^-, Br^-, Cl^-) or cation mixing (MA^+, FA^+, Cs^+)— enables tailored absorption and emission spectra, characteristics that are critical for applications in photovoltaics, LEDs, and photodetectors. Mixed-halide perovskites exemplify this tunability, achieving seamless bandgap adjustments from 1.5 to 3.0 eV, while 2D variants leverage quantum confinement for sharp excitonic features.

However, challenges in stability and environmental sensitivity of perovskites remain significant barriers to their commercialization. Organic–inorganic hybrids such as $MAPbI_3$ decompose at modest temperatures (~100 °C–150 °C) and degrade under moisture, oxygen, and light. Inorganic perovskites (e.g. $CsPbBr_3$) exhibit

Table 3.1. Summary of the energy gap (E_g) and refractive index (n) of some perovskites [33, 34].

Perovskite	Energy gap, E_g (eV)	Refractive index (n), (wavelength, λ) (nm) where applicable
$SrTiO_3$	4.1	2.388 (632.8)
$SrSnO_3$	3.93	≈2.4
$KMgF_3$	10.2	1.404 (632.8)
$CaTiO_3$	3.5	≈2.46
$PbTiO_3$	3.4	2.52
$CsPbF_3$	3.8	2.134 (217.7)
$CsPbI_3$	1.67	2.46 (435)
$KTaO_3$	4.35	2.2 (632.8)
$CsPbBr_3$	2.3	≈2.3 (580)
$CsPbCl_3$	3.0	≈1.91
$LiTaO_3$	4.7	≈2.183 (632.8)
$BaZrO_3$	5.3	2.13
$SrZrO_3$	6.15	2.16
$CaZrO_3$	6.4	2.1
$KCaF_3$	10.86	1.388 (583.9)
$LiBaF_3$	9.8	1.544 (632.8)
$KZnF_3$	7.237	1.53 (583.9)
$RbCaF_3$	10.9	1.46
$CsCaCl_3$	6.89	1.58, 1.603 (583.9)

superior thermal resilience (stable up to ~460 °C) but face phase-instability issues at ambient conditions. Moisture and oxygen synergistically accelerate degradation, with photo-induced superoxide formation exacerbating breakdown of their lattice. Strategies such as compositional engineering (mixed A-site cations, Mn^{2+} doping), surface passivation (sulfonyl chloride hydrolysis, Sb_2Se_3 coating), and encapsulation have mitigated these issues, enhancing their operational lifetimes and humidity resistance.

Mechanically, perovskites are soft (Young's modulus ~10–20 GPa), enabling flexibility but requiring grain-boundary reinforcement to prevent fracture. Strain engineering and 2D/3D heterostructures improve their mechanical durability without compromising optoelectronic performance. Lead-free alternatives (e.g. $Cs_2AgBiBr_6$, $FASnI_3$, GeSe) address toxicity concerns but lag in efficiency due to indirect bandgaps, Sn^{2+} oxidation, or defect-prone surfaces. Recent advances in defect passivation and bandgap optimization, such as Sb_2Se_3-passivated GeSe achieving 5.2% efficiency, highlight progress in balancing sustainability with performance.

Thermal management of perovskite-based devices remains critical, as phase transitions (e.g. the α-to-δ transition of $CsPbI_3$) and interfacial cation segregation (e.g. Cs^+ migration in FA/Cs mixtures) degrade device integrity. Innovations such as acetate ligand coordination homogenize cation distribution, stabilizing perovskite phases under thermal stress. Similarly, Mn^{2+} doping in $CsPbI_3$ nanocrystals suppresses halide vacancy formation, enhancing its thermal and UV stability.

In summary, halide perovskites represent a versatile semiconductor class with unparalleled optoelectronic versatility, yet their commercial viability hinges on resolving the stability–composition–performance trade-offs. Continued research into lead-free materials, defect passivation, and scalable encapsulation will be pivotal. By leveraging compositional diversity, interfacial engineering, and mechanistic insights into degradation pathways, research in this excellent area of materials science and engineering can overcome the current limitations, unlocking the full potential of perovskites for sustainable, high-performance optoelectronics, solar cells, sensors and other devices.

References

[1] Liu S, Biju V P, Qi Y, Chen W and Liu Z 2023 Recent progress in the development of high-efficiency inverted perovskite solar cells *NPG Asia Mater.* **15** 27

[2] Zhang S, Tang M-C, Nguyen N, Anthopoulos T and Hacker C 2021 Wide-band-gap mixed-halide 3D perovskites: electronic structure and halide segregation investigation *ACS Appl. Electron. Mater.* **3** 2277–85

[3] Liu W W, Wu T H, Liu M C, Niu W J and Chueh Y L 2019 Recent challenges in perovskite solar cells toward enhanced stability, less toxicity, and large-area mass production *Adv. Mater. Interfaces* **6** 1801758

[4] Kim N-K *et al* 2017 Investigation of thermally induced degradation in $CH_3NH_3PbI_3$ perovskite solar cells using *in-situ* synchrotron radiation analysis *Sci. Rep.* **7** 4645

[5] Han D *et al* 2023 Thermal and chemical durability of metal halide perovskite $CsPbBr_3$ single crystals *Chem. Eng. J.* **475** 146209

[6] Li B, Wu H and Zheng Y 2025 Mechanical stability analysis of flexible perovskite solar cells via opto-electro-mechanical simulation *Sol. Energy* **287** 113211

[7] Liu W *et al* 2024 Regulation interfacial energy by reducing the Young's modulus of perovskite film in carbon-based HTM free $MAPbI_3$ perovskite solar cell *Appl. Surf. Sci.* **677** 161029

[8] Guo Z, Wang J and Yin W-J 2022 Atomistic origin of lattice softness and its impact on structural and carrier dynamics in three dimensional perovskites *Energy Environ. Sci.* **15** 660–71

[9] Filip M R, Eperon G E, Snaith H J and Giustino F 2014 Steric engineering of metal-halide perovskites with tunable optical band gaps *Nat. Commun.* **5** 5757

[10] Dutta A, Behera R K, Pal P, Baitalik S and Pradhan N 2019 Near-unity photoluminescence quantum efficiency for all $CsPbX_3$ (X = Cl, Br, and I) perovskite nanocrystals: a generic synthesis approach *Angew. Chem. Int. Ed.* **58** 5552–6

[11] Zeidell A M *et al* 2018 Enhanced charge transport in hybrid perovskite field-effect transistors via microstructure control *Adv. Electron. Mater.* **4** 1800316

[12] Steinmann V, Brandt R E and Buonassisi T 2015 Non-cubic solar cell materials *Nat. Photonics* **9** 355–7

[13] Walsh A 2015 Principles of chemical bonding and band gap engineering in hybrid organic-inorganic halide perovskites *J. Phys. Chem.* C **119** 5755–60

[14] Liu S-C *et al* 2021 An antibonding valence band maximum enables defect-tolerant and stable GeSe photovoltaics *Nat. Commun.* **12** 670

[15] Jia Z *et al* 2025 Optoelectronics of lead-free antimony- and bismuth-based metal halides for sensitive and low-noise photodetection *Adv. Funct. Mater.* **35** 2413612

[16] Yang F *et al* 2020 Fully solution processed pure α-phase formamidinium lead iodide perovskite solar cells for scalable production in ambient condition *Adv. Energy Mater.* **10** 2001869

[17] Li Y, Bahnick A, Lohr P J, Raglow S and Printz A D 2024 Enhanced α-phase stability of formamidinium lead iodide with addition of 5-ammonium valeric acid chloride *Energy Adv.* **4** 262–72

[18] Niu G, Li W, Li J, Liang X and Wang L 2017 Enhancement of thermal stability for perovskite solar cells through cesium doping *RSC Adv.* **7** 17473–9

[19] Ji Y *et al* 2021 Improving the stability of α-$CsPbI_3$ nanocrystals in extreme conditions facilitated by Mn^{2+} doping *ACS Omega* **6** 13831–8

[20] Li S *et al* 2024 High-efficiency and thermally stable $FACsPbI_3$ perovskite photovoltaics *Nature* **635** 82–8

[21] Rathore S, Han G, Kumar A, Leong W and Singh A 2021 Elastic modulus tailoring in $CH_3NH_3PbI_3$ perovskite system by the introduction of two dimensionality using (5-AVA) $2PbI_4$ *Sol. Energy* **224** 27–34

[22] Islam M R *et al* 2024 Tuning the optical, electronic, and mechanical properties of inorganic Ca_3PCl_3 perovskite via biaxial strain *J. Phys. Chem. Solids* **184** 111722

[23] Rahman M F *et al* 2023 A computational study of electronic, optical, and mechanical properties of novel Ba_3SbI_3 perovskite using DFT *Opt. Quantum Electron.* **56** 206

[24] Liu S *et al* 2024 Mask-inspired moisture-transmitting and durable thermochromic perovskite smart windows *Nat. Commun.* **15** 876

[25] Deng X and Mani N S 2006 A facile, environmentally benign sulfonamide synthesis in water *Green Chem.* **8** 835–8

[26] Lou T S-B and Willis M C 2022 Sulfonyl fluorides as targets and substrates in the development of new synthetic methods *Nat. Rev. Chem.* **6** 146–62

[27] Liu X *et al* 2024 $CsPbI_3$ perovskite solar module with certified aperture area efficiency >18% based on ambient-moisture-assisted surface hydrolysis *Joule* **8** 2851–62

[28] Aristidou N *et al* 2017 Fast oxygen diffusion and iodide defects mediate oxygen-induced degradation of perovskite solar cells *Nat. Commun.* **8** 15218

[29] Gao P *et al* 2019 Promoted peroxymonosulfate activation into singlet oxygen over perovskite for ofloxacin degradation by controlling the oxygen defect concentration *Chem. Eng. J.* **359** 828–39

[30] Bryant D *et al* 2016 Light and oxygen induced degradation limits the operational stability of methylammonium lead triiodide perovskite solar cells *Energy Environ. Sci.* **9** 1655–60

[31] An introduction to perovskites *Perovskite-info* https://perovskite-info.com/introduction (Accessed: 11 August 2025)

[32] He C and Liu X 2023 The rise of halide perovskite semiconductors *Light Sci. Appl.* **12** 15

[33] Lamichhane A and Ravindra N M 2020 Energy gap—refractive index relations in perovskites *Materials* **13** 1917

[34] Lamichhane A 2021 First-principles density functional theory studies on perovskite materials *PhD thesis* New Jersey Institute of Technology

IOP Publishing

Perovskites

Fundamentals, properties, preparation and applications

N M Ravindra, Priyanka Singh, Leqi Lin and Pankaj Kumar

Chapter 4

Processing methods and materials

This chapter presents a review of the processing techniques for fabricating high-quality perovskite thin films and devices. Both lab-scale routes and industrial methods are discussed. Solution-based approaches, such as spin coating, printing as well as the vapor-based processes such as thermal evaporation, are covered. Scalable fabrication methods, including slot-die coating, blade coating, inkjet printing, and roll-to-roll (R2R) printing, are highlighted as important steps toward the commercialization of perovskites, although challenges such as defects, film uniformity, and stability still need to be addressed. Strategies such as the use of antisolvents, additives, and solvent annealing are also described for improving grain boundaries and charge transport in perovskite films.

4.1 Introduction

Fabrication of uniform and defect-free perovskite thin films is one of the most important steps in achieving efficient and stable devices. Although this book broadly addresses perovskite materials and their diverse applications, this chapter mainly uses perovskite solar cells (PSCs) as examples because they are the most studied and more practical for showing different fabrication methods. Insights gained from PSC processing methods are also relevant to other perovskite-based optoelectronic devices. Over the past decade, researchers have explored a wide range of deposition methods, from simple spin coating in research laboratories to advanced large-area coating techniques suitable for industrial production. Beyond deposition, controlling the crystallization and growth of the perovskite layer is critical, as it strongly influences the morphology, grain size, and stability. The overall efficiency and durability of PSCs are determined not only by the deposition processes but also by the functional roles of individual layers and their interfacial properties. When photons are absorbed by the perovskite layer, electron–hole pairs are generated and subsequently separated through proper energy band alignment with selective transport layers [1, 2]. The conduction band/lowest unoccupied molecular orbital

doi:10.1088/978-0-7503-5427-1ch4

(LUMO) of the electron transport layer (ETL) must be slightly lower than that of the perovskite to enable smooth electron extraction, while the valence band/highest occupied molecular orbital (HOMO) level of the hole transport layer (HTL) should align with the perovskite HOMO level to ensure efficient hole extraction. In this way, ETLs selectively extract electrons and block holes, while HTLs extract holes and block electrons, thereby reducing recombination losses and enhancing the charge carrier transport [1, 2].

In addition to band alignment, processing methods play a decisive role in the overall performance of PSCs. The deposition of one layer should not degrade or dissolve the underlying perovskite or charge-transport layers, which necessitates careful selection of deposition techniques and solvent systems. Thus, the relationship between material selection, processing techniques, and interfacial compatibility ultimately governs the efficiency, reproducibility, and commercial viability of PSCs. Advances in scalable fabrication approaches are therefore central to the development of high-performance, stable, and commercially feasible perovskite-based devices [2].

4.2 Thermal evaporation technique

Thermal evaporation is less common than solution-based methods in PSC fabrication. In most research and commercial labs, spin coating and other solution-processing techniques dominate because of their simplicity, and lower cost. However, thermal evaporation offers distinct advantages such as high purity, precise control over film thickness, and the ability to form uniform, pinhole-free layers. It is particularly useful for depositing charge-transport layers (such as C_{60}, bathocuproine, or metal oxides) and metal electrodes such as gold (Au), silver (Ag), or aluminum (Al). Thermal co-evaporation has also been explored for directly depositing perovskite films, enabling better stoichiometric control and improved reproducibility, although this method is more complex and expensive. As a result, while thermal evaporation is not the dominant technique, it remains an important approach in high-performance device fabrication and is often used in tandem with solution processing to achieve optimized perovskite solar cells. The following section is provided to give readers a broader understanding of the fabrication techniques beyond the commonly used solution-based methods.

Thermal evaporation is a simple method in which the material to be deposited (the evaporant) is placed in a heated source until it reaches its sublimation or melting point. The source can be a ceramic crucible wrapped with tungsten wire, a tantalum (Ta), molybdenum (Mo), or graphite boat, or a tungsten coil. The source is kept inside an evaporation chamber that is first evacuated to high vacuum before the process begins. Every material has a vapor pressure that increases with temperature. Some solids evaporate directly without melting (sublimation), while others first melt and then evaporate. For example, organic semiconductors usually sublime, whereas metals typically melt before evaporation. Once vaporized, the molecules travel in straight lines from the source to the substrate surface, provided they do not collide with other gas molecules. Condensation of the evaporated molecules on the relatively cold surface forms a thin film. A high vacuum is essential to ensure high-quality films, as it prevents unwanted reactions or scattering with gas molecules, increases the mean free path of the vapor, and minimizes impurities from residual gases [3].

Thermal evaporation can also be explained using a Knudsen cell (or K-cell). This cell consists of a small container (the source) enclosed within an isothermal chamber that has a tiny opening. The evaporant material is heated inside the container, producing vapors that maintain an equilibrium pressure in the enclosure. Since the opening is very small compared to the surface area of the evaporant, the vapor escapes in a controlled way through an orifice (process known as effusion) and such a cell is known as Knudsen cell, effusion cell or K-cell. The vapor exiting the orifice follows a Lambertian angular distribution. If A_e is the area of the opening, $P^*(T)$ is the vapor pressure of the evaporant at temperature T, and P is the ambient pressure, then the evaporation rate from a Knudsen cell is proportional to $(P^*(T) - P)$.

$$\frac{dN_e}{dt} = A_e \frac{1}{\sqrt{2\pi m k_B T}} (P^*(T) - P). \tag{4.1}$$

Here, N_e represents the number of evaporated atoms, k_B is the Boltzmann constant, and m denotes the molecular or atomic mass of the evaporant. Accordingly, the mass evaporation rate through the orifice can be expressed as

$$\Gamma = m \frac{d}{dA_e}\left(\frac{dN_e}{dt}\right) = \sqrt{\frac{m}{2\pi k_B T}} (P^*(T) - P). \tag{4.2}$$

And the total mass of evaporated material (M_e) can be calculated from

$$M_e = \int_0^t \int_0^{A_e} \Gamma dA_e dt. \tag{4.3}$$

Considering the Lambertian angular distribution of the evaporated material, the amount of material emitted into a solid angle $d\omega = \frac{dA_e}{r^2}$ at an angle ϕ with respect to the surface normal can be expressed as

$$dM_e = M_e \cos\phi \frac{d\omega}{\pi}. \tag{4.4}$$

And the spatial distribution of vapor current density $\Phi(\phi)$ will now be given by

$$\Phi(\phi) = M_e \cos\phi \frac{1}{\pi}. \tag{4.5}$$

Assuming that the substrate is placed at a distance r from the source and oriented at a tilt angle θ with respect to it, the deposition rate at any arbitrary point on the substrate can be expressed as

$$R(\phi) = \Phi(\phi)\cos\theta \frac{1}{r^2}, \tag{4.6}$$

or

$$R(\phi) = M_e \cos\phi \cos\theta \frac{1}{\pi r^2}. \tag{4.7}$$

This relation, known as the Hertz–Knudsen law, allows the determination of the rate distribution for any given geometry. In order to illustrate its application, three representative cases of substrate types and positions are considered, as shown in figure 4.1.

Case I represents an infinite plane substrate as shown in figure 4.1(a). Here, the substrate S is positioned at a normal distance D from the source Q. For an arbitrary point B located at a distance x from the normal on the substrate and at an angle ϕ, the relations $\phi = \theta$ and $r^2 = x^2 + D^2$ hold, leading to the expression for the deposition rate distribution,

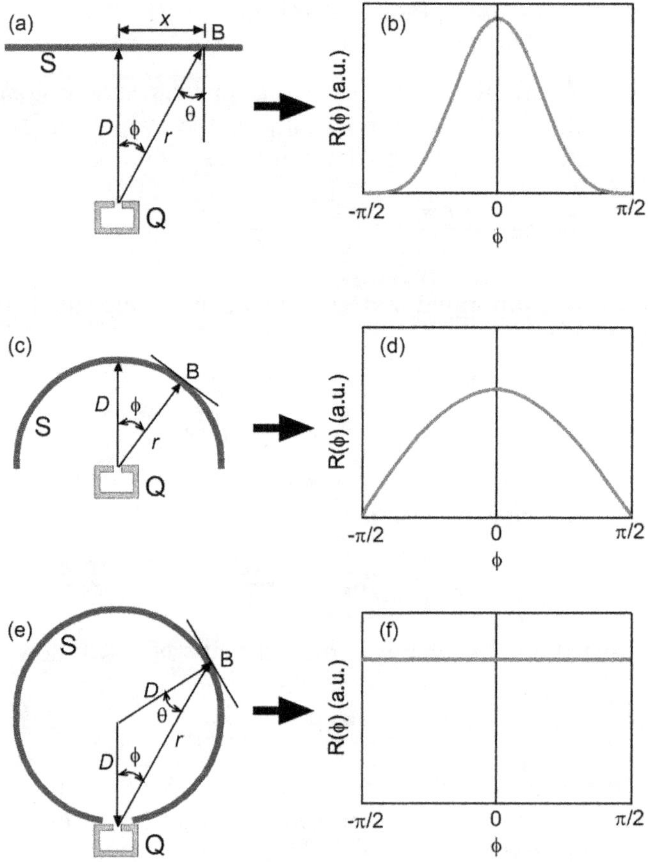

Figure 4.1. Schematic diagrams showing different substrate shapes for film deposition: (a) plane infinite surface, (c) hemispherical and (e) spherical. Panels (b), (d), and (f) show the rate distribution (or film thickness if the evaporation rate is stable) for each corresponding substrate shape. (Reproduced with permission from [3]. Copyright 2016 Taylor and Francis.)

$$R(\phi) = M_e \cos^4 \phi \frac{1}{\pi D^2}. \tag{4.8}$$

The corresponding rate distribution over the infinite plane substrate is illustrated in figure 4.1(b), which clearly indicates that the deposition rate (and consequently the film thickness, assuming a stable evaporation rate), is non-uniform across the substrate surface.

In case II, a hemispherical substrate S with the source Q located at its center is considered, as illustrated in figure 4.1(c). Where, D represents the hemisphere radius. For a point B situated at an angle ϕ and a distance r from the source, the conditions are $\theta = 0$ and $r = D$, leading to the deposition rate distribution

$$R(\phi) = M_e \cos \phi \frac{1}{\pi D^2}. \tag{4.9}$$

The corresponding distribution on the hemispherical substrate, shown in figure 4.1 (d), indicates that the deposition rate (or film thickness) is non-uniform across the substrate surface.

In case III, the configuration involves a spherical substrate S with the source Q positioned at one of its poles, as depicted in figure 4.1(e). For this geometry, D is the sphere radius. At a point B, located at an angle ϕ from the normal to the source and at a distance r, the relations are $\phi = \theta$ and $r = 2D\cos\phi$, yielding the deposition rate distribution

$$R(\phi) = M_e \frac{1}{4\pi D^2}. \tag{4.10}$$

Notably, this distribution is independent of the angle, resulting in a constant rate in all directions. The corresponding distribution, shown in figure 4.1(f), demonstrates that this arrangement produces uniform film thickness over the spherical substrate. Such a configuration is widely employed in industrial applications where highly uniform thin films are essential.

In practice, substrates are generally smaller in size and planar in shape, unlike the idealized geometries discussed earlier. A schematic of a typical thermal evaporation chamber is shown in figure 4.2. The chamber is evacuated using suitable vacuum pumps, typically achieving pressures better than 10^{-5} torr.

If the evaporation source is made of a conductive material such as Mo, Ta, or W, it is filled with the target material and heated directly by passing an electric current through it. Alternatively, if a ceramic crucible is used, it is wrapped with tungsten wire, and the current is passed through the wire, providing indirect heating of the source. Heating causes the material inside to evaporate, and the evaporation rate is controlled by adjusting the source temperature through precise regulation of the current.

The film deposition rate and thickness on the substrate are commonly measured with a quartz crystal thickness monitor, which operates based on the piezoelectric effect. The monitor uses a quartz crystal sensor positioned close to the substrate. An RF voltage excites the crystal, causing it to vibrate at its natural frequency (~5–6 MHz). The vibration frequency of the quartz crystal depends on its mass and

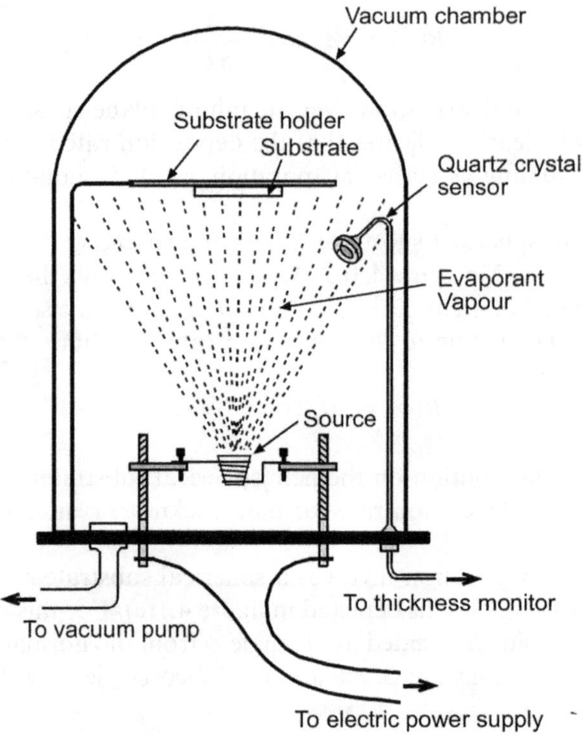

Figure 4.2. Schematic diagram of a standard thermal evaporation system. (Reproduced with permission from [3]. Copyright 2016 Taylor and Francis.)

temperature. When a material is deposited on the crystal surface, its added mass reduces the frequency. This frequency shift is detected by an oscillator circuit connected to the crystal. By using the known Z-factor of the depositing material, the frequency change is converted into the deposited mass per unit area, and with the material's density, the film thickness can be determined. The sensor directly provides the deposition rate and film thickness on the crystal, and through appropriate calculations, the corresponding values on the substrate are obtained [3].

An important parameter in these calculations is the tooling factor (TF), which corrects the difference in film thickness between the sensor and the substrate due to their relative positions. Accurate determination of the TF is essential for reliable thickness measurements. It is usually determined experimentally: an arbitrary TF value (e.g. 1.0) is entered in the quartz crystal monitor, and a film of thickness T_i is deposited. The actual thickness T_m is then measured using a profilometer. The TF is calculated as T_m/T_i. Now this value of TF is fed into the quartz crystal monitor and the above experiment is repeated again to obtain the new value of TF. This process is repeated until the difference between T_m and T_i becomes negligible. Although the quartz crystal monitor does not yield the exact film thickness on the substrate, its accuracy is generally acceptable when the error is below 2%. In general, TF is expressed as

$$\text{TF}_{\text{actual}} = \text{TF}_{\text{approx}} \frac{T_m}{T_i}. \tag{4.11}$$

The TF depends on the relative positions of the substrate and sensor and varies for different geometries. When multiple substrates are deposited simultaneously, they are usually mounted on a rotating substrate holder placed at equal distances from the center, ensuring uniform deposition. In this case, TF must be recalculated, though it remains independent of the substrate holder's rotation speed.

In thermal evaporation, substrates are generally kept at room temperature. However, for improved film quality, substrate heating may be employed. Both evaporation rate and substrate temperature strongly influence the resulting film morphology: higher evaporation rates tend to produce rough films, whereas slower rates yield smoother surfaces [3].

Thermal evaporation in perovskite fabrication typically involves co-evaporating PbI_2 and CH_3NH_3I in a vacuum chamber, allowing precise control over film stoichiometry and thickness. In this process, PbI_2 and CH_3NH_3I are heated in separate crucibles, with deposition rates monitored by quartz crystal microbalances to maintain a 1:1 molar ratio and avoid non-stoichiometric phases. Hwang *et al* [4] demonstrated sequential vapor deposition of $CH_3NH_3PbI_3$ within high-aspect-ratio on 250 nm wafers, achieving conformal coverage unattainable by solution methods. Their vapor-deposited perovskite memory devices exhibited low-voltage operation, 200 ns switching speed, stable retention ($>10^5$ s), and scalability in cross-point arrays, attributes linked to suppressed ion migration due to reduced grain-boundary density [4]. In order to address dual-source calibration challenges, Fan *et al* [5] developed a single-source vapor deposition (SSPVD) approach using pre-reacted $CH_3NH_3PbI_3$ powder as the evaporation source. This room-temperature method produces high-quality perovskite films with 10.9% efficiency, full coverage, phase purity, and improved moisture stability, while eliminating post-annealing. Its compatibility with heat-sensitive substrates makes it particularly attractive for flexible and wearable devices [5]. Another advancement, hybrid chemical vapor deposition (HCVD), introduced by Leyden *et al* [6] combines thermal evaporation of PbI_2 with carrier-gas-assisted delivery of MAI. This technique achieved 11.8% efficiency and demonstrated large-area deposition on 30×30 cm^2 glass substrates with $<5\%$ thickness variation, highlighting its scalability and reproducibility for industrial applications [6].

Hybrid methods, such as vapor-assisted solution processing (VASP), offer further improvements. Zhou *et al* [7] reported a low-temperature VASP strategy, in which a spin-coated PbI_2 film is exposed to MAI vapor at 150 °C, enabling solid-state transformation into perovskite. This method achieved 16.8% efficiency, with *in situ* XRD confirming the transformation mechanism [7]. Chen *et al* [8, 9] further demonstrated VASP films with micrometer-sized grains (1–2 μm), reducing grain-boundary density by $\sim$90% compared to spin-coated films. These films exhibited smooth surfaces, full coverage, and enhanced device performance in planar hetero-junction architectures [8].

Vapor deposition is a simple technique but suffers from low precursor utilization, with up to 70% of material condensing on chamber walls rather than the substrate. This inefficiency, along with non-uniform film thickness on large-area substrates, high vacuum requirements ($<10^{-5}$ Torr), and size limitations of evaporation chambers, restricts its scalability and increases cost. Recent strategies such as spatial confinement, substrate rotation, and localized heating have improved utilization efficiency to over 40%, although atmospheric-pressure CVD variants, with plasma or aerosol-assisted precursors, are being explored to further reduce energy demand.

4.3 Solution-processed techniques

(Parts of this section have been reproduced with permission from [3]. Copyright 2016 Taylor & Francis Ltd, http://www.tandfonline.com.)

Solution-based methods are widely used for fabricating perovskite thin films, as the precursor materials can be readily dissolved in suitable solvents and deposited using low-cost processes. A variety of coating techniques have been developed, each with specific advantages. Spin coating remains the most common method that is utilized in research laboratories due to its simplicity and ability to produce high-quality films on small substrates. Other solution-processing techniques such as dip coating, doctor blading (knife-over-edge coating), and screen printing provide alternative approaches for film deposition. For scalable manufacturing, methods such as inkjet printing and slot-die coating have received considerable attention, particularly when integrated with roll-to-roll (R2R) processing. These scalable approaches are highly attractive for the production of PSCs, as they minimize material wastage, improve uniformity over large areas, and reduce the overall manufacturing costs [3, 10, 11].

4.3.1 Spin coating

Spin coating is a simple, low-cost, and widely adopted technique for thin-film fabrication. It has been employed for several decades to produce smooth and uniform coatings on planar substrates [12–14]. High-efficiency perovskite devices have been fabricated by spin coating of active layers on small-area substrates under inert conditions, typically in a N_2 glovebox [15–17]. The stepwise procedure of spin coating is illustrated in figure 4.3. The process is performed in a spin coater equipped with a rotatable vacuum chuck, on which the substrates are placed and held securely by vacuum. This ensures that the substrate remains fixed during rotation. A solution of the desired material is prepared in a suitable solvent at the required concentration and dispensed onto the substrate positioned on the chuck (figure 4.3(a)). The solution is applied on the substrate in an amount greater than what is necessary for film formation. The chuck is then rotated at a defined speed for a specific duration (figure 4.3(b)). During rotation, centrifugal force spreads the solution across the entire substrate surface (figure 4.3(c)), while an opposing viscous force acts radially inward on the liquid. Excess solution is expelled from the substrate in the form of droplets. At high rotational speeds, the centrifugal force is large enough to produce a thin and uniform coverage of the material on the substrate. Rapid spinning also accelerates solvent evaporation, leading to the solidification of the spreading film

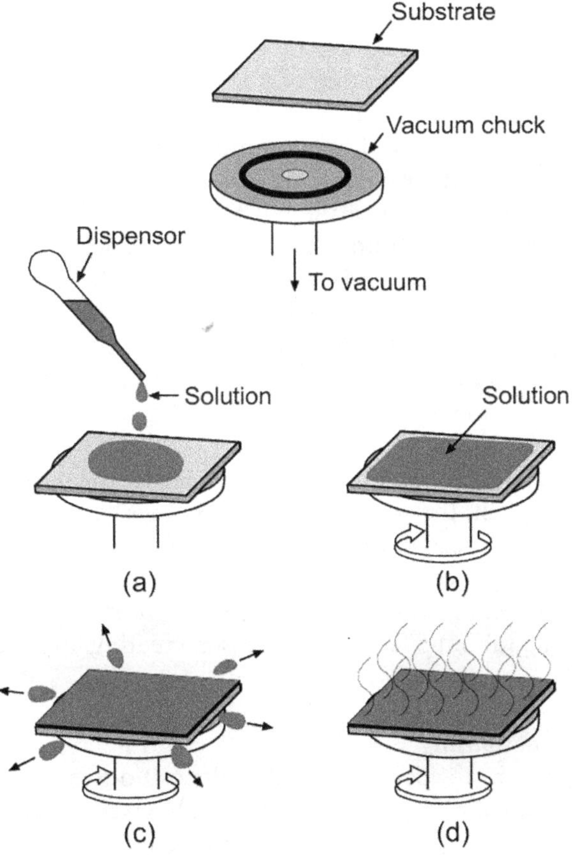

Figure 4.3. Schematic illustration of the spin-coating process: (a) solution dispensing, (b) spin-up stage, (c) spin-off stage, and (d) solvent evaporation. (Reproduced with permission from [3]. Copyright 2016 Taylor and Francis.)

(figure 4.3(d)). As the film becomes thinner, the removal rate of excess solution slows down, since thinner films offer greater resistance to flow. After spinning for a certain time, a uniform solid thin film of the deposited material is formed on the substrate. However, residual solvent may remain trapped in the solidified layer, which can be eliminated by heating the film at an appropriate temperature for a specified time. The film thickness is typically measured using a profilometer. According to Meyerhofer [18], the film thickness (d) shows $d \propto f^{-2/3}\eta^{1/3}e^{1/3}$, where f is the rotation speed, η is the viscosity of the solution and e is the solvent evaporation rate that is proportional to $f^{1/2}$. Based on this, the spin-coating process can be categorized into four main stages: (a) solution dispensing, (b) spin-up, (c) spin-off, and (d) solvent evaporation. Among these, the spin-off stage plays the most crucial role in determining the quality of the film.

Variations in the spin-coating parameters result in significant changes in the morphology of the films. Factors such as spin speed, rotation time, and substrate temperature strongly affect film quality. In addition, the solution concentration and the choice of solvent also influence the final morphology and film thickness. For instance, increasing the concentration of the solution and/or decreasing the spin speed generally results in the formation of a relatively thicker film. To achieve high-quality films with the desired thickness, it is essential to carefully optimize all these parameters for a given solvent and temperature. Spin coating has therefore proven to be an important method for preparing high-quality perovskite thin films. This process has led to the fabrication of the most efficient PSCs [3, 19].

Spin coating of perovskite thin films starts by placing a precursor solution containing usually lead iodide (PbI_2) and methylammonium iodide (CH_3NH_3I), mixed in a polar solvent like dimethylformamide (DMF) or dimethyl sulfoxide (DMSO), onto a glass or ITO/FTO coated substrate after ETL/HTL deposition. The substrate then spins at very high speeds (~2000–6000 rpm), which spreads the solution evenly due to the spinning force [20]. For obtaining high-quality films, choosing the right solvent is very important. Solvents that evaporate quickly such as diethyl ether can speed up crystal formation but might make the film less uniform. As the substrate spins fast, the solvent evaporates quickly, creating thin and smooth films. After this, the film needs to be heated (usually 100 °C–150 °C) to remove the leftover solvents and improve the growth of perovskite crystals. The kinetics of annealing, including heating ramp rate time, strongly influence grain growth. For instance, slow heating (about 5 °C min^{-1}) helps in creating larger crystals with fewer defects.

4.3.1.1 One-step spin coating

The simplest approach to perovskite film deposition is one-step spin-coating technique, in which all the required materials are mixed together in one solution as shown in figure 4.4(a) [21]. Min *et al* [22] reported epitaxial heterodimensional Sn perovskite films fabricated using one-step spin-coating, which yielded efficient light emitting diodes (LEDs) with an external quantum efficiency (EQE) of 11.6%. However, the method often suffers from incomplete surface coverage and pinhole defects, which serve as recombination centers and severely restrict the device performance. To mitigate these issues, antisolvent engineering has been developed. In this approach, a nonpolar solvent such as toluene or chlorobenzene is dropped during spinning, which extracts the excess polar solvent out and induces rapid crystallization, resulting in improved film quality. The antisolvent removes the polar solvents (DMF/DMSO) while leaving the perovskite precursors intact, creating a supersaturated medium that enhances homogeneous nucleation. Jeon *et al* [23] demonstrated that chlorobenzene treatment significantly enhances the PCE by reducing defects by more than 50%.

4.3.1.2 Two-step spin coating

Two-step spin-coating method has been proposed as a more reliable route for fabricating high-efficiency PSCs while providing better insight into crystallization dynamics through controlled phase formation [24]. In the two-step spin-coating process for perovskite films, the deposition is carried out in two stages as shown in figure 4.4(b) [21]. In the first stage, a solution of a PbI_2 precursor is dispensed onto

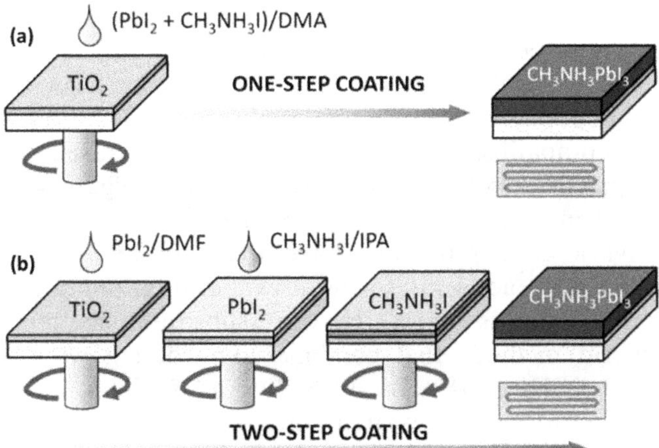

Figure 4.4. One-step and two-step spin-coating methods for making $CH_3NH_3PbI_3$ perovskite films. (a) One-step method: PbI_2 and CH_3NH_3I are mixed together in DMA solvent, then spin-coated and heated. (b) Two-step method: PbI_2 dissolved in DMF is first spin-coated and dried, followed by a second spin-coating step using CH_3NH_3I dissolved in IPA solution and then dried. (Reproduced from [21].)

the substrate and spread evenly by spinning, resulting in the formation of a uniform precursor layer. In the second stage, a solution of an organic halide such as CH_3NH_3I (MAI) or formamidinium iodide (FAI) is dropped onto the spinning substrate, where it reacts with the lead halide to form perovskite films. During this step, an antisolvent is often introduced to promote rapid crystallization and improve the film uniformity. Finally, the solvent evaporates, leaving behind a smooth and compact perovskite film that is suitable for device fabrication. By incorporating DMF into FAI/MAI solutions, perovskite conversion is enhanced during multicycle spin coating, yielding compact films with minimized defects. DMF functions as a coordination modulator, weakening Pb–I bonds and promoting MA^+/FA^+ insertion into the PbI_2 lattice, as confirmed by *in situ* grazing-incidence x-ray diffraction (GIXRD). With optimized DMF content and coating cycles, this method has achieved solar cells with a PCE of 20.1% [24] and a steady-state efficiency of 19.1%. Similarly, Cheng *et al* [25] introduced a modified two-step spin-coating approach to address the poor solubility of cesium iodide (CsI) in conventional isopropyl alcohol (IPA) based solution. By replacing IPA with a tunable ethanol/methanol mixture (7:3), CsI, FAI, and methylammonium chloride (MACl) were co-dissolved, achieving a CsI solubility of 1.2 M, six times higher than in pure IPA. This enabled stoichiometric control in mixed-cation perovskites, leading to champion PCEs of 21.17% (0.1 cm^2 device area) and 14.65% (0.25 cm^2 device area). The ethanol/methanol blend allowed controlled precursor layering, while annealing optimized the microstructure, thereby enhancing light absorption and charge-transport properties of the perovskite films. Time-resolved photoluminescence (TRPL) measurements further confirmed superior carrier dynamics, with lifetimes exceeding 800 ns in optimized two-step films compared to < 200 ns in one-step films. Controlled annealing at 150 °C for 10 min promotes α-phase perovskite formation with minimal δ-phase impurities, as validated by temperature-dependent XPS [26].

Despite its popularity at the laboratory scale, spin coating faces serious challenges for large-area fabrication and commercial applications. The process is inherently non-continuous, incompatible with flexible substrates, and wasteful, with more than 90%–98% of the precursor solution lost during spinning (figure 4.3(c)) [27–29]. To overcome these challenges, advanced strategies such as gas quenching, solvent-vapor annealing, and hybrid coating methods have been explored. Gas quenching using nitrogen flow at 50 ms^{-1} accelerates solvent evaporation uniformly, yielding 95% thickness uniformity across 15 × 15 cm^2 substrates in pilot-scale studies. Furthermore, hybrid methods that combine spin coating with slot-die coating have demonstrated PCEs of 18% on 30 × 30 cm^2 modules, offering a practical bridge between laboratory-scale research and industrial-scale perovskite manufacturing. Therefore, while spin coating remains indispensable for small-area, high-performance research devices, it is unsuitable for R2R and industrial-scale production. The problem of material wastage is completely resolved in printing processes, which are therefore considered to be the most cost-effective techniques for the commercial fabrication of PSCs.

4.3.2 Scalable manufacturing and R2R processing

As discussed in section 4.3.1, spin coating is the most common method in research due to its simplicity and ability to obtain high-quality films. However, it is not suitable for high-volume manufacturing, as it involves significant material wastage and process variability that can compromise device performance. To overcome these limitations, scalable deposition and printing techniques have been developed for cost-effective large-area fabrication. These include slot-die coating, blade coating, gravure coating, spray coating, inkjet printing, pad printing, and screen printing, each offering advantages in terms of material utilization, scalability, and compatibility with industrial processes [27].

Such techniques are particularly attractive for continuous production lines, paving the way for commercial deployment of PSCs. Other coating methods, such as curtain coating, multilayer slot-die coating, slide coating, flexographic printing, and offset printing, are also available. Further details of these techniques can be found in reference [27]. Scalable fabrication approaches, particularly R2R processing, are crucial for transitioning PSC from lab-scale prototypes to commercially viable large-area devices. This section provides a concise review of R2R processing.

R2R processing represents one of the most promising large-scale manufacturing routes, particularly for flexible perovskite devices on plastic or metal foil substrates [27, 30, 31]. Unlike discrete batch methods, continuous R2R systems shorten production time, reduce energy consumption, and allow high-throughput fabrication. In R2R process, a flexible substrate, commonly polyethylene terephthalate (PET) or polyethylene naphthalate (PEN), is unwound, sequentially coated or printed, and then rewound. Intermediate steps such as annealing, ultraviolet curing,

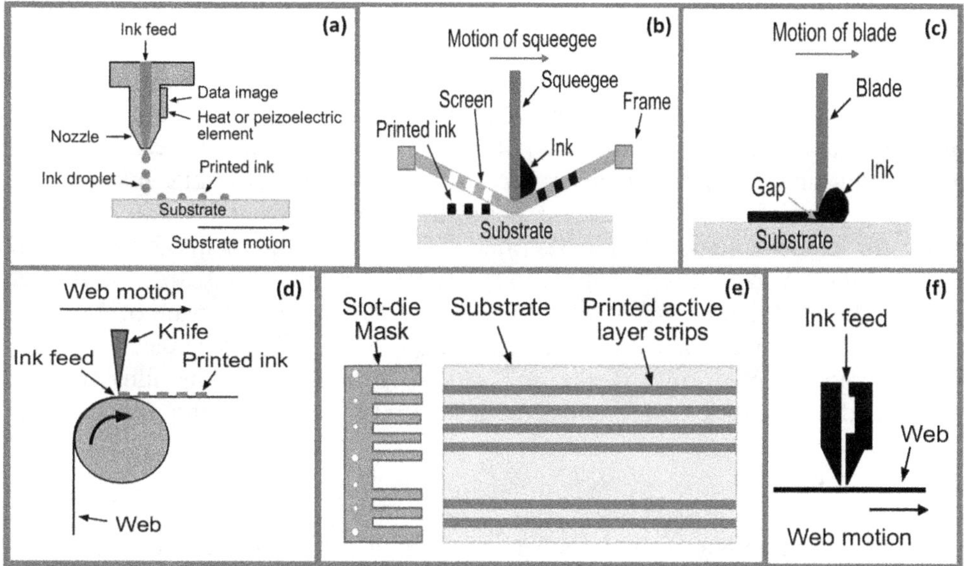

Figure 4.5. Schematic illustration of (a) inkjet printing, (b) screen printing, (c) doctor blading, (d) knife-over-edge coating, (e) slot-die coating with mask and printed substrate, and (f) slot-die coating head. (Reproduced with permission from [3]. Copyright 2016 Taylor and Francis.)

drying, and encapsulation are integrated into the production line to build complete PSC modules [28, 29, 31, 32].

As illustrated in figure 4.5, common R2R-compatible techniques include, inkjet printing, screen printing, doctor-blade coating, knife-over edge coating, slot-die coating, spray coating, gravure and meniscus coatings. These can be classified as one-dimensional (substrate motion only, e.g. doctor blade, slot-die, screen printing) or two-dimensional (substrate and coating-head motion, e.g. spray coating or inkjet printing). Selection of the method depends on the desired device structure; for example, screen printing [33] offers straightforward electrode and interconnect patterning [27, 30, 33–36].

Among flexible substrates, PET provides the most practical balance of thermal stability, strength, and flexibility, while alternatives such as Kapton (thermally stable but costly) and polylactic acid (PLA-biodegradable but solvent-sensitive) see limited use. Integrated R2R deposition lines can continuously produce complete PSC modules, greatly increasing throughput. Although early printed PSCs were less efficient than their spin-coated counterparts, the advances in scalable coating and printing methods have filled much of this gap. Continued optimization of R2R-compatible methods is expected to accelerate the transition from laboratory prototypes to industrial-scale PSC production.

4.3.2.1 Inkjet printing

Inkjet printing is a high-resolution, mask-free, and computer-controlled deposition method that allows direct digital patterning of functional materials, using solvent-resistant print heads and inks formulated from organic or hybrid systems [37–39]. It has emerged as an effective way for fabricating large-area PSCs, providing a cost-effective, scalable, and maskless deposition technique, that is compatible with meter-scale production. Operating under ambient conditions, it reduces equipment complexity and energy consumption compared to vacuum-based methods. Unlike spin coating, inkjet printing delivers picoliter droplets of perovskite ink with high precision, ensuring material-efficient deposition, large-area uniformity, and suitability for tandem devices and integrated circuits [37–39]. In inkjet process, the substrate moves in a forward direction while the print head scans transversely, releasing droplets that are generated either by mechanical compression of the ink via the piezoelectric effect or by thermal excitation, where the ink is electrostatically charged and directed toward the substrate (figure 4.5(a)). The film thickness deposited through inkjet printing can be estimated by the following equation,

$$d = N_d V_d \frac{c}{\rho}, \tag{4.12}$$

where N_d is the number of droplets per unit area, V_d is the droplet volume, c is the solute concentration, and ρ is the density of the dried material [27]. Successful inkjet printing requires inks of low viscosity and electrostatic chargeability, typically achieved using multisolvent mixtures with sufficient surface tension, although this complicates ink formulation and stability.

To improve film quality and stability, Lu *et al* [40] developed a cosolvent system combining high-boiling-point ethylene glycol (197 °C) with low-viscosity ethanol, which balances ink stability and jetting reliability through viscosity modulation. This approach enables scalable deposition of TiO_2, $SrTiO_2$, and SnO_2 ETLs, yielding SnO_2-based devices with 17.37% efficiency and low hysteresis, while $SrTiO_3/TiO_2$ heterostructures further enhance current density by aligning energy levels (TiO_2: –4.2 eV; $SrTiO_3$: −3.9 eV versus vacuum), facilitating electron extraction from the perovskite layer (−3.9 eV) [40].

Despite these advances, most studies on inkjet-printed PSCs rely on toxic solvents such as dimethylformamide (DMF) and chlorobenzene, which pose serious environmental and health risks due to their high toxicity and bioaccumulation potential. To address this, Chalkias *et al* [41] proposed a low-toxicity, inkjet-printable perovskite ink (0.8 M precursors) processed under ambient air without post-annealing, using cyrene (dihydrolevoglucosenone) as a green solvent alternative to DMF. This formulation not only suppresses coffee-ring defects, but also delivers carbon-based devices with >13% efficiency, long-term stability ($T_{95} = 1000$ h), and scalable mini-modules (100 cm^2) showing minimal degradation [41]. The use of conductive and hydrophobic carbon electrodes further enhances device stability by blocking moisture ingress and eliminating the need for costly noble metals. Beyond ink formulation, post-deposition processing plays a critical role in film quality. Rapid solvent evaporation during droplet deposition often triggers chaotic crystallization, leading to discontinuous films with point defects such as iodide vacancies. Controlled thermal annealing (100–120 °C for 10–20 min) facilitates Ostwald ripening, enlarging grain sizes from <100 nm in as-printed films to >500 nm in annealed layers, thereby reducing the defect density [42]. Strategies such as additive-assisted solvent engineering, exemplified by Gu *et al* with 1,8-diiodooctane, have been shown to suppress coffee-ring effects and enhance crystallinity [43]. More recently, Li *et al* introduced a mixed-cation perovskite ink system capable of retarding crystallization kinetics, thereby enabling smoother, pinhole-free films with improved optoelectronic properties [44].

Inkjet printing has emerged as a scalable and precise technique for fabricating PSCs, offering additive manufacturing capabilities with fine control over droplet placement and spacing at micrometer resolution [37–39]. By optimizing ink formulations, solvent engineering, and thermal treatment, this method enables the possibility to obtain uniform films with reduced defect formation, while also addressing the environmental concerns and large-area integration challenges. Its ability to sequentially deposit active layers, charge-transport layers, and electrodes facilitates streamlined fabrication of complete multilayered devices. Although first applied in polymer solar cells [37–39], the adaptability and compatibility of inkjet printing with solution-processable materials make it a promising route for PSC technologies.

4.3.2.2 Screen printing

Screen printing is a well-established technique for fabricating solar cells, particularly for depositing electrodes and interfacial layers due to its simplicity, scalability, and cost-effectiveness [37, 45–48]. In this method, a viscous ink, generally a precursor dissolved in suitable solvents with additives, is deposited through a patterned mesh screen onto the substrate. The screen, made of polyester or stainless steel, is coated with a stencil that defines the printable pattern, while impermeable areas block deposition. During printing, a squeegee spreads the ink across the stencil and forces it through the open mesh areas onto the substrate (figure 4.5(b)). Multiple functional layers, such as ETLs, HTLs and Ag/carbon electrodes, can be deposited sequentially using separate stencils and ink formulations.

The wet film thickness depends on parameters such as screen mesh volume, ink viscosity, squeegee force, speed, and snap-off distance, and the final dry film thickness can be estimated from the following equation [27]:

$$d = V_{\text{screen}} k_p \frac{c}{\rho}, \tag{4.13}$$

where V_{screen} is the volume of screen per open area, c is the concentration of ink, ρ is the density of the material in the dry film and k_p is the pick-out ratio.

Compared to spin coating, screen printing minimizes material loss and is versatile for producing different patterns with uniform coverage. Rotary screen printers are widely integrated into R2R manufacturing, where both the cylindrical screen and substrate web move synchronously while the squeegee remains fixed inside the screen figure 4.5(b). This configuration enables continuous large-area deposition and high-throughput fabrication of solar modules. Although screen printing may be less suitable for active layer deposition due to thickness control challenges, it is highly effective for large-scale, low-cost manufacturing of interfacial and electrode layers, making it an essential technique for scalable PSC production [37, 45–49].

4.3.2.3 Doctor blade coating

Doctor blade coating, also known as knife-over-roll coating, is a simple yet effective method for fabricating PSCs, well-suited to depositing thin, uniform layers of perovskite precursors and charge-transport materials on glass or flexible substrates [37, 38, 50–52]. In this process, a rigid metal/plastic blade is held at a controlled gap above the substrate; the perovskite solution is dispensed ahead of the blade and spread as the substrate (or blade) tanslates, producing a uniform wet film (figure 4.5 (c)). After drying/annealing, the final dry thickness is determined by [27]

$$d = \frac{1}{2}\left(g\frac{c}{\rho}\right), \tag{4.14}$$

where g is the blade-substrate gap, c is the solution concentration, and ρ is the density of the solid in the dry film.

The wet-film thickness typically ranges from 0.5–5 μm and is controlled by parameters such as the blade-substrate gap (10–100 μm), coating speed (1–10 cm s^{-1}), ink viscosity (10–1000 mPa.s), surface tension, substrate surface energy, and shear rate during coating [53]. Deng *et al* achieved 15% PCE on flexible substrates using low-cost doctor-blade coating process which can be compatible with the roll-to-roll fabrication process for the large-scale production of perovskite solar cell panels [53]. Compared with spin coating, it minimizes material wastage but requires careful solvent evaporation and crystallization control; otherwise, issues such as ink aggregation, premature crystallization, or rougher films may arise. Despite these challenges, blade coating is well established for scalable PSCs, including deposition of nanoparticle-based transport layers (e.g. TiO_2, ZnO), and is compatible with continuous R2R processes for large-area device manufacturing [37, 38, 50–52].

4.3.2.4 Knife-over edge coating

Knife-over edge coating is a technique similar to doctor blading, with the key difference that the knife remains stationary while the substrate moves. As illustrated in figure 4.5(d), the set-up includes an ink bath positioned in front of the knife, where the ink is delivered into the gap between the knife and the substrate. This method produces a smooth and uniform film, with film thickness and coating area determined primarily by the knife-to-web distance. The edge quality depends on both the knife-to-web gap and the substrate speed. Ink flow towards the knife is regulated using either a piston pump or a pressure tank. Notably, this process minimizes material loss, as all ink supplied to the knife is deposited onto the substrate.

4.3.2.5 Slot-die coating

Slot-die coating is a straightforward and highly efficient technique that enables material deposition in strips with virtually no material loss, making it particularly well-suited for multilayer solar cells and R2R production of perovskite solar panels where cells are serially interconnected [37, 46, 47, 54]. In this method, different layers can be aligned sequentially and coated in the same pattern by vertically adjusting the coating head relative to the web motion. The ink is delivered to the coating head using a piston pump or pressure tank. The coating head, which is the most critical yet complex component of the set-up, contains several parts, with the mask being the most essential. The mask defines the slots and voids for ink deposition and is typically fabricated from stainless steel, resistant to ink solvents and chemicals. A schematic representation of the slot-die mask and the corresponding substrate pattern, after coating, is shown in figure 4.5(e), while figure 4.5(f) depicts the overall slot-die coating-head design.

For polymer or perovskite precursor solutions with viscosities ranging between 1–20 cP, masks of 20–50 μm thickness are generally employed [27]. However, excessively high web speeds make coating unstable or impractical; therefore, the substrate must be translated within an optimized speed range to ensure uniform films. Importantly, slot-die coating allows easy pattern modification by simply replacing the mask. The film thickness is primarily determined by the web speed and ink feeding rate, and the dry film thickness can be calculated by [27]

$$d = \frac{f}{Sw}\frac{c}{\rho}, \tag{4.15}$$

where f is the ink flow rate, S is the web speed, w is the coating width, c is the solid content of the ink, and ρ is the density of the dried film.

Optimization of coating parameters, solution flow rate, slot geometry, substrate translation speed, and slot–substrate gap-enable uniform wet film formation. The viscosity of the precursor solution strongly influences the continuity of the ink meniscus and thus affects both wet and dry film thickness. Additionally, parameters such as pump pressure, web rotation speed, and meniscus stability can be finely adjusted to tune coating uniformity [37, 46, 47, 54]. Overall, slot-die coating combines high precision, excellent film uniformity, and compatibility with R2R processes, making it one of the most promising scalable techniques for fabricating large-area perovskite solar modules.

4.3.2.6 Spray coating

Spray coating is a widely explored technique for fabrication of large-area PSCs [37, 46, 47, 54, 55]. In this method, a perovskite precursor solution or nanoparticle dispersion is deposited onto a substrate (rigid or flexible) using a spray nozzle or atomizer, as illustrated in figure 4.6(a). The precursor solution must be well-mixed and homogeneous to ensure uniform film formation. During deposition, the solution is atomized into fine droplets, which are directed towards the substrate surface under controlled conditions. As the droplets impinge on the substrate, they spread, coalesce, and subsequently form a continuous thin film. The film quality strongly depends on parameters such as spray rate, distance, angle, and substrate motion, which must be optimized carefully to obtain smooth, defect-free layers with the

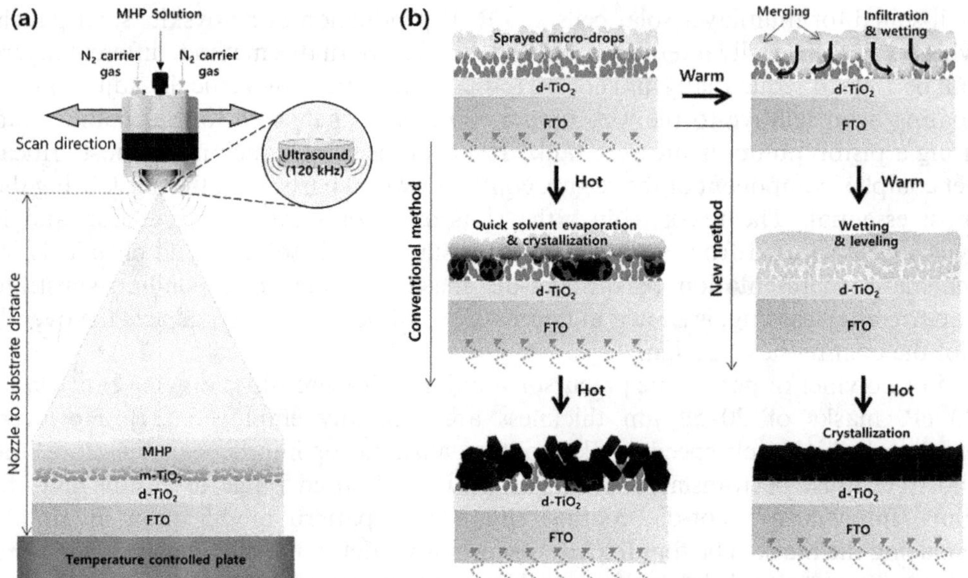

Figure 4.6. (a) Schematic of the ultrasonic spray coater used for depositing the MHP solution and (b) schematic illustration of MHP/m-TiO_2 film morphologies obtained via the conventional method compared with the new wetting-induced infiltration approach. (Reproduced from [58]. CC BY 3.0.)

desired thickness. Bishop *et al* reported spray-coated PSCs achieving 19.4% efficiency through controlled crystallization and film uniformity for high-performance devices [56]. Liang *et al* introduced formamidine acetate (FAAc) pseudo-halogen engineering to lower the annealing temperature to 150 °C, enabling spray/evaporation-coated wide-bandgap PSCs with 19.46% efficiency and improved stability (80% after 600 h at 60 °C; 90% after 1500 h under 50% RH) [57]. Kim *et al* [58] developed a green ultrasonic spray-coating technique for fabricating void-free metal halide perovskite (MHP) films on mesoscopic TiO_2 (m-TiO_2) scaffolds, providing a scalable and environmentally friendly alternative to conventional deposition processes (figure 4.6). By optimizing important processing parameters, including a substrate temperature of ~30 °C, a spray flow rate of ~11 ml h^{-1}, a nozzle–substrate distance of ~8 cm, and a metal halide perovskite (MHP) solution concentration of ~0.6 M, the authors achieved highly uniform perovskite layers. The mesoscopic PSCs, with active areas ranging from 0.096 to 100 cm^2, successfully fabricated using this approach, exhibited power conversion efficiencies of 17.14%, 16.03%, 12.93%, and 10.67%. This work demonstrates that wetting-induced ultrasonic spray coating enables efficient large-area PSC fabrication with reduced environmental impact. Despite challenges such as overspray and solvent splashing, advances in ultrasonic spray systems have enhanced droplet size control [58]. Using this approach, Zhao *et al* demonstrated high-quality $MAPbI_3$ films without anti-solvent treatment, where (2-(3,6-dimethoxy-9H-carbazol-9-yl)ethyl)phosphonic acid self assembled monolayers (MeO-2PACz SAMs) facilitated efficiencies of 20.3%

(small-area) and 18.83% (100 cm^2), highlighting the potential of ultrasonic spray coating for scalable PSCs with improved crystallization and carrier extraction [59].

Spray coating provides distinct advantages for perovskite photovoltaics, including its simplicity, compatibility with large-area and flexible substrates, and high-throughput potential. Nonetheless, precise control of deposition conditions is essential to minimize film inhomogeneity, pinholes, or roughness that can adversely affect device performance [37, 46, 47, 54, 55].

4.3.2.7 Gravure and meniscus coatings

In gravure coating, two rollers are used; the coating roller, which has engraved patterns, and a support roller that directs the web. The coating roller is partially immersed in an ink bath, where the engraved cells are filled with ink. Excess ink is then removed using a doctor blade before the roller contacts the web. Upon contact, the ink contained in the engraved cells is transferred to the web, creating a patterned coating layer. A limitation of this method is that changing the coating pattern requires replacing the entire engraved roller, making the process costly. However, gravure coating offers the advantage of producing high-quality films even with low-viscosity inks and at high web speeds.

Meniscus coating also employs two rollers, one to support and guide the web, and the other to meter the ink. The ink is supplied between the two rollers, forming a meniscus in the gap. This meniscus enables deposition of a continuous film, where the final thickness is controlled by adjusting the gap and the web speed. The rollers can rotate either in the same direction or opposite to the web movement. Meniscus coating can achieve uniform films that are comparable to those obtained by knife-over-edge coating.

4.3.3 R2R solar modules

R2R processing has emerged as a promising method for the scalable production of perovskite solar modules. This continuous manufacturing technique involves the sequential deposition of functional layers, including ETLs, perovskite active layers, HTLs, and electrodes, onto flexible substrates such as PET or PEN. These layers are typically applied using scalable techniques such as slot-die coating, doctor blading, or screen printing, which are compatible with R2R systems.

A recent study [60] reports a breakthrough in the scalable fabrication of PSCs by demonstrating the first fully ambient R2R production of modules without vacuum-based techniques. Using slot-die coating for perovskite layers and a specially formulated perovskite-compatible carbon ink, upscaled through a three-roll mill, the team fabricated and optimized several devices on an automated R2R platform. The approach achieved 15.5% efficiency for individual small-area cells and 11.0% for serially interconnected large-area modules, marking a record for fully R2R-processed PSCs. Crucially, the process shows potential for ultra-low-cost production (~\$0.70 W^{-1}) at gigawatt scale, highlighting R2R manufacturing as a viable route for commercialization of PSCs. As illustrated in figure 4.7, the process begins with tailored carbon inks that are developed for reliable automated testing (figure 4.7(a)),

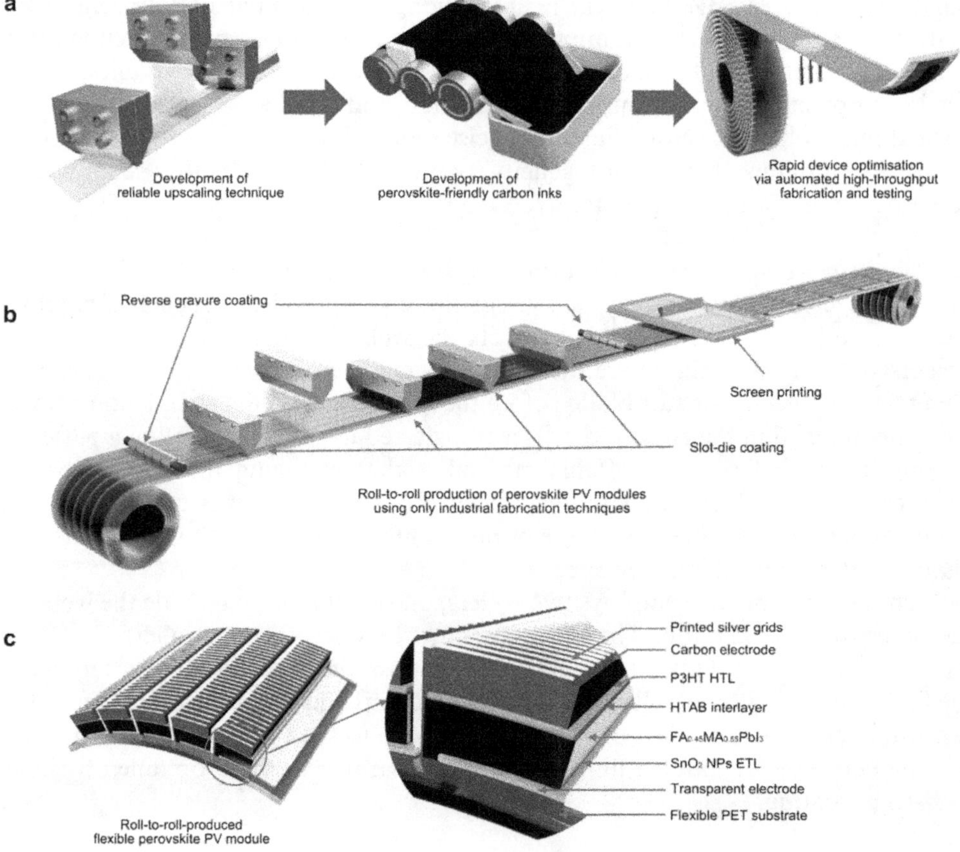

Figure 4.7. (a) A reliable slot-die coating process and development of perovskite-compatible carbon ink using a three-roll mill. (b) A schematic representation of R2R production of perovskite modules using slot-die coating, reverse gravure coating, and screen printing. (c) The detailed structure of a series-connected module, fully fabricated through R2R processing on commercially available transparent electrodes. (Reproduced from [60]. CC BY 4.0.)

followed by continuous layer deposition on flexible poly(ethylene terephthalate) (PET) substrates using reverse gravure, slot-die, and screen-printing techniques (figure 4.7(b)). The resulting module comprises a multilayer stack of transparent electrode, SnO_2 ETL, $FA_{0.45}MA_{0.55}PbI_3$ absorber, N-hexyltrimethylammonium bromide (HTAB) interlayer, poly(3-hexylthiophene) (P3HT) HTL, carbon electrode, and printed silver grids (figure 4.7(c)).

4.4 Crystallization and growth mechanisms

(Parts of this section have been reproduced with permission from [3]. Copyright 2016 Taylor & Francis Ltd, http://www.tandfonline.com.)

The crystallization process in MHPs plays a decisive role in determining device performance, as it directly affects film quality, defect density, charge-transport pathways, and interfacial contact with transport layers. In solution-processed methods, the relationship between solvent coordination, precursor reactivity, and nucleation dynamics governs film morphology and optoelectronic properties. In the widely used one-step spin-coating method, solvent evaporation progressively increases precursor concentration, eventually reaching supersaturation and triggering nucleation. However, rapid solvent escape (commonly from DMF (dimethylformamide) or DMF/DMSO (dimethyl sulfoxide) mixtures) often results in uncontrolled nucleation, heterogeneous grain sizes, and pinholes in the perovskite layer. To address this, the antisolvent method is commonly employed, where a nonpolar solvent such as toluene or chlorobenzene (CB) is dripped within a narrow time window (5–20 s post-spin). This induces abrupt supersaturation, enabling uniform nucleation and improved film coverage, although precise timing is critical to avoid film disruption.

Zheng *et al* [61] investigated the influence of precursor compositions and annealing conditions on the crystallization and growth direction of perovskite thin films. The redistribution of elements during spin-coating process of perovskite film is illustrated in figure 4.8 [61]. During deposition, the combined effect of CB dripping and extended spin time leads to an accumulation of residual solvent in the upper portion of the film after coating. In contrast, the lower region undergoes stronger precursor crystallization, resulting in relative lead enrichment. The introduction of CB facilitates the migration of Pb species within the film and enhances the formation of the intermediate precursor phase. This behavior reflects a general trend observed during the spin-coating step in the preparation of perovskite layers and plays an important role in controlling subsequent crystal growth.

The two-step sequential deposition method provides enhanced control over crystallization. The PbI_2 layer often forms a porous framework, whose crystallographic spacing can be tuned by solvent coordination. For example, DMSO coordinates with Pb^{2+} to form PbI_2.DMSO, which expands the interlayer distance from ~6.3 Å (pristine PbI_2) to ~9.4 Å. This intermediate phase facilitates cation intercalation, promoting uniform perovskite formation. Using *in situ* grazing-incidence x-ray diffraction (GIXRD), Zheng *et al* demonstrated that PbI_2.DMSO intermediates act as structural templates, aligning the perovskite ⟨110⟩ planes parallel to the substrate and reducing lattice strain during MA^+/FA^+ incorporation. This templating mechanism yields vertically aligned grains with fewer interfacial defects [62]. Crystallization kinetics is further modulated by solvent choice, annealing conditions, and additives. Slow-crystallizing solvents such as DMSO promote larger grains, whereas DMF accelerates nucleation, yielding smaller crystallites [63].

Additives such as MACl have been shown to regulate crystallization rates, passivate trap states, and improve film morphology [64, 65]. Environmental conditions such as humidity also play a role by stabilizing intermediate hydrates that can later convert to perovskite phases. Beyond solution-based routes, vapor-assisted and gas–solid methods allow highly ordered perovskite crystal growth. In these methods, precursors adsorb sequentially onto the substrate and react under controlled temperature and pressure, producing dense, oriented films with reduced grain-boundary defects [66]. However, the requirement for precise process control

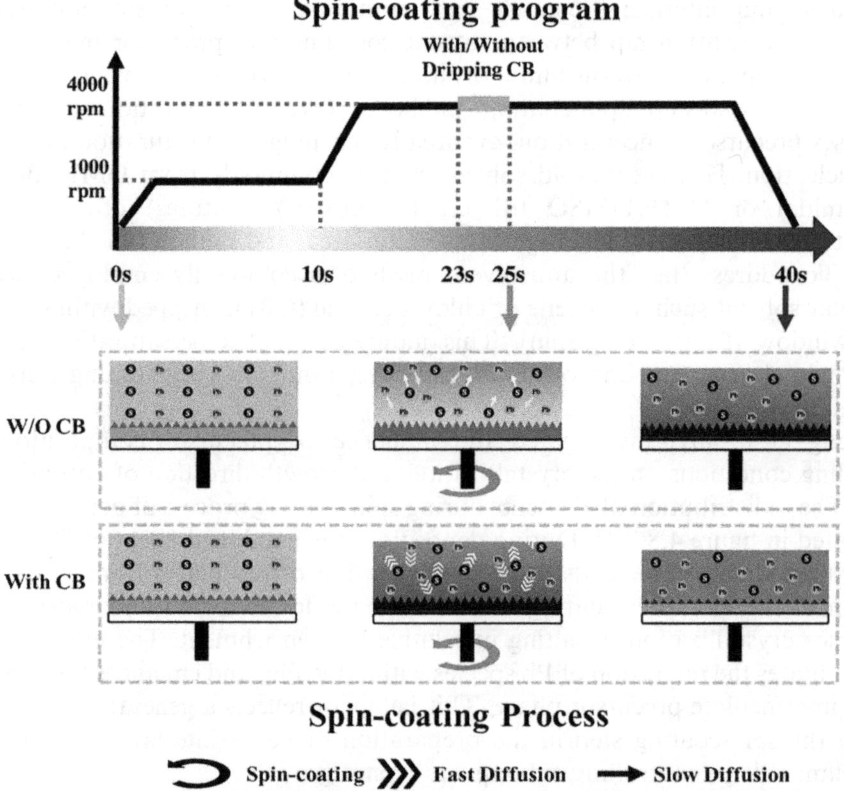

Figure 4.8. Element redistribution during spin coating. Top: Sequence of the spin-coating process. Bottom: Schematic representation of Pb and S distributions at 0, 25, and 40 s, comparing conditions without and with CB dripping. (Reproduced from [61]. CC BY 4.0.)

limits scalability compared to solution processing. Schematic representation of the crystallization mechanism in organic–inorganic halide perovskites is represented in figure 4.9 [67]. The figure schematically depicts the system at important stages. Initially, the solution is depleted of essential components: Pb^{2+} and halide ions remain confined within colloids, while MAH^+ concentration is reduced due to partial deprotonation. As proton activity increases, either through deliberate acid addition or thermal dissociation of existing acids, three major processes are triggered. First, lead halide colloids dissolve, releasing ionic species that are necessary for crystal nucleation. Second, protonation of MA enhances the concentration of MAH^+. When proton activity increases by deliberate addition to the system or by dissociation of existing acid with increasing temperatures, the following three changes occur: first, lead halide colloids are dissolved and they produce a higher concentration of the ion species required for crystal formation; second, the concentration of MAH^+ is increased because of protonation of the MA; and, third, the solvent becomes weaker because of a lower concentration of the MA. In other words, an increase in acidic medium raises the concentration of all solutes, while

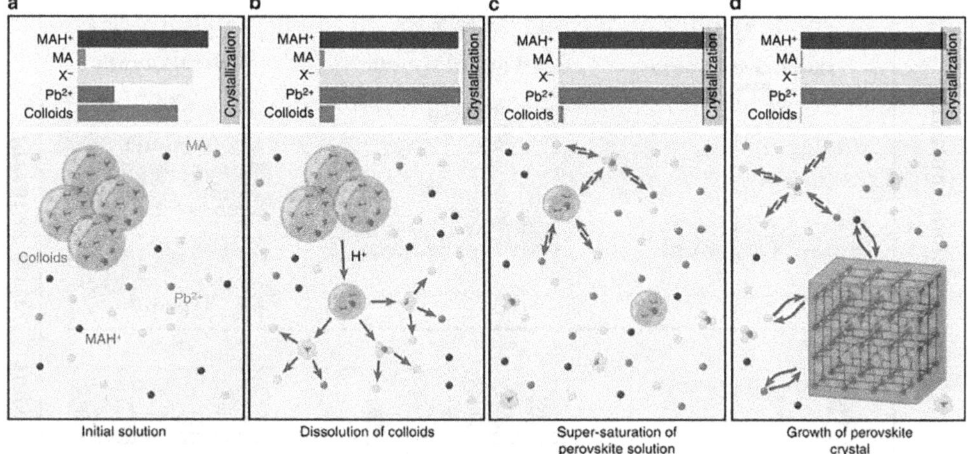

Figure 4.9. Top horizontal bars illustrate the relative concentrations of Pb^{2+}, X^-, MAH^+, MA, and colloidals (not to scale), with spheres of corresponding colors depicting each species in the lower panels. The vertical bar in the upper panel marks the supersaturation regime where crystallization occurs. (a)–(d) depict the sequential stages of crystal growth. (Reproduced from [67]. CC BY 4.0.)

simultaneously decreasing the strength of the solvent; this results in supersaturation and the onset of crystallization.

Zheng *et al* [61] also provided an overview of the stages and mechanisms of perovskite film growth during thermal annealing, as illustrated schematically in figure 4.10. The study identifies two distinct stages in this process. Stage 1 begins with the thermal decomposition of precursor crystallites consisting of solvent molecules, cations, lead, and halides. As the solvent evaporates, a rearrangement and recrystallization process occurs, leading to the formation of perovskite crystals that grow in multiple directions. The location and density of the initial nuclei or seeds, determined by the microstructure and composition of the precursor layer, play a decisive role in subsequent growth. Specifically, seeds concentrated at the top, middle, or bottom of the layer drive growth in downward, lateral, or upward directions, respectively (figure 4.10, middle panel). Moreover, the density of these initial crystallites influences the final grain size: sparsely distributed nuclei favor the formation of larger grains, which are beneficial for device performance.

In stage 2, the film growth behavior depends on the composition of the precursor and processing system. Additives are shown to influence film formation from the very beginning, tuning the size and composition of precursor aggregates and the density of the initial nuclei [68, 69]. For example, the gradual evaporation of volatile salts such as MACl or NH_4Cl can self-regulate crystal growth and enable a slower, more uniform conversion [70]. In general, additives help reduce nucleation site density, promote the formation of larger precursor crystallites, control solubility and intermediate phases, and regulate both solvent elimination and overall growth rate. During this stage, a transition toward directional grain growth occurs (figure 4.10),

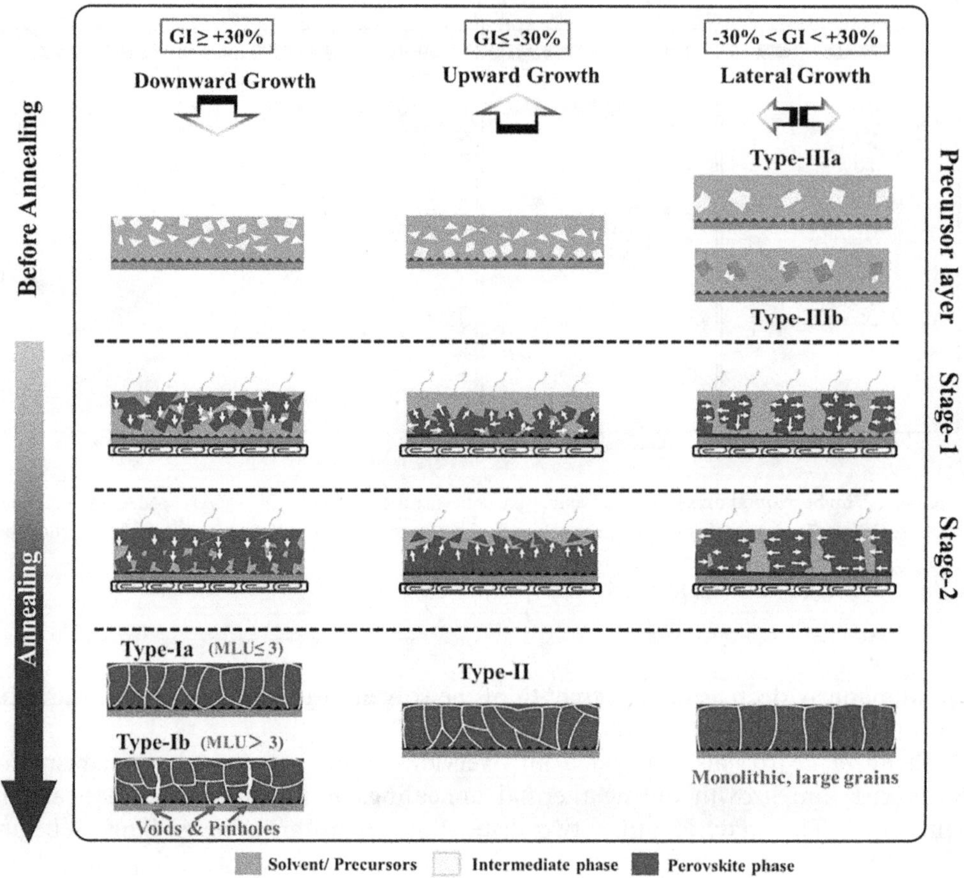

Figure 4.10. Growth models of perovskite layers: parameters and typologies. Film formation during annealing: grain growth direction, mechanisms, and growth parameters. White arrows indicate the growth directions. (Reproduced from [61]. CC BY 4.0.)

with the growth direction and speed ultimately defining the final properties of the perovskite layer.

The authors have proposed a universal typology of growth modes: type I (downward), type II (upward), and type III (lateral). If the film is dense with a rather slow growth, it is type-Ia. A special case is encountered when a surface crust is formed. The inner solvent is especially difficult to eliminate then and is entrapped in the inner part of the film. Its escape leads to the formation of voids at the bottom contact interface and pinholes (type-Ib). The second case is when the growth direction is upward (type II). Both type I and type II lead to rather small grains with multiple grain boundaries. Lateral growth (type III) was identified as the most favorable, producing large monolithic grains with smooth morphology, reduced grain boundaries, and improved charge transport. Their findings highlight the crucial role of controlled crystallization and growth direction in achieving high-quality, efficient PSCs [61].

4.5 Conclusions

Deposition techniques form the backbone of perovskite device development, from lab-scale spin coating to industrially relevant slot-die, blade, spray, inkjet, and roll-to-roll processing. While spin coating remains valuable for research, scalable methods are essential for commercialization, each offering trade-offs in cost, precision, and film quality. Vapor-phase techniques provide excellent control but face scalability and cost limitations. The future of perovskite manufacturing lies in combining scalable deposition with eco-friendly formulations and advanced crystallization control to achieve uniform, stable, and durable devices.

References

[1] Zhang C and Park N-G 2024 Materials and methods for cost-effective fabrication of perovskite photovoltaic devices *Commun. Mater.* **5** 194

[2] Krishna B G, Ghosh D S and Tiwari S 2023 Hole and electron transport materials: a review on recent progress in organic charge transport materials for efficient, stable, and scalable perovskite solar cells *Chem. Inorg. Mater.* **1** 100026

[3] Kumar P 2016 *Organic Solar Cells: Device Physics, Processing, Degradation, and Prevention* (CRC Press)

[4] Hwang B and Lee J-S 2017 A strategy to design high-density nanoscale devices utilizing vapor deposition of metal halide perovskite materials *Adv. Mater.* **29** 1701048

[5] Fan P *et al* 2016 High-performance perovskite $CH_3NH_3PbI_3$ thin films for solar cells prepared by single-source physical vapour deposition *Sci. Rep.* **6** 29910

[6] Leyden M R *et al* 2014 High performance perovskite solar cells by hybrid chemical vapor deposition *J. Mater. Chem.* A **2** 18742–5

[7] Swartwout R, Hoerantner M T and Bulović V 2019 Scalable deposition methods for large-area production of perovskite thin films *Energy Environ. Mater.* **2** 119–45

[8] Chen Q *et al* 2013 Planar heterojunction perovskite solar cells via vapor-assisted solution process *J. Am. Chem. Soc.* **136** 622–5

[9] Zhou H, Chen Q and Yang Y 2015 Vapor-assisted solution process for perovskite materials and solar cells *MRS Bull.* **40** 667–73

[10] Chen J *et al* 2025 Solution-processed coating methods for perovskite films towards large-scale photovoltaics *Sol. Energy Mater. Sol. Cells* **293** 113872

[11] Park N-G and Zhu K 2020 Scalable fabrication and coating methods for perovskite solar cells and solar modules *Nat. Rev. Mater.* **5** 333–50

[12] McCullough R D *et al* 1993 Design, synthesis and control of conducting polymer architectures: structurally homogeneous poly(3-alkylthiophenes) *J. Org. Chem.* **58** 904

[13] Li Y *et al* 2022 Recent progress in organic solar cells: a review on materials from acceptor to donor *Molecules* **27** 1800

[14] Facchetti A 2013 Polymer donor–polymer acceptor (all-polymer) solar cells *Mater. Today* **16** 123–32

[15] Hashmi G *et al* 2011 Review of materials and manufacturing options for large area flexible dye solar cells *Renew. Sustain. Energy Rev.* **15** 3717–32

[16] Service R F 2011 Outlook brightens for plastic solar cells *Science* **332** 293–3

[17] Jung E H *et al* 2019 Efficient, stable and scalable perovskite solar cells using poly (3-hexylthiophene) *Nature* **567** 511–5

[18] Meyerhofer D 1978 Characteristics of resist films produced by spinning *J. Appl. Phys.* **49** 3993–7

[19] Bahtiar A, Rahmanita S and Inayatie Y 2017 Pin-hole free perovskite film for solar cells application prepared by controlled two-step spin-coating method *IOP Conf. Ser.: Mater. Sci. Eng.* **196** 012037

[20] Wang F *et al* 2018 Materials toward the upscaling of perovskite solar cells: progress, challenges, and strategies *Adv. Funct. Mater.* **28** 1803753

[21] Im J-H, Kim H-S and Park N-G 2014 Morphology-photovoltaic property correlation in perovskite solar cells: one-step versus two-step deposition of $CH_3NH_3PbI_3$ *APL Mater.* **2** 081510

[22] Min H *et al* 2024 Spin coating epitaxial heterodimensional tin perovskites for light-emitting diodes *Nat. Nanotechnol.* **19** 632–7

[23] Jeon N J *et al* 2014 Solvent engineering for high-performance inorganic–organic hybrid perovskite solar cells *Nat. Mater.* **13** 897–903

[24] Wu J *et al* 2017 DMF as an additive in a two-step spin-coating method for 20% conversion efficiency in perovskite solar cells *ACS Appl. Mater. Interfaces* **9** 26937–47

[25] Cheng N *et al* 2023 A modified two-step sequential spin-coating method for perovskite solar cells using CsI containing organic salts in mixed ethanol/methanol solvent *Sol. Energy Mater. Sol. Cells* **250** 112107

[26] Aigbodion V S, Ako P A and Mbohwa C 2024 Two-step spin coating method: revealing the effect of thermal annealing temperature on the properties of hybrid perovskite solid state solar cells *2024 1st Int. Conf. on Smart Energy Systems and Artificial Intelligence (SESAI)*

[27] Krebs F C 2009 Fabrication and processing of polymer solar cells: a review of printing and coating techniques *Sol. Energy Mater. Sol. Cells* **93** 394–412

[28] Shen X, Hu W and Russell T P 2016 Measuring the degree of crystallinity in semicrystalline regioregular poly (3-hexylthiophene) *Macromolecules* **49** 4501–9

[29] Sahu N, Parija B and Panigrahi S 2009 Fundamental understanding and modeling of spin coating process: a review *Indian J. Phys.* **83** 493–502

[30] Rubio Arias J J *et al* 2021 Solution processing of polymer solar cells: towards continuous vacuum-free production *J. Mater. Sci., Mater. Electron.* **32** 11367–92

[31] Krebs F C 2009 All solution roll-to-roll processed polymer solar cells free from indium-tin-oxide and vacuum coating steps *Org. Electron.* **10** 761–8

[32] Zuo C *et al* 2018 One-step roll-to-roll air processed high efficiency perovskite solar cells *Nano Energy* **46** 185–92

[33] Krebs F C *et al* 2009 A complete process for production of flexible large area polymer solar cells entirely using screen printing—first public demonstration *Sol. Energy Mater. Sol. Cells* **93** 422–41

[34] Hall D B, Underhill P and Torkelson J M 1998 Spin coating of thin and ultrathin polymer films *Polym. Eng. Sci.* **38** 2039–45

[35] Du X *et al* 2019 Efficient polymer solar cells based on non-fullerene acceptors with potential device lifetime approaching 10 years *Joule* **3** 215–26

[36] Zhang Y *et al* 2018 Thermally stable all-polymer solar cells with high tolerance on blend ratios *Adv. Energy Mater.* **8** 1800029

[37] Søndergaard R R, Hösel M and Krebs F C 2013 Roll-to-roll fabrication of large area functional organic materials *J. Polym. Sci.* **B51** *16–34*

[38] Søndergaard R *et al* 2012 Roll-to-roll fabrication of polymer solar cells *Mater. Today* **15** 36–49

[39] Corzo D *et al* 2019 Digital inkjet printing of high-efficiency large-area nonfullerene organic solar cells *Adv. Mater. Technol.* **4** 1900040

[40] Lu D *et al* 2021 Inkjet-printed electron transport layers for perovskite solar cells *Materials* **14** 7525

[41] Chalkias D A *et al* 2023 Development of greener and stable inkjet-printable perovskite precursor inks for all-printed annealing-free perovskite solar mini-modules manufacturing *Small Methods* **7** 2300664

[42] Zhang Z *et al* 2023 Progress on inkjet printing technique for perovskite films and their optoelectronic and optical applications *ACS Photonics* **10** 3435–50

[43] Gu X *et al* 2018 The meniscus-guided deposition of semiconducting polymers *Nat. Commun.* **9** 534

[44] Li Z *et al* 2020 Ink engineering of inkjet printing perovskite *ACS Appl. Mater. Interfaces* **12** 39082–91

[45] Larsen-Olsen T T *et al* 2012 Roll-to-roll processed polymer tandem solar cells partially processed from water *Sol. Energy Mater. Sol. Cells* **97** 43–9

[46] Gilot J, Wienk M M and Janssen R A 2007 Double and triple junction polymer solar cells processed from solution *Appl. Phys. Lett.* **90** 143512

[47] Tanenbaum D M *et al* 2012 Edge sealing for low cost stability enhancement of roll-to-roll processed flexible polymer solar cell modules *Sol. Energy Mater. Sol. Cells* **97** 157–63

[48] Angmo D *et al* 2013 Roll-to-roll inkjet printing and photonic sintering of electrodes for ITO free polymer solar cell modules and facile product integration *Adv. Energy Mater.* **3** 172–5

[49] Krebs F C *et al* 2007 Large area plastic solar cell modules *Mater. Sci. Eng.* B **138** 106–11

[50] Kim J H *et al* 2015 Enhanced environmental stability of planar heterojunction perovskite solar cells based on blade-coating *Adv. Energy Mater.* **5** 1401229

[51] Yang Z *et al* 2015 High-performance fully printable perovskite solar cells via blade-coating technique under the ambient condition *Adv. Energy Mater.* **5** 1500328

[52] Ji G *et al* 2019 12.88% efficiency in doctor-blade coated organic solar cells through optimizing the surface morphology of a ZnO cathode buffer layer *J. Mater. Chem.* A **7** 212–20

[53] Deng Y *et al* 2015 Scalable fabrication of efficient organolead trihalide perovskite solar cells with doctor-bladed active layers *Energy Environ. Sci.* **8** 1544–50

[54] Zhong W *et al* 2019 *In situ* structure characterization in slot-die-printed all-polymer solar cells with efficiency over 9% *Sol. RRL* **3** 1900032

[55] La Notte L *et al* 2018 Fully-sprayed flexible polymer solar cells with a cellulose-graphene electrode *Mater. Today Energy.* **7** 105–12

[56] Bishop J E, Smith J A and Lidzey D G 2020 Development of spray-coated perovskite solar cells *ACS Appl. Mater. Interfaces* **12** 48237–45

[57] Liang C *et al* 2024 Low temperature method-based evaporation/spray-coating technology for wide bandgap perovskite solar cells *Mater. Today Energy.* **44** 101612

[58] Kim S S, Heo J H and Im S H 2020 Wetting-induced formation of void-free metal halide perovskite films by green ultrasonic spray coating for large-area mesoscopic perovskite solar cells *RSC Adv.* **10** 33651–61

[59] Zhao J *et al* 2024 Performance amelioration of spray-coated perovskite solar cells utilizing a self-assembled monolayer *ACS Appl. Energy Mater.* **7** 10

[60] Weerasinghe H C *et al* 2024 The first demonstration of entirely roll-to-roll fabricated perovskite solar cell modules under ambient room conditions *Nat. Commun.* **15** 1656

[61] Zheng D *et al* 2022 Control of perovskite film crystallization and growth direction to target homogeneous monolithic structures *Nat. Commun.* **13** 6655

[62] Zheng X *et al* 2017 Defect passivation in hybrid perovskite solar cells using quaternary ammonium halide anions and cations *Nat. Energy* **2** 17102

[63] Gallant B M *et al* 2024 A green solvent enables precursor phase engineering of stable formamidinium lead triiodide perovskite solar cells *Nat. Commun.* **15** 10110

[64] Zhou B *et al* 2024 Unlocking the potential of antisolvent-free perovskite solar cells: modulating crystallization and intermediates through a binary volatile additive strategy *Nano Energy* **124** 12

[65] Hoang M T *et al* 2025 Lead (II) fluoride additive modulating grains growth of water-processed metal halide perovskites for enhanced efficiency in solar cells *Mater. Futures.* **4** 025103

[66] Chen H *et al* 2017 A solvent- and vacuum-free route to large-area perovskite films for efficient solar modules *Nature* **550** 92–5

[67] Nayak P K *et al* 2016 Mechanism for rapid growth of organic–inorganic halide perovskite crystals *Nat. Commun.* **7** 13303

[68] Abdelsamie M *et al* 2021 Mechanism of additive-assisted room-temperature processing of metal halide perovskite thin films *ACS Appl. Mater. Interfaces* **13** 13212–25

[69] Song T-B *et al* 2020 Revealing the dynamics of hybrid metal halide perovskite formation via multimodal *in situ* probes *Adv. Funct. Mater.* **30** 1908337

[70] Stone K H *et al* 2018 Transformation from crystalline precursor to perovskite in $PbCl_2$-derived $MAPbI_3$ *Nat. Commun.* **9** 3458

IOP Publishing

Chapter 5

Characterization techniques

This chapter presents an overview of several important characterization techniques for determining the properties of perovskite-based materials and devices for a variety of applications. The techniques discussed include structural analysis methods such as x-ray diffraction (XRD), transmission electron microscopy (TEM), and scanning electron microscopy (SEM). Next, optical measurement techniques are discussed, focusing on UV–vis absorption spectroscopy for bandgap determination and tunability, as well as steady-state and time-resolved photoluminescence (PL, TRPL) for quantifying the radiative efficiency, trap-state densities, and carrier lifetimes. The chapter then focuses on electrical testing techniques, e.g. four-point probe and Hall effect measurements for determining the electrical conductivity and carrier mobility, respectively, electrochemical impedance spectroscopy (EIS) and space-charge-limited current (SCLC) analysis for interface and trap characterization in perovskites. For performing thermal evaluation, thermogravimetric analysis (TGA) and differential scanning calorimetry (DSC), for assessing thermal stability and transport properties under operational conditions, are briefly introduced. Finally, advanced spectroscopy methods such as Fourier transform infrared (FTIR) spectroscopy, Raman spectroscopy, transient absorption spectroscopy (TAS), x-ray photoelectron spectroscopy (XPS), and terahertz (THz) spectroscopy, which probe the strength of hydrogen bonding, dimensionality and connectivity, individual nanograins, ultrafast carrier dynamics, surface chemical evolution, and real-time crystallization and degradation phenomena during film formation and operation, respectively, are explored. Studies of real-time *in situ* characterization techniques are discussed in brief to monitor the changes in the properties of perovskites under the influence of various stimuli such as humidity, light and temperature. Representative examples from the literature illustrate how each technique contributes to critical insights into the design, optimization, and stabilization of high-performance perovskite optoelectronic devices.

doi:10.1088/978-0-7503-5427-1ch5

5.1 Introduction

From the perspective of process–properties–performance correlations, perovskites represent an excellent family of materials in which characterization plays a critical role to be able to understand, analyse and optimize such relationships. The ability to identify the degradation mechanisms and enhance the stability of perovskites will be the outcome of such studies. In addition to advancing the research and development efforts on perovskites, these studies will also lead to better understanding of perovskites from a materials and device application point-of-view and transitioning them from prototyping to manufacturing.

As discussed in chapter 4, in general, there are two main approaches to the synthesis of perovskite thin films. These methods include the following: (a) solution-based techniques such as spin-coating, spray-coating and slot-die coating and (b) vacuum-based methods such as thermal evaporation, sputtering and chemical vapor deposition (CVD). The CVD methods include low-pressure CVD (LPCVD), ambient pressure CVD (APCVD), tubular CVD, aerosol-assisted CVD (AACVD), hybrid physical CVD (HPCVD) and plasma-enhanced CVD (PECVD).

5.2 Structural analysis

Structural characterization techniques provide the critical link between the process parameters during synthesis and the resulting crystallographic, defect, and morphological features that govern the optoelectronic properties and stability of perovskites [1, 2]. Three complementary techniques, x-ray diffraction (XRD), transmission electron microscopy (TEM), and scanning electron microscopy (SEM), span length scales from meters to angstroms, enabling a full multiscale analysis.

5.2.1 X-ray diffraction

Since the discovery of x-rays by Wilhelm Conrad Röntgen on 8 November 1895, there have been numerous developments in their applications. Walter Friedrich and Paul Knipping reported the first x-ray diffraction measurements, under the direction of Max von Laue, in 1912. These measurements laid the foundation for the modern x-ray crystallography. X-ray diffraction (XRD), as shown in figure 5.1, is a versatile, non-destructive analytical method to analyse the material properties such as phase, composition, structure, texture and many more characteristics of solid samples, powder samples or even liquid samples [3, 4]. In a θ–2θ geometry, the positions of diffracted peaks obey Bragg's law ($n\lambda = 2d\sin\theta$), where λ is the wavelength of the incident x-rays, θ is the angle between the incident x-ray beam and the crystallographic planes of the sample (θ, the Bragg angle, is the specific angle of incidence at which x-rays are diffracted by a crystal lattice, that results in constructive interference and a strong reflected beam), n is an integer representing the order of diffraction, and d is the lattice spacing or the spacing between the crystal planes. When x-rays are incident onto the surface of a sample, a unique diffraction pattern that acts like a 'fingerprint' for the material is produced. This pattern is compared with reference databases (standards, e.g. The International Centre for Diffraction

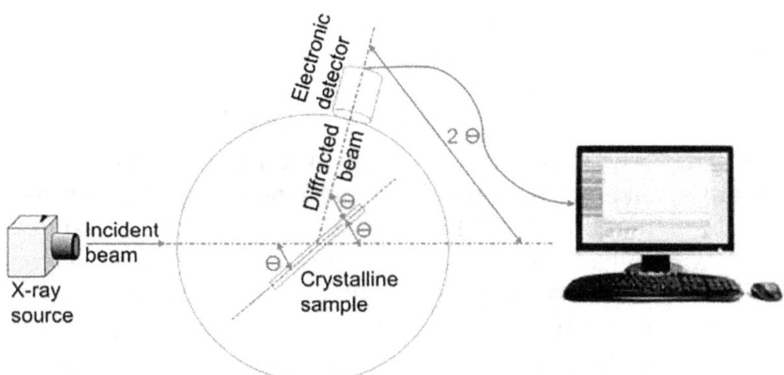

Figure 5.1. XRD instrument schematic. An incident x-ray beam shines on the surface and a film or electronic detector captures the signal as it completes an arc. (Reproduced from [16] with permission from John Wiley & Sons. Copyright 2020 Canadian Society for Chemical Engineering.)

Data (ICDD)) to identify phases, much like matching fingerprints in forensics. Basically, XRD can be categorized into the following: (a) single-crystal x-ray diffraction, in which a single crystal is rotated around various axes in order to find a large number of different reflections; (b) x-ray powder diffraction, in which the sample consists of an infinitely large number of small crystallites, ideally randomly oriented with respect to each other; and (c) special x-ray techniques such as small-angle x-ray scattering (SAXS), pair distribution function (PDF) analysis, high-resolution XRD and x-ray reflectometry (XRR).

Single-crystal x-ray diffraction serves as an advanced analytical technique for identifying the 3D atomic and molecular structure of crystalline materials. This method is utilized to determine the influence of structural distortions or phase changes in perovskite materials on carrier lifetimes. Examples include methylammonium lead iodide ($MAPbI_3$) films which exhibit sharp peaks at 14.1°, 28.4°, and 31.8°, assignable to the (110), (220), and (310) planes of the room-temperature tetragonal phase [2]. Single-crystal XRD is indispensable for understanding perovskites at the atomic level. It enables researchers to decode the intricate structures, optimize their functional properties, and design the next-generation perovskite materials for applications in energy, electronics, and catalysis.

For performing XRD of powders, the Debye and Scherrer method [5, 6] provides a quick qualitative and quantitative analysis of pure and multi-component mixtures and requires minimal sample preparation. The method involves the interaction between x-rays and the tiny crystals that are present in the powders, with no preferred orientation. The result is the form of diffraction patterns that reflect the structural physico-chemical characteristics of the tiny crystals. XRD based techniques, including conventional laboratory-based XRD and synchrotron-based grazing-incidence wide-angle x-ray scattering, are widely used to probe the microstructure of perovskite thin films for applications in photovoltaics. Beyond qualitative phase assignment, Rietveld refinement of powder diffraction data enables quantitative lattice parameter determination, site occupancy factors in mixed-cation systems, and identification of minor secondary phases such as PbI_2 or δ-phase perovskite in lead containing perovskites such as δ-$FAPbI_3$ and δ-$CsPbI_3$ [7].

The principle advantage of grazing-incidence XRD (GI-XRD) is its surface sensitivity by fixing the angle of incidence just below the critical angle for total reflection, ensuring penetration depths of only a few tens of nanometers [8, 9]. By varying this angle, one can perform depth profiling of texture, strain, and phase composition through the film thickness—even on reflective substrates in which standard θ–2θ is dominated by substrate peaks. This geometry enhances surface sensitivity, allowing precise probing of crystallographic order while minimizing bulk contributions. Key features of this technique include reciprocal space mapping, detection of Bragg rods/crystal truncation rods, and depth-dependent structural analysis via controlled penetration depth [10].

Ghadiri *et al* used XRD to investigate the moisture-induced degradation and dark-resting recovery of perovskite solar cells (PSCs) with reduced graphene oxide (rGO) as a hole-transporting layer (HTL). For instance, XRD revealed light-induced irreversible orthorhombic dominance and dynamic defect/self-healing processes, guiding future designs for stable perovskites [11]. Luo and Daoud [12] used XRD to distinguish the cubic (high-temperature) and tetragonal (low-temperature) phases of $MAPbI_3$. The cubic phase shows single peaks (e.g. (100), (200)), while the tetragonal phase exhibits peak splitting (e.g. (100) → (002)/(110)) due to lattice distortion. In addition, the incorporation of Cl^- reduces lattice defects (e.g. interstitial PbI_2) and improves crystallinity, which could be evidenced by sharper XRD peaks [12]. Pereñiguez *et al* proposed an advanced system of XRD in combination of operando XRD and x-ray absorption near edge structure (XANES) to track real-time structural dynamics during PrOx (CO oxidation in H_2-rich gas). XRD provides critical insights into the redox-driven structural dynamics of $LaCo_{0.8}Cu_{0.2}O_3$, highlighting its reversible phase transitions and the necessity of oxidation treatments to maintain catalytic functionality. The integration with XANES elucidates the interplay between bulk structure (XRD) and local metal coordination (XANES), underscoring the role of exsolution/dissolution processes in perovskite-based catalysts [13]. Common wavelengths that are utilized in XRD are summarized in table 5.1 [14, 15]. Generally, the wavelengths listed are a weighted average of the $K\alpha_1$ and $K\alpha_2$ lines; these lines are typically close to each other. $K\alpha_1$,

Table 5.1. Common XRD wavelengths and $K\beta$ filters [14, 15].

	Wavelength (Å)			
Anode	$K\alpha_1$	$K\alpha_2$	$K\beta$	$K\beta$-filter
Cr	2.289 70	2.293 61	2.084 87	V
Fe	1.936 04	1.939 98	1.756 61	Mn
Co	1.788 97	1.792 85	1.620 79	Fe
Cu	1.540 56	1.544 39	1.392 22	Ni
Mo	0.709 30	0.713 59	0.632 29	Zr

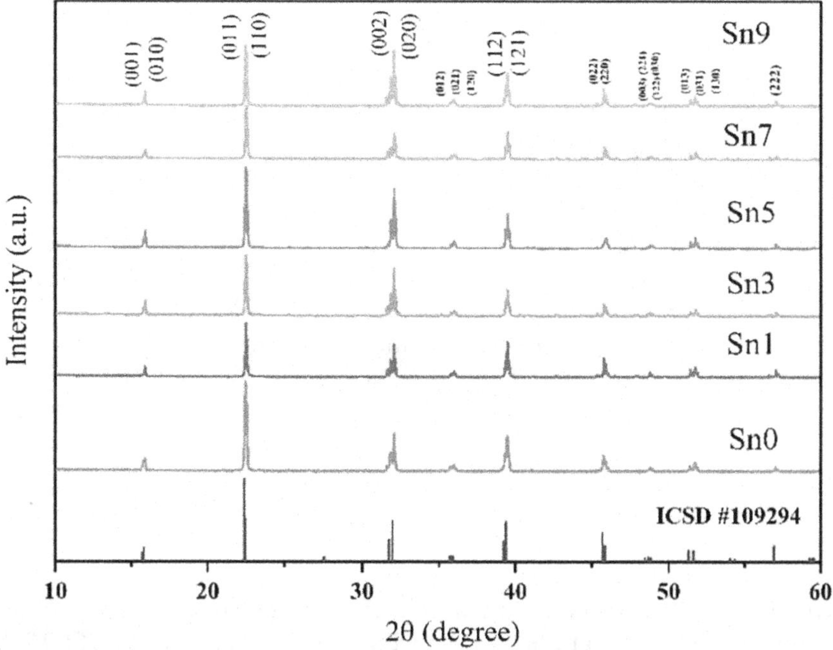

Figure 5.2. XRD patterns of tin doped $CsPbCl_3$. (Reproduced with permission from [17]. Copyright 2025 IOP Publishing Ltd. All rights reserved.)

$K\alpha_2$ and $K\beta$ represent specific characteristic x-ray emissions that are produced by the metal target (copper is commonly used; molybdenum or tungsten are also utilized) in an x-ray tube.

An illustration of the XRD spectra of Sn^{2+} doping in $CsPbCl_3$ is provided in figure 5.2 [17]. The objective of the doping of tin in this halide perovskite is to improve the charge transport and light emitting properties of $CsPbCl_3$. As can be seen in this figure, there is strong agreement of the experimentally obtained XRD spectra with that of the standard. The results demonstrate the successful incorporation of Sn^{2+} ions in the host $CsPbCl_3$ lattice [17].

5.2.2 Transmission electron microscopy

The synthesized films of perovskites exhibit differing microstructures which generally lead to the following situations: (a) varying concentrations of defects depending on the degree of crystallinity, (b) differing interfaces as function of the polycrystalline nature of the films and the underlying substrate, and (c) surface chemistry. By utilizing high energy electron beams that pass through an extremely thin sample in high vacuum or ultrahigh vacuum, TEM offers atomic-resolution imaging and diffraction capabilities that enable the study of crystal domains, dislocations, and grain boundaries. TEM provides real-space imaging down to atomic resolution ($\leq$1 Å) and complementary selected-area electron diffraction (SAED) for local crystallography [18]. High-resolution TEM (HRTEM) can resolve lattice fringes, while SAED patterns provide complementary structural information. TEM has been particularly useful in identifying degradation pathways in perovskite materials [19]. For instance, grain boundary-assisted degradation in $MAPbI_3$ under humidity can be directly visualized via TEM imaging, offering insights into the role of microstructure on stability.

In 2013, Walther *et al* [20] investigated the structure of perovskites using an aberration-corrected scanning transmission electron microscopy (STEM) to reveal a 0.8 nm mixed-layer interface in La–Ca–Mn–O/$SrTiO_3$. This was attributed to cation interdiffusion, demonstrating the role of TEM in uncovering atomic-scale interfacial phenomena. This STEM system enables precise correlation of the atomic structure (e.g. epitaxial relationships, super-periodicities) with functional properties (magnetism, conductivity). Moreover, STEM is critical for optimizing thin-film synthesis (e.g. pulsed laser deposition) by identifying interfacial diffusion, segregation, and strain effects [20]. Orum *et al* [21] investigated the topochemical conversion interface between layered perovskite $La_2Ti_2O_7$ and the ordered perovskite $Li_{0.16}La_{0.62}TiO_3$ (LLTO) using high-resolution microscopy and diffraction. These experiments revealed that the semi-reacted $La_2Ti_2O_7$ particles (heated at 1273 K with $Li_4Ti_5O_{12}$ and TiO_2) form a direct regular perovskite phase above 1173 K without intermediates [21]. The crystallographic alignment at the interface was determined as orthorhombic LLTO 001 parallel to monoclinic $La_2Ti_2O_7$ 001, with STEM showing La-rich/poor planes in LLTO oriented perpendicular to the interface, aligned with the $La_2Ti_2O_7$ (200) interlayers. In recent years, advanced *in situ* TEM technologies have emerged as an excellent characterization technique for performing real-time, nanoscale observations of the dynamic behavior of perovskite thin films. In 2023, Duan *et al* [22] reported an on-chip light-incorporated *in situ* TEM (LI^2ST) approach for probing metal halide perovskites (MHPs). The novel LI^2ST methodology allows real-time tracking of light-triggered processes, such as intragrain degradation and strain evolution, under controlled illumination. These studies reveal the dynamic mechanisms in perovskites such as bromine extraction and $PbBr_2$ formation [22].

While standard TEM (using SAED/dark-field imaging) provides crystallographic and phase distribution data, it lacks atomic-scale resolution with nanoscale features due to projection artifacts. HRTEM directly images atomic lattices in real space, resolving sub-nanometer ordered regions, defects, or interfaces that are critical in perovskites. The high resolution of HRTEM's imaging and computational analysis (e.g. Fourier filtering) enable the precise detection of localized structural distortions, strain, or domain boundaries, which standard TEM cannot reliably achieve. Pan *et al* [23] utilized HRTEM, paired with image processing, to enable precise detection and size measurement of nanoscale ordered regions in La-doped lead magnesium niobate perovskites. This approach overcomes the projection artifacts from SAED and dark-field imaging, and resolves sub-nanometer structural order [23]. Xu *et al* [24] proposed HRTEM for characterizing the structural and interfacial properties of Ce-bearing perovskite within the $CaCeTi_2O_7$-based waste-form system. The technique revealed a semi-coherent boundary between the pyrochlore and perovskite phases, indicating partial crystallographic alignment that minimizes the interfacial strain while maintaining structural compatibility between these phases. Additionally, HRTEM confirmed the absence of glassy (amorphous) phases at grain boundaries, both between pyrochlore and perovskite and between CeO_2 and pyrochlore, demonstrating phase purity and robust interfacial cohesion that is critical for long-term stability in nuclear waste forms [24].

However, halide perovskites are beam sensitive; prolonged electron irradiation induces decomposition into PbI_2, requiring low-dose protocols, cryogenic cooling stages, or pulsed-beam TEM to preserve native structures. Recent operando TEM studies, performed under controlled humidity or bias, have facilitated visualizing grain boundary migration, phase transitions, and degradation pathways in real time. Operando TEM bridges the gap between *ex situ* characterization and real-world device behavior, offering unparalleled insights into perovskite degradation. By correlating structural changes with operational stressors, it guides the design of durable perovskite-based optoelectronic devices [7].

An illustration of TEM studies of $La_{0.72}Sr_{0.18}Fe_{0.9}Ni_{0.1}O_{3-\delta}$ sample, after reduction by forming gas, is presented in figure 5.3 [25]. The FeNi particle size is within the range of 120–240 nm and is highly crystalline. Based on Fourier transform (FT) patterns, a layer of Fe_3O_4 is observed—this oxide is due to the partial oxidation of metal particles upon exposure to air [25].

5.2.3 Scanning electron microscopy

SEM is widely used to analyse the surface morphology and grain size of perovskite thin films. SEM bridges the gap between bulk XRD and atomic TEM by imaging surface topography and mesoscale morphology (10 nm–10 μm) with high throughput. High-quality SEM images reveal film uniformity, pinholes, cracks, and the distribution of grains—factors that are crucial to device performance [26]. Low-voltage SEM (1–5 kV) minimizes charging and beam damage in perovskite films, resolving grain boundaries, pinholes, and surface roughness down to a few nanometers [26]. Coupled with energy-dispersive x-ray spectroscopy (EDX), SEM maps elemental distributions, revealing halide segregation, unreacted precursors, or impurity phases that degrade performance. Cross-section SEM of focussed ion beam (FIB)-milled lamellae measures film thickness (300–500 nm for one-step $MAPbI_3$) and visualizes interfaces with transport layers, that are critical for understanding the charge extraction and recombination at buried junctions. An illustration of the SEM images of $CaMnO_3$ perovskite, as function of thermal annealing, at temperatures of 700 °C, 800 °C, 900 °C and 1000 °C, are presented in figure 5.4 [27]. With an increase in annealing temperature, the SEM images show the presence of lesser voids.

Liu *et al* [28] have utilized SEM to leverage charging effects to map phase distributions on quasi-2D perovskite surfaces, overcoming limitations of prior techniques that only identified phase *n*-values (for materials such as Ruddlesden–Popper perovskites, different inorganic layer thicknesses, called *n*-values, result in multiple phases within the film). This insight has guided compositional and antisolvent engineering, yielding uniform films with optimal phase distribution,

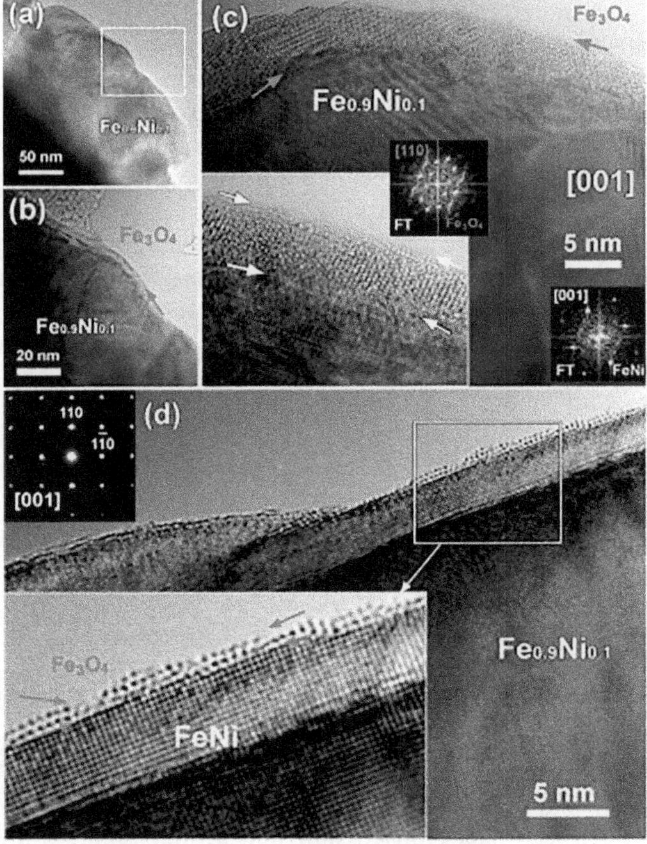

Figure 5.3. (a), (b) Low-magnification and (c), (d) high-resolution high-angle annular dark-field-scanning transmission electron microscopy (HAADF-STEM) images of singular exsolved FeNi particles along the [001] zone axis in a $La_{0.72}Sr_{0.18}Fe_{0.}$ $Ni_{0.1}O_{3-\delta}$ sample after reduction by forming gas at 800 °C for 4 h, together with the corresponding selected-area electron diffraction (SAED) and Fourier transform (FT) data. The observed Fe_3O_4 surface oxide layer is indicated by arrows. (Reproduced from [25]. Copyright The Author(s). Published by IOP Publishing Ltd. CC BY 4.0.)

enabling green perovskite light emitting diodes (PeLEDs) with >20% external quantum efficiency. SEM has proven to be vital for optimizing perovskite morphology [28].

5.3 Optical measurements

Optical characterization techniques are fundamental to understanding the electronic structure, carrier dynamics, and defect states in perovskite materials. Given the tunable bandgap (1.2–2.3 eV), strong light absorption, and exceptional photocarrier generation of perovskites, techniques such as UV–vis absorption spectroscopy and photoluminescence (PL) spectroscopy are widely used for evaluating the perovskite material quality and predicting their device performance.

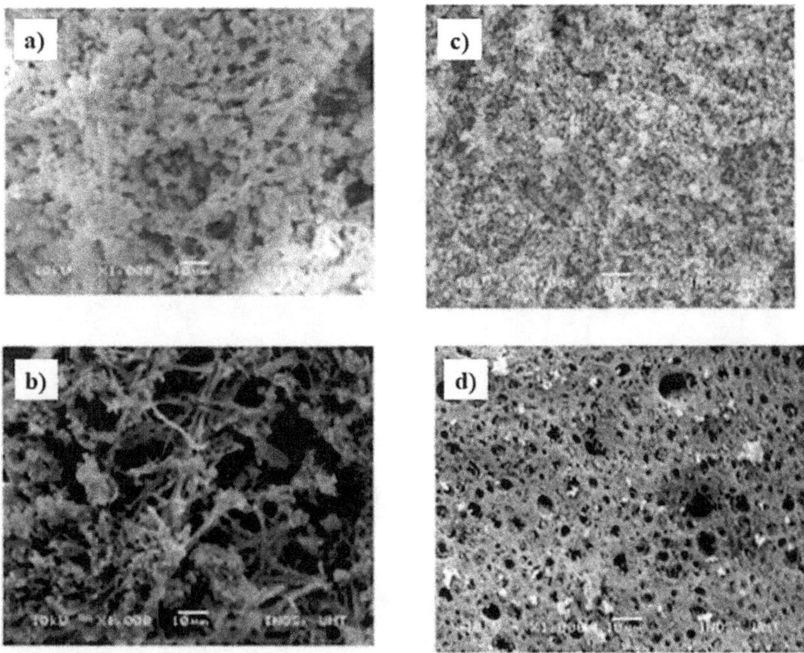

Figure 5.4. SEM images of $CaMnO_3$ after heat treatment at (a) 700 °C, (b) 800 °C, (c) 900 °C and (d) 1000 °C. (Reproduced from [27]. Copyright 2020 IOP Publishing Ltd. CC BY 3.0.)

5.3.1 UV–vis absorption spectroscopy

UV–vis absorption spectroscopy provides direct insight into the optical bandgap and absorption behavior. By analysing the absorption edge using Tauc plots, researchers can determine whether the material has a direct or indirect bandgap and estimate the bandgap energy accordingly [29]. For example, $CH_3NH_3PbI_3$ ($MAPbI_3$) typically shows an absorption onset around 750 nm, corresponding to a direct bandgap of approximately 1.55 eV. By partially substituting halides (e.g. replacing I^- with Br^-), the bandgap can be tuned to match tandem solar cell architectures. Kuznetsov *et al* [30] have utilized UV–vis diffuse reflectance spectroscopy to reveal temperature-dependent charge-transfer transitions in 1D pyridinium lead triiodide perovskite. Cooling to 100 K induces excitonic resonance (3.20 eV) and fine-structured absorption edges (3.05, 2.92, 2.78 eV), that are attributed to organic cation-involved charge-transfer transitions, highlighting its role in probing electronic interactions in hybrid perovskites [30]. Tauc plots of undoped and Sn^{2+} doped $CsPbCl_3$ is presented in figure 5.5 [17]. The optical bandgap of these perovskites is in the range of 2.865–3.040 eV [17].

Lei *et al* [31] reported that under UV–vis light, sulfate-modified perovskite (SO_4^{2-}/TBBFS) efficiently catalyzes Cr(VI) reduction. Citric acid enhances electron–hole separation, boosting kinetic rates by 9.2×. The synergy between UV–vis activation, perovskite photocatalyst, and citric acid highlights the critical role of UV–vis spectroscopy in driving Cr(VI) removal via optimized charge carrier dynamics [31].

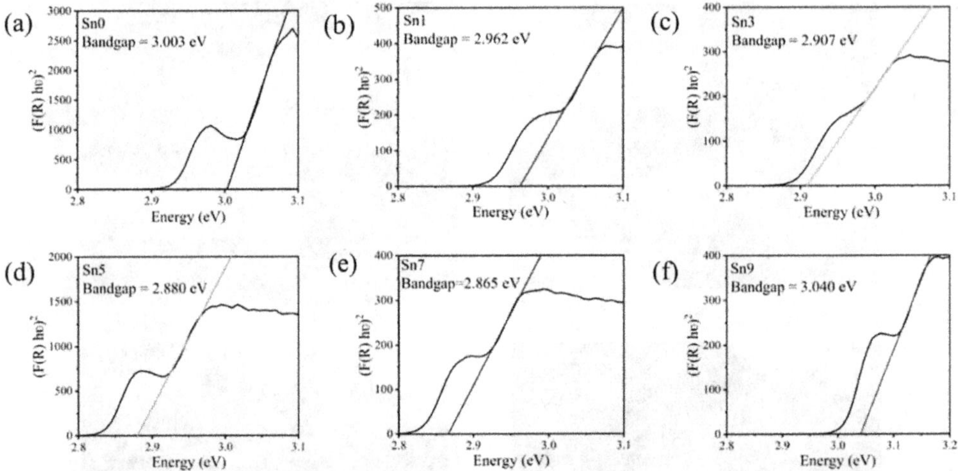

Figure 5.5. Tauc plots of undoped and Sn^{2+} doped $CsPbCl_3$. (Reproduced with permission from [17]. Copyright 2025 IOP Publishing Ltd. All rights reserved.)

Harb *et al* [32] use UV–vis diffuse reflectance spectra to validate DFT/HSE06 (density functional theory/Heyd–Scuseria–Ernzerhof screened hybrid functional version 06, a computational method used to accurately calculate the electronic and optoelectronic properties of materials)—calculated bandgap narrowing in $(Na_{2-x}Cu_x)Ta_4O_{11}$ perovskite solid solutions, revealing a red-shift from 4.58 to 2.76 eV with Cu doping. Cu-induced localized 3d states above the valence band reduces the gap, which is confirmed by UV–vis spectroscopy, but limits the hole mobility due to electronic localization [32].

5.3.2 Photoluminescence spectroscopy

PL spectroscopy is a sensitive probe for characterizing carrier recombination and defect states. A strong and sharp PL peak typically indicates low nonradiative recombination and high crystal quality. Conversely, a red-shifted or broadened PL peak often signals the presence of surface traps or defect states. Thus, PL has been used widely to differentiate different types of perovskite samples such as $CsPbBr_3$, $CsPbI_3$ and $FAPbBr_3$, especially as a function of temperature [33]. Time-resolved PL (TRPL) and temperature-dependent PL can provide further insights into carrier lifetime, trap densities, and exciton behavior, making PL a powerful tool for evaluating perovskite films and interfaces. Yamada [34] utilize TRPL spectroscopy to reveal carrier dynamics in $CH_3NH_3PbI_3$ crystals. One-photon PL red-shifts over time due to carrier diffusion from surface to interior, while two-photon PL remains static at lower energy, indicating localized states. Excitation-dependent dynamics highlight radiative recombination and diffusion effects, elucidating the optoelectronic properties of perovskites [34].

In Singh's work [35], as shown in figure 5.6, PL was pivotal in optimizing Dy^{3+}-doped $CaTiO_3$ perovskite nanophosphors for solid-state lighting applications.

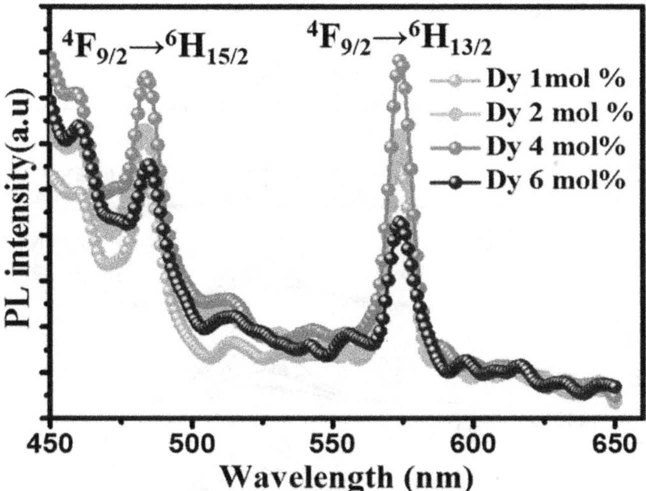

Figure 5.6. Emission spectra of Dy^{3+} doped $CaTiO_3$ nanophosphors for different concentrations of Dy^{3+} under the 386 nm excitation. (Reproduced with permission from [35]. Copyright 2017 Springer Nature.)

By analysing the excitation and emission spectra, PL identified the characteristic blue (484 nm, $^4F_9/_2 \rightarrow {}^6H_{15}/_2$) and yellow (575 nm, $^4F_9/_2 \rightarrow {}^6H_{13}/_2$) emissions of Dy^{3+} under 386 nm UV excitation, with the hypersensitive yellow transition confirming the low-symmetry sites of Dy^{3+} in the host lattice. PL intensity peaked at 4 mol% Dy^{3+}, beyond which concentration quenching occurred due to dipole–dipole interactions, as evidenced by the decay kinetics and a critical Dy^{3+}–Dy^{3+} distance of ~21.72 Å. The technique also enabled photometric evaluation, yielding Commission internationale de l'éclairage (International Commission on Illumination (CIE)) coordinates (0.28, 0.32) and a correlated color temperature (9222 K) indicative of near-white, cold light emission ideal for LEDs. Temperature-dependent PL further demonstrated thermal stability, retaining 87% intensity at 425 K. Additionally, PL excitation spectra and nephelauxetic analysis revealed covalent Dy^{3+}–O^{2-} bonding, linking structural symmetry to emission efficiency. Collectively, PL spectroscopy guided doping optimization, emission tuning, and performance validation, establishing $CaTiO_3$:Dy^{3+} as a promising, energy-efficient phosphor for advanced lighting applications [35].

In a study by Segovia *et al* [36] PL served as a critical tool for probing the humidity-induced evolution of $CH_3NH_3PbI_3$ perovskite microwires (MWs). As shown in figure 5.7, PL spectra revealed a progressive 21 nm red-shift in emission wavelength (from ~759 to ~780 nm) over 11 weeks of humidity exposure, corresponding to a reduction in the optical bandgap from 1.63 to 1.59 eV. This red-shift was attributed to structural disorder caused by H_2O infiltration into the perovskite lattice and radiative recombination via moisture-induced shallow trap states near the band edges. Concurrently, PL intensity enhancement in the initial degradation phase suggested passivation of nonradiative deep traps by water molecules, suppressing carrier recombination losses [36].

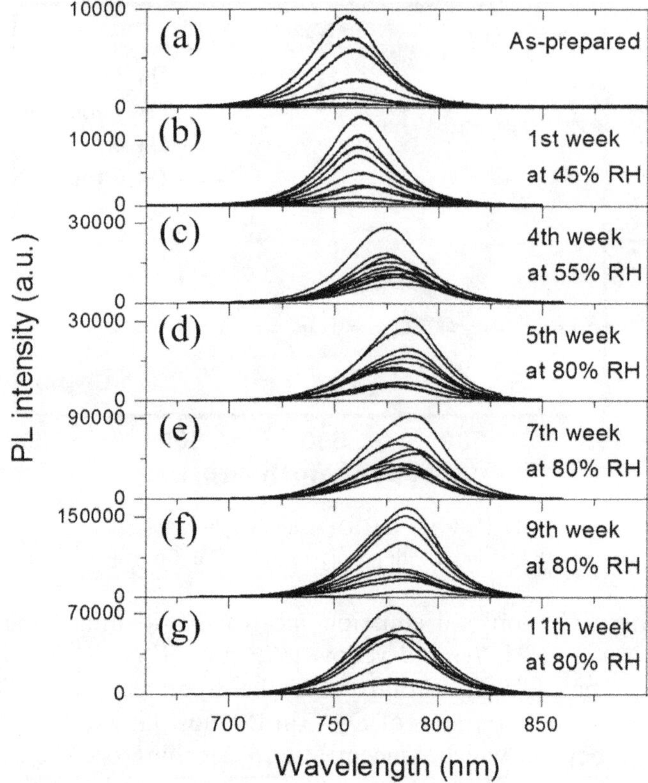

Figure 5.7. Photoluminescence spectra of $MAPbI_3$ MWs at different stages of humidity exposure. (a) As-prepared sample, after the (b) first week at 45% RH, (c) fourth week at 55% RH, and (d) fifth, (e) seventh, (f) ninth, and (g) eleventh weeks at 80% RH. All spectra taken in ambient conditions, with an excitation wavelength of 633 nm, a laser power $\sim$5 μW, an acquisition time of 10 s, and a laser spot diameter of $\sim$1 μm on the sample. (Reproduced from [36]. CC BY 4.0.)

Examples of the optical absorption edge, PL emission peak and bandgap of some perovskites are summarized in table 5.2.

5.4 Electrical and thermal testing

The understanding of the electrical and thermal properties of perovskite materials is vital for evaluating their potential in photovoltaic and optoelectronic applications. These properties are directly related to charge transport, energy efficiency, and device stability. In this section, the principal techniques that are employed to measure the electrical conductivity, carrier mobility, and thermal behavior in perovskites are discussed, offering insights into their structure–property relationships.

5.4.1 Electrical testing

The four-point probe method is a widely used technique (industry standard in semiconductors) for measuring the sheet resistance of thin films. Sheet resistance is a

Table 5.2. A summary of the optical parameters of perovskite single crystals reported in the literature [37].

	The optical absorption edge (nm)	PL emission peak (nm)	The bandgap (eV)
$MAPbCl_3$	435	440	2.88
$Cs_2AgInCl_6$	384	595	3.2
$(PEA)_2PbBr_4$	428	412	2.91
$CsPbCl_3$	–	412	3.01
BA_2PbBr_4	400	411	3.1
PMA_2PbCl_4	340	350	3.63
$MAPbBr_3$	553	545	2.24
$MAPbBr_3$	566	535	2.19
$CsPbBr_3$	561	556	2.21
$MAPbI_3$	850	781	1.45
$(FAPbI_3)_{0.85}(MAPbBr_3)_{0.15}$	825	775	1.50
$Cs_3Bi_2I_9$	550	654	2.12
$Cs_2AgBiBr_6$	625	550	1.98
$MAPb_{0.5}Sn_{0.5}I_3$	950	–	1.3
$Cs_3Bi_2I_9$	660	660	1.89
$FA_{0.6}MA_{0.4}PbI_3$	850	805	1.49
$FAPbI_3$	870	812	1.49
$DABCO\text{-}NH_4Br_3$	250	–	5.25
$MDABCO\text{-}NH_4I_3$	263	609	4.95

critical parameter for evaluating the electrical conductivity. This method reduces contact resistance errors by separating the current-carrying and voltage-sensing electrodes. In perovskite thin films, this technique helps to assess the homogeneity of the material and optimize doping levels or annealing conditions. For instance, in Ftikos *et al*'s study [38], the four-point probe method was central to evaluating the mixed electronic/ionic conductivity of Sr- and Ni-doped $LaCoO_3$ perovskites ($La_{1-x}Sr_xCo_{1-\gamma}Ni_\gamma O_{3-\delta}$). By measuring the electrical conductivity (10^2–10^3 S cm^{-1}) in air in the temperature range of 500 °C–1000 °C, the technique revealed metallic-like behavior for most compositions, with conductivity influenced by Sr/Ni codoping interdependencies. Combined with isotopic exchange ($^{18}O/^{16}O$) and SIMS analysis, oxygen self-diffusion coefficients ($D^* \sim 10^{-6}$–10^{-8} cm^2 s^{-1}) and surface-exchange kinetics ($k \sim 10^{-6}$–10^{-7} cm s^{-1}) were quantified, demonstrating that oxygen flux was surface-exchange-limited. The four-point probe data highlighted optimal doping ranges (e.g. Sr = 0.5, Ni = 0.4), balancing high electronic conductivity and oxygen mobility, properties that are critical for applications in solid oxide fuel cell cathodes. This approach linked dopant chemistry to transport properties, establishing these perovskites as efficient mixed conductors [38].

In a study performed by Cabral *et al* [39] the four-point probe method was instrumental in optimizing Ta–Si–N thin films as oxidation-resistant diffusion barriers for perovskite-integrated DRAM structures. By systematically measuring sheet resistance before and after aggressive 650 °C/30 min oxygen annealing, the technique identified compositions with post-annealling resistivities below 260 $\mu\Omega$·cm, a critical threshold to ensure functional conductivity (<10 kΩ resistance) in scaled 30 nm thick, 0.1 μm-diameter device geometries. This analysis excluded nitrogen-rich (>60 at.%) or Ta-deficient (<20 at.%) films, which exhibited excessive resistivity or oxidation-driven insulation, while highlighting optimal ternary compositions (Ta: 20–25 at.%, Si: 20–45 at.%, N: 35–60 at.%) that balanced electrical performance and oxidation resistance. The four-point probe data correlated with structural characterization, revealing that the most effective barriers retained quasi-amorphous/nanocrystalline microstructures—key to suppressing oxygen diffusion pathways—even after high-temperature processing. Additionally, through-film resistance measurements on patterned Pt/Ta–Si–N/Si stacks confirmed minimal SiO_2 formation at the barrier–substrate interface, validating the method's role in assessing interfacial stability. Temperature-dependent resistance trends further aligned with XRD and Rutherford back-scattering (RBS) findings, demonstrating suppressed Pt–Si interdiffusion and oxidation-induced phase changes. Collectively, the four-point probe enabled rigorous screening of composition–property relationships, guiding the development of Ta–Si–N barriers that met stringent requirements for perovskite dielectric integration, including thermal stability, electrical reliability, and compatibility with CMOS manufacturing processes [39].

In a related experiment [40], electrical resistivity measurements were performed on $LaNiO_3$ by utilizing platinum net electrodes that were attached to the ends of the rod samples with platinum paint. The results of the electrical resistivity of $LaNiO_3$, as function of temperature, are presented in figure 5.8 [40] The measured room temperature resistivity of $LaNiO_3$ is ~9×10^{-3} Ω cm. It increases with temperature with a slight kink at 300 °C followed by continued increase up to 800 °C. At 860 °C, the resistivity attains a maximum value of 1.5×10^{-2} Ω cm, followed by a decrease

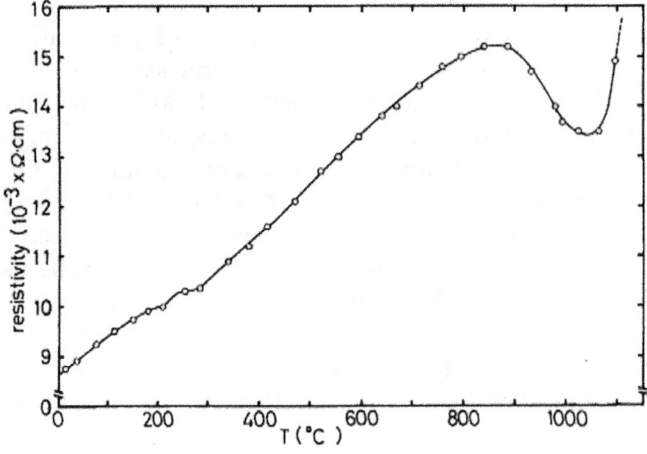

Figure 5.8. The effect of temperature on the resistivity of $LaNiO_3$. (Reproduced with permission from [40]. Copyright 1975 The Japan Society of Applied Physics. All rights reserved.)

and attaining a minimum at 1050 °C. Subsequently, the resistivity exhibits a sharp increase.

The Hall effect method provides comprehensive information on carrier concentration, mobility, and the type of dominant charge carriers (n- or p-type, electrons or holes). In perovskites, Hall measurements have revealed relatively high mobilities (~10–60 cm^2 V^{-1}·s^{-1}), which are essential for efficient solar energy conversion. For example, Shi *et al* [41] demonstrated electron mobilities of 40–60 cm^2 V^{-1}·s^{-1} in methylammonium lead iodide using this technique [41]. Ideue *et al* [42] explore the magnon (spin wave)-induced thermal Hall effect in ferromagnetic insulators, emphasizing the critical role of lattice geometry, particularly in perovskite structures. Hall effect measurements reveal that pyrochlore materials (class of inorganic compounds, general formula: AB_2O_7, e.g. $Lu_2V_2O_7$, $Ho_2V_2O_7$, $In_2Mn_2O_7$) exhibit significant thermal Hall conductivity below their Curie temperatures, driven by the Dzyaloshinskii–Moriya (DM) (antisymmetric exchange) interaction, which generates Berry curvature (intrinsic band geometric quantity—plays a critical role in anomalous transport properties) in momentum space. This effect arises from the inequivalent loops of pyrochlore lattice, enabling non-cancellation of the fictitious magnetic flux. In contrast, perovskite systems such as La_2NiMnO_6 and $YTiO_3$—with distorted structures and four transition metal ions per unit cell—show negligible or no thermal Hall signal due to symmetry-induced cancellation of Berry curvature. However, $BiMnO_3$, a perovskite with a larger unit cell (16 Mn sites) and orbital ordering, breaks this symmetry, allowing a detectable magnon Hall effect. The Hall effect measurements thus highlight how lattice geometry dictates the presence or absence of magnon-mediated thermal transport, underscoring the interplay between structural topology, DM interactions, and Berry phase effects in perovskites [42]. Hall and electrical properties of Sn^{2+} doped $CsPbCl_3$ perovskites are summarized in table 5.3

Electrochemical impedance spectroscopy (EIS) is an AC technique used to analyse charge transport, recombination, and dielectric behavior. By applying a small sinusoidal voltage and measuring the resulting current, researchers can extract equivalent circuit parameters. For PSCs, EIS helps in evaluating the interfacial resistance, ion migration, and capacitive effects, which are critical for understanding

Table 5.3. Hall and electrical parameters of pure and Sn^{2+} doped $CsPbCl_3$ perovskites [17].

Perovskite	Hall coefficient (cm^3 C^{-1}) (× 0^2)	Charge carrier mobility (cm^2 V^{-1} s^{-1})	Resistance (Ω) (× 10^5)	Resistivity (Ω cm) (× 10^7)
Sn0	−3.7	11.3	23.62	3.18
Snl	−1.8	14.8	14.25	1.92
Sn3	−1.1	29.3	11.85	1.59
Sn5	−7.02	33.4	3.56	0.47
Sn7	−15.9	56.3	2.66	0.35
Sn9	−5.5	21.2	7.94	1.07

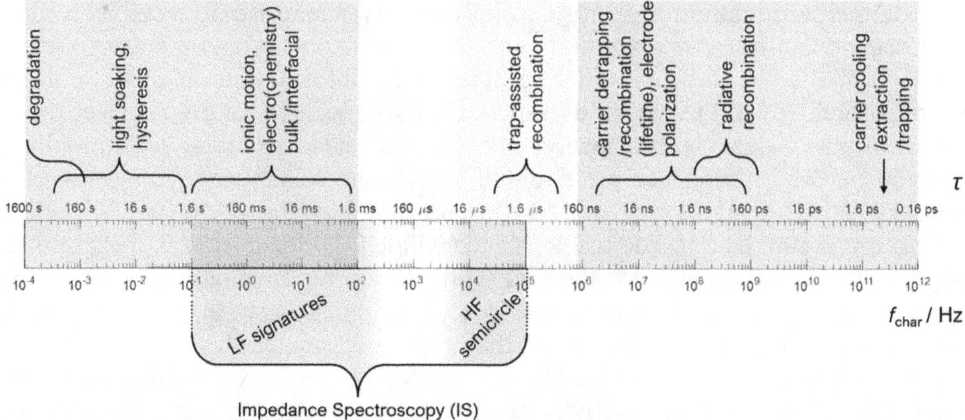

Figure 5.9. Timescales of dynamics in PSC. The top scale bar indicates the relaxation time τ of each process. The bottom scale bar indicates the corresponding frequency. The maximum measurement frequency possible in IS measurements without high-frequency compensation bridges is between 1 and 10 MHz. For the reliable characterization of dynamics, the measurement frequency should be about an order of magnitude higher than the characteristic frequency. This limits characterization at the high-frequency end of the spectrum. At the lower end, the characterization of slow processes is limited by more practical aspects such as measurement time, signal-to-noise, and device stability. (Reproduced with permission from [43]. Copyright 2022 the Royal Society of Chemistry.)

the hysteresis phenomena. Figure 5.9 illustrates the different electronic, ionic, and (electro-)chemical dynamics in PSCs and how these compare to the IS measurement window (shaded in dark blue) [43]. Torres *et al* [44] reported using EIS in unraveling the degradation mechanisms and recombination dynamics of triple-cation PSCs under ambient aging conditions. By modeling the impedance response with an equivalent circuit, EIS provided critical insights into how varying formamidinium lead iodide (FAPI, $FAPbI_3$) concentrations influenced stability and performance. The study revealed that reduced FAPI content correlated with lower recombination resistance and higher ideality factors ($n > 3.9$ for 74.5% FAPI), indicating increased defect-mediated recombination pathways due to the formation of multiple energy states. Additionally, EIS identified a decline in dielectric relaxation resistance with lower FAPI concentrations, pointing to enhanced ion vacancy formation and accelerated ionic migration, which exacerbated degradation. These findings underscore EIS as an indispensable tool for linking compositional changes to physical degradation processes, such as nonradiative recombination and ion vacancy accumulation [44].

5.4.2 Thermal testing

Thermal gravimetric analyser (TGA) is employed to study the thermal stability of perovskite materials by monitoring mass loss under controlled heating. This helps to identify decomposition temperatures and volatile species, which are critical factors for establishing the device longevity. In figure 5.10, the results of the TGA measurements of $CaMnO_3$ [27] are presented. In the temperature range of 30 °C–

200 °C, an endothermic reaction occurs leading to loss of water; the weight loss between 200 °C and 400 °C is attributed to the decomposition of excess citric acid that is present in the sample; the results exhibit slow decomposition of the removal of organic impurities from 400 °C to 600 °C. Further heating from 600 °C to 900 °C leads to the removal of the surface hydroxyls and evaporation of chemisorbed water.

Differential scanning calorimetry (DSC) measures the heat flow associated with phase transitions. In perovskites, it helps identify polymorphic transformations such as the orthorhombic-to-tetragonal transition in $MAPbI_3$ near 160 K. These transitions impact material properties such as bandgap and carrier mobility. Utilizing DSC, Lee *et al* [45] proposed a study to evaluate the thermal stability of Li–La–Ti–O-coated Li $[Ni_{0.3}Co_{0.4}Mn_{0.3}]O_2$ cathodes, emphasizing its critical role in assessing perovskite-based coatings. DSC analysis revealed that the Li–La–Ti–O coating, particularly the amorphous phase formed at 400 °C, significantly enhanced thermal stability by delaying exothermic reactions and reducing heat generation compared to uncoated electrodes. While pristine samples exhibited intense exothermic peaks at ~262 °C, coated samples showed shifted peaks (e.g. ~280 °C for amorphous coatings) and lower heat flow, indicating suppressed electrolyte–electrode reactivity. The study highlighted the utility of DSC in correlating coating phase (amorphous versus crystalline perovskite) with stability, demonstrating that amorphous Li–La–Ti–O layers effectively mitigate thermal degradation, a characteristic that is crucial for the use of high-voltage lithium battery applications [45]. Hashimoto *et al* [46] proposed a study utilizing DSC to analyse the structural phase transition of perovskite $LaCrO_3$. Simultaneous DSC–XRD measurements revealed that the orthorhombic-to-rhombohedral phase transition occurred between 250 °C and 268 °C (heating rate: 5 °C min^{-1}), accompanied by heat absorption. DSC confirmed the first-order nature of the transition, with no intermediate phases observed. The volume compression during transition was primarily attributed to shrinkage of $[CrO_6]$ octahedra, as validated by neutron diffraction [46].

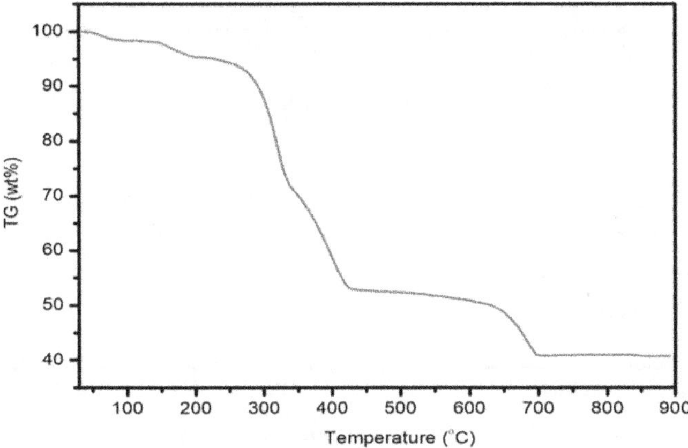

Figure 5.10. TGA Curves of the synthesized $CaMnO_3$. (Reproduced with permission from [27]. Copyright 2020 IOP Publishing Ltd. CC BY 3.0.)

5.5 Advanced spectroscopic techniques

5.5.1 Fourier transform infrared spectroscopy

Fourier transform infrared (FTIR) spectroscopy is a powerful technique that can provide details of chemical composition, degradation pathways and structural modifications in hybrid organic–inorganic perovskites. Examples of such studies include monitoring the hydration/dehydration, identifying the chemical changes, stability etc, in perovskites by detecting their vibrational modes. In providing the molecular fingerprint of the perovskite, one of the main advantages of FTIR spectroscopy is its non-destructive nature. Figure 5.11 presents an FTIR spectra of $CaMnO_3$, as function of temperature in the range of 700 °C–1000 °C. The observed band at 873 cm^{-1} is due to $CaCO_3$ and the band at 1419 cm^{-1} is due to $MnCO_3$. The strong band of $CaMnO_3$ appears at 1010 cm^{-1} and 1417 cm^{-1}. Such a study is useful for the determination of the onset of calcination of materials.

5.5.2 Raman spectroscopy

Similar to FTIR, Raman spectroscopy also provides a technique to study molecular vibrations in materials. Unlike FTIR which involves the measurement of light absorption due to a change in the dipole moment of the molecule, Raman spectroscopy involves the measurement of light scattering due to a change in its polarizability. Raman spectroscopy has been extensively utilized to characterize the crystal structure, phase, quality and stability of perovskites. A schematic of a micro-Raman set-up for polarization-resolved measurements is presented in figure 5.12 [47].

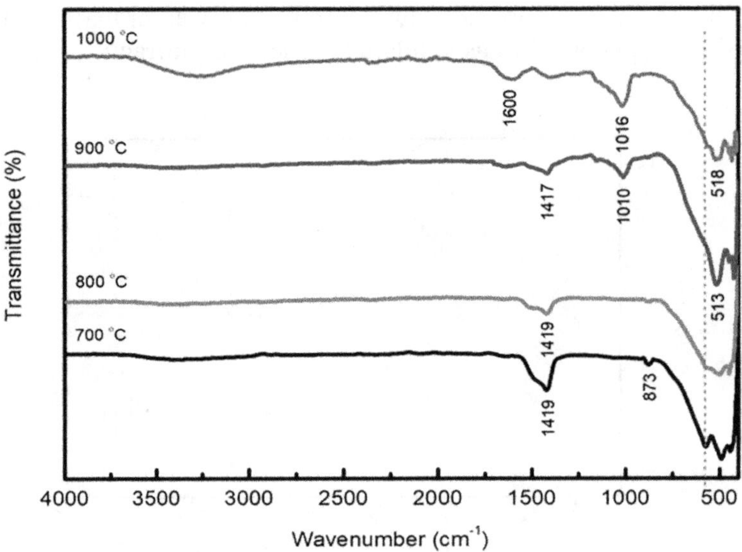

Figure 5.11. FTIR spectra of the synthesized $CaMnO_3$. (Reproduced with permission from [27]. Copyright 2020 IOP Publishing Ltd. CC BY 3.0.)

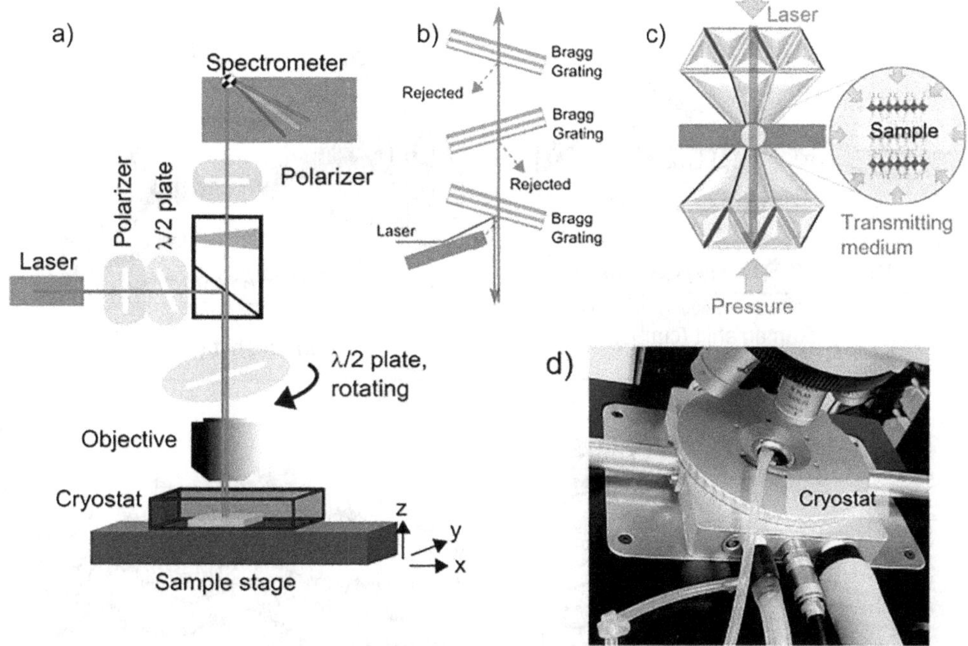

Figure 5.12. (a) Scheme of a micro-Raman set-up for polarization-resolved measurements. (b) Arrangement of filters for low Raman shift measurements. (c) Sketch of a diamond anvil for measurement as a function of pressure. (d) Photo of a liquid-nitrogen flow cryostat with optical access under an optical microscope coupled to a Raman instrument. (Reproduced with permission from [47]. Copyright 2022 The Author(s). Published by IOP Publishing Ltd. CC BY 4.0.)

An illustration of the measured Raman spectra of $La_{0.7}Sr_{0.3}Mn_{1-x}M_xO_3$ (a) Fe, (b) Co, (c) Ti, and (d) Ni is presented in figure 5.13 [48]. As can be seen in this figure, 'with increasing concentration of the metal, a strong mode appears at 670 cm^{-1}; the mode at 427 cm^{-1} gradually disappears while two broad bands appear at $\sim$500 cm^{-1} and 620 cm^{-1}. For Fe and Co doped samples, with increasing doping, the peak at 600 cm^{-1} up-shifts, whereas the peak downshifts in case of Ti and Ni' [48].

A summary of Raman spectroscopy studies on 3D hybrid organic–inorganic metal halide perovskites (HOIPs) indicating the compound used, details of the technique and available information regarding the assignment of modes and phase transitions, in the literature, is presented in table 5.4 [47].

5.5.3 Transient absorption spectroscopy

Transient absorption spectroscopy (TAS) probes ultrafast carrier dynamics, such as hot-carrier cooling, exciton dissociation, and charge-transfer processes. By tracking transient absorption signals across picosecond to millisecond timescales, TAS quantifies carrier mobilities and trap densities.

Wang *et al* proposed a study leveraging TAS to elucidate charge carrier dynamics in $CH_3NH_3PbI_3$ perovskite films deposited with and without TiO_2 under visible light

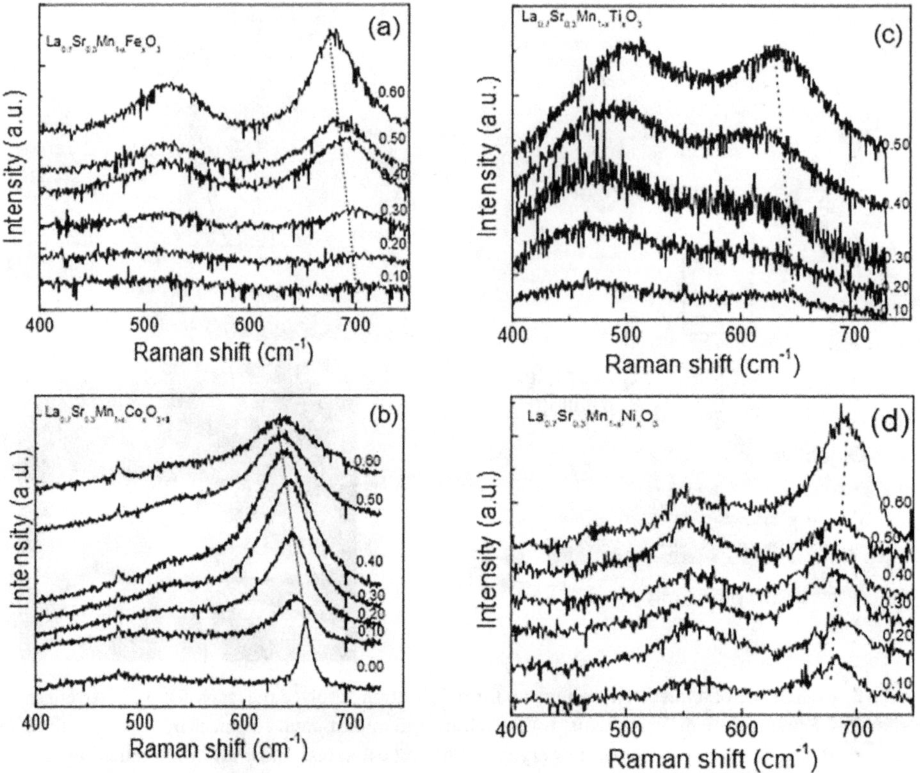

Figure 5.13. Raman spectra of $La_{0.7}Sr_{0.3}Mn_{1-x}M_xO_3$ (a) Fe, (b) Co, (c) Ti, and (d) Ni. (Reproduced with permission from [48]. Copyright 2009 The Author(s). Published by IOP Publishing Ltd. CC BY 4.0.)

(450–800 nm). TAS revealed electron injection from perovskite to TiO_2 via transient bleaching at ~750 nm, while distinguishing this signal from red-shifted fluorescence at ~780 nm. A distinct bleach feature at ~510 nm identified the presence of PbI_2, with its amplitude correlating to PbI_2 content. Films with less PbI_2 exhibited faster carrier relaxation due to grain boundary trapping, whereas PbI_2-rich samples showed slower dynamics, attributed to defect passivation. TAS thus critically uncovered structural and compositional influences on charge behavior, aligning with mechanisms of perovskite performance degradation and stabilization [49]. Jo *et al* [50] elucidate the role of atmospheric degradation on ultrafast carrier dynamics in $Cs_{0.05}(FAPbI_3)_{0.85}(MAPbBr_3)_{0.15}$ perovskite films by using TAS. TAS revealed three decay modes—carrier trapping (tens of ps), extraction (hundreds of ps), and recombination (ns), by analysing transient bleaching at band-edge transitions. Atmospheric exposure slowed carrier extraction and increased recombination, attributed to defect formation at grain boundaries. Furthermore, it demonstrated that higher carrier densities accelerated extraction rates, partially counteracting degradation effects. By correlating decay lifetimes and coefficients with exposure time, TAS identified reduced extraction efficiency and enhanced trapping as the key degradation mechanism [50].

Table 5.4. Summary of Raman spectroscopy studies on 3D HOIPs indicating the compound used, details of the technique and available information about assignment of modes and phase transitions [47].

Compound	Details of Raman measurement	Available information
$MAPbI_3$	Room temperature	MA mode assignment, comparison with DFT
$MAPbI_3$ $MAPbI_{3-x}Br_x$ ($x = 0$–1)	Room temperature low frequency	Inorganic peaks, effect of composition, comparison with Pb salts
$MAPbI_3$	Different excitation laser	Deconvolution of modes associated to organic and inorganic part; effect of laser exposure
$MAPbI_3$	Pressure dependent	Phase transition with pressure
$MAPbI_3$	Temperature dependent Polarization	Detailed mode assignment by comparison with DFT
$MAPbI_3$	Temperature dependent Polarization angle dependent	Mode assignment from polarization dependence
$MAPbX_3$ (X = I, Br, Cl)	Temperature dependent	Detailed mode assignment and effect of halide, comparison with DFT
$MAPbBr_3$ $CsPbBr_3$	Low frequency temperature dependent	Phase transition
$FAPbI_3$ $FAPbBr_3$	Low frequency	Detailed mode assignment and effect of composition, comparison with DFT
$Cs_2AgBiBr_6$	Low temperature	Mode assignment
$Cs_2AgBiBr_6$ $CsPbBr_3$	Low frequency temperature dependent Polarization angle dependent	Phase transition and mode symmetry, comparison single and double perovskites

5.5.4 X-ray photoelectron spectroscopy

X-ray photoelectron spectroscopy (XPS) is essential for analysing surface chemistry and elemental composition. By probing core-level electron binding energies, XPS identifies oxidation states and chemical degradation products.

Ning *et al* [51] proposed that coating $CH_3NH_3PbI_3$ with 4,4′-dibromotriphenylamine (DBTPA) enhances thermal stability, validated via XPS. Near-ambient pressure XPS revealed that DBTPA suppresses iodide/methylamine loss under heat (150 °C) and humidity by forming Lewis acid–base and cation−π interactions with perovskite surfaces. This passivation improved device stability (96% efficiency retention after 1000 h at 85 °C) and efficiency (20.6%), highlighting the role of XPS in probing degradation mechanisms and interfacial chemistry [51]. Reo *et al* [52] proposed a vapor-deposited $CsSnI_3$-based perovskite transistor using XPS to elucidate the role of chlorine in film optimization. XPS depth profiling revealed uniform Pb distribution and surface Cl residue, indicating partial sublimation during annealing. This facilitated solid-state reactions, enhancing perovskite crystallinity and reducing defects. The optimized $CsSnI_3$:$PbCl_2$ films achieved high hole mobility (33.8 $cm^2 V^{-1} s^{-1}$) and stability, demonstrating the critical role of XPS in correlating Cl dynamics with improved device performance and large-scale uniformity [52]. Moreover, in contrast to traditional XPS that measures static chemical states after sample exposure to environmental factors, *in situ* XPS facilitates real-time analysis during x-ray exposure or environmental treatment within a controlled ultrahigh vacuum (UHV) system. Lin *et al* [53] utilized *in situ* XPS, under ultrahigh vacuum, to probe real-time x-ray-induced degradation of $CH_3NH_3PbI_3$. Unlike conventional XPS, which captures static post-exposure states (e.g. $Pb^{2+} \rightarrow Pb^{\circ}$), *in situ* XPS dynamically revealed rapid nitrogen loss and Pb° formation during irradiation, highlighting x-ray-driven decomposition [53].

5.5.5 Ultrafast terahertz spectroscopy

Ultrafast terahertz (THz) spectroscopy measures low-frequency lattice vibrations and carrier mobilities with sub-picosecond resolution. Luo *et al* [54] proposed that THz enables direct observation of excitonic Rydberg states and electronic coherence dynamics in organometal halide perovskites (e.g. $CH_3NH_3PbI_3$). By probing the internal quantum transitions (e.g. 1s-2p, 1s-3p), this technique revealed an intermediate exciton binding energy (~13.5 meV) and a delayed build-up of incoherent excitons under resonant and off-resonant pumping. THz spectroscopy identified coexisting excitons, carriers, and discrete THz phonons, distinguishing electronic coherence loss (~1 ps dephasing time) from hot-state cooling processes. The study highlighted perovskite-specific interactions, such as efficient electron–phonon scattering and suppressed long-range charge separation, offering insights into quantum dynamics critical for light-harvesting and energy conversion applications [54]. Gong *et al* [55] proposed that THz serves as a pivotal tool for unraveling carrier dynamics in perovskite optoelectronics. By probing hot-carrier cooling, interfacial transport, and spin relaxation with sub-picosecond resolution, THz spectroscopy captures real-time interactions between carriers, phonons, and lattice vibrations, as illustrated in figure 5.14. It quantifies key parameters such as carrier mobility, recombination rates, and phonon scattering, revealing mechanisms such as the hot phonon bottleneck that delays energy loss. This technique provides critical insights into perovskite-specific phenomena, such as efficient charge extraction and spin coherence preservation, guiding the design of high-performance solar cells and spintronic devices [55].

In figure 5.15, a schematic describing the correlation between perovskite single crystal, its physical properties and applications is presented. The methods for the growth of single crystals of perovskites are generally classified along the following lines: (a) temperature cooling cystallization, (b) inverse temperature crystallization and (c) antisolvent vapor-assisted crystallization.

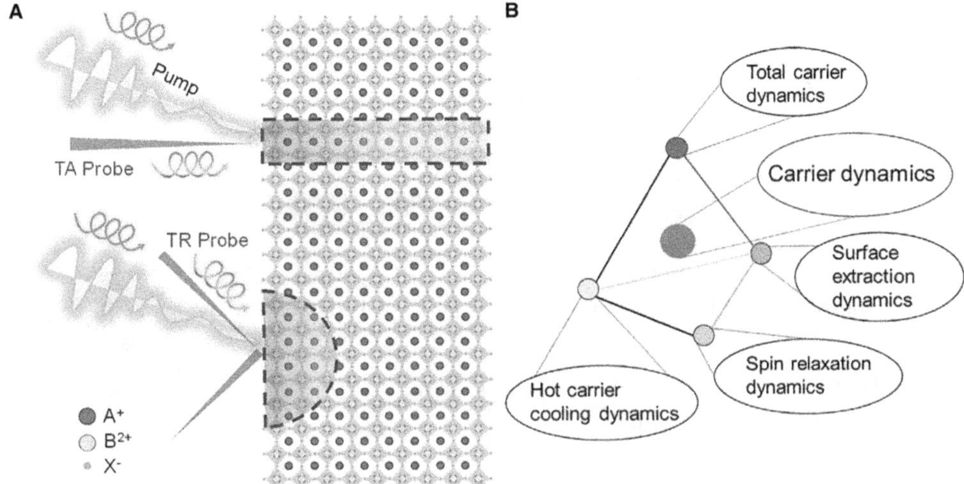

Figure 5.14. TAS measurement and carrier dynamics for perovskite-based optoelectronic device. (A) General scheme for two modes of transient spectroscopic techniques. (B) The carrier dynamics include the total carrier dynamics, hot-carrier cooling dynamics, surface extraction dynamics, and spin relaxation dynamics. The above contents correspond to the bulk carrier recombination lifetime, hot-carrier cooling mechanism, interfacial carrier transport configuration between charge transport layer and perovskite layer, and carrier spin relaxation lifetime, respectively. (Reproduced with permission from [55]. Copyright 2023 Elsevier.)

5.6 Conclusion

Characterization techniques for determining the properties of perovskite materials are indispensable for unraveling the intricate relationships between synthesis, structure, and functional properties. Structural methods such as XRD, TEM, and SEM provide multiscale insights into crystallographic phases, atomic-scale defects, and mesoscale morphology, forming the foundation for understanding the optoelectronic performance and degradation pathways in perovskites.

Optical techniques decode the electronic band structures, carrier dynamics, and defect-mediated recombination while providing the strategies for bandgap engineering and defect passivation. Electrical and thermal analyses bridge material properties to device functionality by quantifying charge transport, ion migration, and thermal stability under operational stressors.

Advanced spectroscopies probe ultrafast carrier dynamics, surface chemistry, and real-time structural evolution, offering unprecedented resolution of degradation mechanisms and interfacial phenomena. Integrating these techniques enables a holistic approach to material optimization, in which atomic-scale insights inform macroscopic device design. For instance, correlating *in situ* TEM with impedance spectroscopy reveals how grain boundaries govern both ion migration and recombination losses.

Ultimately, the synergy of structural, optical, electrical, and advanced characterization techniques accelerates the development of robust, high-efficiency perovskites for solar cells, LEDs, and beyond. By coupling multiscale analysis with dynamic

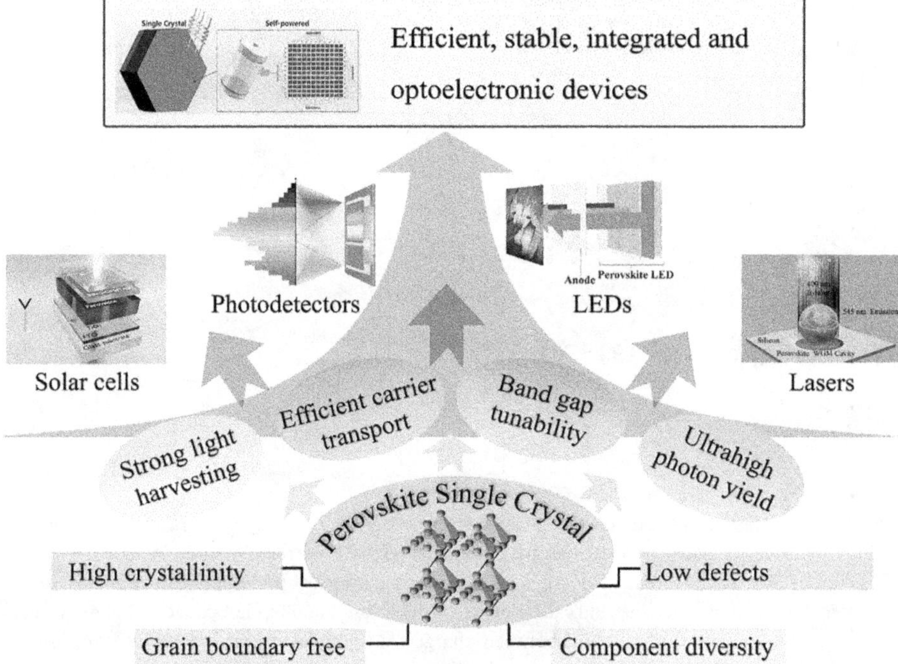

Figure 5.15. Schematic demonstrating how the fundamental physical properties of single crystals affect the optoelectronic application. ('Solar cells' reproduced from [9]. CC BY-NC. 'Photodetectors' reproduced from [11]. CC BY 4.0. 'LEDs' reproduced with permission from [15]. Copyright Springer Nature. 'Lasers' reproduced with permission from [18]. Copyright 2017 American Chemical Society. 'Efficient, stable, integrated and micro-nano optoelectronic devices' reproduced from [12]. CC BY 4.0. Reproduced from [37]. Copyright 2023 The Author(s). Published by IOP Publishing Ltd on behalf of the Songshan Lake Materials Laboratory. CC BY 4.0.)

operando studies, researchers can address stability challenges and unlock new frontiers in perovskite-based optoelectronics.

References

[1] Gupta S, Deepshikha , Chaudhary R, Singh S and Singh 2022 Review: Perovskite structure, characterization, & synthetic methods *Int. J. Mech. Eng.* **7** 51–9

[2] Zhou Y, Zhou H, Deng J, Cha W and Cai Z 2020 Decisive structural and functional characterization of halide perovskites with synchrotron *Matter* **2** 360–77

[3] Quinn P and Benzonelli A 2018 XRD and materials analysis *The Encyclopedia of Archaeological Sciences* (John Wiley and Sons)

[4] Tan W L and McNeill C R 2022 X-ray diffraction of photovoltaic perovskites: principles and applications *Appl. Phys. Rev.* **9** 021310

[5] Holzwarth U and Gibson N 2011 The Scherrer equation versus the 'Debye–Scherrer equation' *Nat. Nanotechnol.* **6** 534–4

[6] Debye P and Scherrer P 1916 Interference of x-rays, employing amorphous substances *Phys. Z.* **17** 277–83

[7] He B *et al* 2023 *In situ* and *operando* characterization techniques in stability study of perovskite-based devices *Nanomaterials* **13** 1983

[8] Hsieh C-M, Yu Y-L, Chen C-P and Chuang S-C 2017 Effects of the additives: N-propylammonium or n-butylammonium iodide on the performance of perovskite solar cells *RSC Adv.* **7** 55986–92

[9] Qureshi M W, Ma X, Tang G, Miao B and Junbo N 2021 Fabrication and mechanical properties of Cr_2AlC MAX phase coatings on TiBw/Ti_6Al_4V composite prepared by HiPIMS *Materials* **14** 826

[10] Werzer O *et al* 2024 X-ray diffraction under grazing incidence conditions *Nat. Rev. Methods Primers* **4** 15

[11] Ghadiri M, Kang A K and Gorji N E 2020 XRD characterization of graphene-contacted perovskite solar cells: moisture degradation and dark-resting recovery *Superlatt. Microstruct.* **146** 106677

[12] Luo S and Daoud W A 2016 Crystal structure formation of $CH_3NH_3PbI_{3-x}Cl_x$ perovskite *Materials* **9** 123

[13] Pereñíguez R and Ferri D 2024 *In situ* XRD and operando XRD-XANES study of the regeneration of $LaCo_{0.8}Cu_{0.2}O_3$ perovskite for preferential oxidation of CO *Mater. Today Sustain.* **27** 100867

[14] Kabekkodu S, Dosen A and Blanton T 2024 PDF-5+ a comprehensive Powder Diffraction File™ for materials characterization *Powder Diffrac.* **39** 47–59

[15] Wilson A J C 2013 *International Tables for Crystallography* vol C (International Union of Crystallography)

[16] Khan H *et al* 2020 Experimental methods in chemical engineering: x-ray diffraction spectroscopy—XRD *Can. J. Chem. Eng.* **98** 1255–66

[17] Sharma D and Sharma S K 2025 Influence of Sn^{2+} doping to improve the charge transport and light-emitting properties of $CsPbCl_3$ perovskites *J. Phys. Condens. Matter* **37** 075704

[18] Villa-Bracamonte M F, Montes-Bojorquez J R and Ayon A A 2024 Optical properties study of a perovskite solar cell film stack by spectroscopic ellipsometry and spectrophotometry *Results Opt.* **15** 100640

[19] Kumar A, Gupta S K, Dhamaniya B P, Pathak S K and Karak S 2023 Understanding the origin of defect states, their nature, and effects on metal halide perovskite solar cells *Mater. Today Energy* **37** 101400

[20] Walther T and Ross I M 2013 Aberration corrected high-resolution transmission and scanning transmission electron microscopy of thin perovskite layers *Phys. Proc.* **40** 49–55

[21] Orum A *et al* 2016 Transmission electron microscopy of topochemical conversion interface between $La_2Ti_2O_7$ reactive template and perovskite product $Li_{0.16}La_{0.62}TiO_3$ electrolyte *Solid State Ion.* **296** 78–84

[22] Duan T *et al* 2023 On-chip light-incorporated *in situ* transmission electron microscopy of metal halide perovskite materials *ACS Energy Lett.* **8** 3048–53

[23] Pan X, Kaplan W D, Rühle M and Newnham R E 1998 Quantitative comparison of transmission electron microscopy techniques for the study of localized ordering on a nanoscale *J. Am. Ceram. Soc.* **81** 597–605

[24] Xu H, Wang Y, Putnam R L, Gutierriez J and Navrosky A 1999 Microstructure and composition of synroc samples crystallized from a $CaCeTi_2O_7$ chemical system: HRTEM/EELS investigation *MRS Proc.* **608** 461

[25] Alizadeh S M *et al* 2025 Investigating reduction and oxidation of $La_{0.72}Sr_{0.18}Fe_{0.9}Ni_{0.1}O_{3-\delta}$ as a Co-free perovskite electrode for symmetrical solid oxide cells *J. Phys.: Energy* **7** 035010

[26] Srivastava S *et al* 2023 Advanced spectroscopic techniques for characterizing defects in perovskite solar cells *Commun. Mater.* **4** 52

[27] Nurhaziqah A M S *et al* 2020 Microstructure study of calcium managanese oxide ($CaMnO_3$) as perovskite materials *J. Phys. Conf. Ser.* **1535** 012024

[28] Liu W *et al* 2022 Perovskite phase analysis by SEM facilitating efficient quasi-2D perovskite light-emitting device designs *Adv. Opt. Mater.* **10** 2200518

[29] Tauc J 1974 *Amorphous and Liquid Semiconductors* (Springer)

[30] Kuznetsov V N *et al* 2023 Absorption edge fine structure in the $PyPbI_3$ perovskite revealed by variable-temperature UV–vis diffuse reflectance spectroscopy *J. Phys. Chem.* C **127** 11

[31] Lei X F and Xue X X 2010 Photocatalytic reduction of Cr(VI) in the presence of citric acid over perovskite type SO_4^{2-}/TBBFS under UV–vis light irradiation *Adv. Mater. Res.* **113–116** 1632–8

[32] Harb M, Masih D, Ould-Chikh S, Sautet P and Takanabe K 2013 Determination of the electronic structure and UV–vis absorption properties of $(Na_{2-x}Cu_x)Ta_4O_{11}$ from first-principle calculations *J. Phys. Chem.* C **117** 17477–84

[33] Yamada T *et al* 2024 *Photoluminescence Properties of Lead Halide Perovskite Nanocrystals Revealed by Single-Dot Spectroscopy* (IOP Publishing)

[34] Yamada Y, Yamada T, Phuong L Q, Maruyama N and Kanemitsu Y 2015 Dynamic optical properties of $CH_3NH_3PbI_3$ single crystals as revealed by one- and two-photon excited photoluminescence measurements *J. Am. Chem. Soc.* **137** 10456–9

[35] Singh D K and Manam J 2017 Investigation of structural, spectral and photometric properties of $CaTiO_3:Dy^{3+}$ nanophosphors for the lighting applications *Electron. Mater. Lett.* **13** 292–301

[36] Segovia R *et al* 2018 Evolution of photoluminescence, Raman, and structure of $CH_3NH_3PbI_3$ perovskite microwires under humidity exposure *Nanoscale Res. Lett.* **13** 79

[37] Li C *et al* 2023 Perovskite single crystals: physical properties and optoelectronic properties *Mater. Futures* **2** 101

[38] Ftikos Ch *et al* 1993 Mixed electronic/ionic conductivity of the solid solution $La_{(1-x)}Sr_x Co_{(1-y)}Ni_yO_{3-\delta}$ (x: 0·4, 0·5, 0·6 and y: 0·2, 0·4, 0·6) *J. Eur. Ceram. Soc.* **12** 79–86

[39] Cabral C, Saenger K L, Kotecki D E and Harper J M E 2000 Optimization of TaSiN thin films for use as oxidation-resistant diffusion barriers *J. Mater. Res.* **15** 194–8

[40] Obayashi H and Kudo T 1975 Some crystallographic, electric and thermochemical properties of the perovskite-type $La_{1-x}M_xNiO_3$ (M: Ca, Sr and Ba) *Jpn. J. Appl. Phys.* **14** 330

[41] Shi D *et al* 2015 Low trap-state density and long carrier diffusion in organolead trihalide perovskite single crystals *Science* **347** 5

[42] Ideue T *et al* 2012 Effect of lattice geometry on magnon Hall effect in ferromagnetic insulators *Phys. Rev.* B **85** 1903–11

[43] Hauff E V and Klotz D 2022 Impedance spectroscopy for perovskite solar cells: characterisation, analysis, and diagnosis *J. Mater. Chem.* C **10** 742–61

[44] Torres J *et al* 2022 Degradation analysis of triple-cation perovskite solar cells by electrochemical impedance spectroscopy *ACS Appl. Energy Mater.* **5** 12545–52

[45] Lee H J, Park K S and Park Y J 2010 Surface modification of $Li[Ni_{0.3}Co_{0.4}Mn_{0.3}]O_2$ cathode by Li–La–Ti–O coating *J. Power Sources* **195** 6122–9

[46] Hashimoto T *et al* 2000 Analysis of crystal structure and phase transition of $LaCrO_3$ by various diffraction measurements *Solid State Ionics* **132** 181–8

[47] Spirito D *et al* 2022 Raman spectroscopy in layered hybrid organic–inorganic metal halide perovskites *J. Phys. Mater.* **5** 034004

[48] Minh N V 2009 Raman scattering study of perovskite manganites *J. Phys. Conf. Ser.* **187** 012011

[49] Wang L, Mccleese C, Kovalsky A, Zhao Y and Burda C 2014 Femtosecond time-resolved transient absorption spectroscopy of $CH_3NH_3PbI_3$ perovskite films: evidence for passivation effect of PbI_2 *J. Am. Chem. Soc.* **136** 12205–8

[50] Jo C, Kim H and Yoon D K 2024 Ultrafast photo-induced carrier dynamics of perovskite films being degraded by atmospheric exposure *J. Korean Phys. Soc.* **84** 17–23

[51] Ning S *et al* 2020 Ambient pressure x-ray photoelectron spectroscopy investigation of thermally stable halide perovskite solar cells via post-treatment *ACS Appl. Mater. Interfaces* **12** 9

[52] Reo Y *et al* 2025 Vapour-deposited high-performance tin perovskite transistors *Nat. Electron.* **8** 403–10

[53] Lin W-C *et al* 2021 *In situ* XPS investigation of the x-ray-triggered decomposition of perovskites in ultrahigh vacuum condition *NPJ Mater. Degradn* **5** 13

[54] Luo L *et al* 2017 Ultrafast terahertz snapshots of excitonic Rydberg states and electronic coherence in an organometal halide perovskite *Nat. Commun.* **8** 15565

[55] Gong S *et al* 2023 Ultrafast dynamics in perovskite-based optoelectronic devices *Cell Rep. Phys. Sci.* **4** 101580

Chapter 6

Computational and analytical simulation techniques

This chapter provides an in-depth overview of the analytical and numerical simulation-based methods for the assessment of the performance of perovskite solar cells (PSCs). It emphasizes the analysis of the device parameters such as short-circuit current (I_{sc}), open-circuit voltage (V_{oc}), fill factor (FF), and power conversion efficiency (η), along with the diode parameters including photocurrent (I_{ph}), ideality factor (n), reverse saturation current (I_0), series resistance (R_s), and shunt resistance (R_{sh}), by utilizing the single- and double-diode models. This chapter further explores the spectral response, quantum efficiency, and charge carrier lifetime in PSCs. A comprehensive review of the optical, electrical, and Multiphysics simulation techniques, including density functional theory (DFT), is presented. Multiscale modeling approaches from the atomistic to device level, that enable the design and optimization of efficient and stable PSCs, are also discussed.

6.1 Introduction

The photovoltaic (PV) effect enables the efficient conversion of sunlight into electricity. In recent years, PV technology has been playing a crucial role in meeting the growing global energy demands [1]. Metal–halide perovskites have emerged as a breakthrough class of materials in photovoltaics and optoelectronics. With a general formula of ABX_3, these materials feature the crystal structure as shown in figure 6.1 (a) where A is a monovalent cation (e.g. MA^+, FA^+, Cs^+), B is a metal cation (e.g. Pb^{2+}, Sn^{2+}), and X is a halide anion (I^-, Br^-, Cl^-) [2]. Their exceptional optoelectronic properties, including strong light absorption, tunable bandgaps, and solution processability, have rapidly advanced solar cell research across the world. Today, perovskites represent a convergence of interests in the chemistry, device engineering, materials science and the physics community.

doi:10.1088/978-0-7503-5427-1ch6

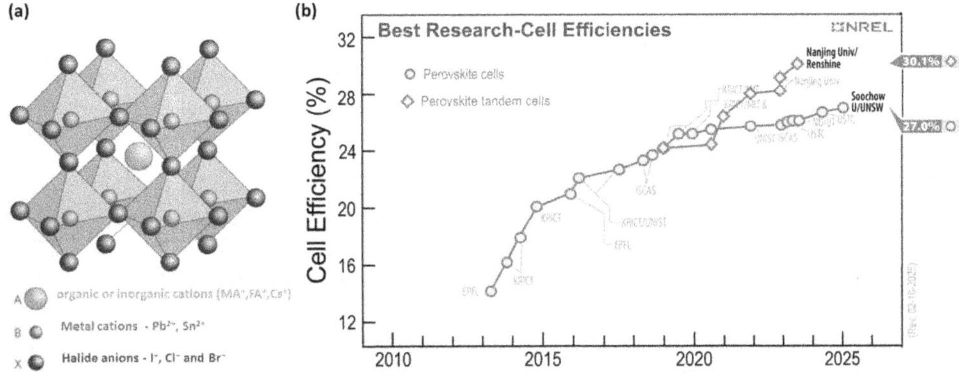

Figure 6.1. (a) Crystal structure of a perovskite (ABX_3). (b) Year-wise progress in the solar cell efficiencies for perovskite single-junction and tandem cells. ((a) Reproduced with permission from [2]. Copyright 2014 Springer Nature. (b) Adapted from [3], courtesy of the National Renewable Energy Laboratory, Golden, CO.)

PSC efficiencies have increased significantly from 3.8% in 2009 to over 25% in 2022 driven by the advances in materials, device architectures, and tandem configurations. As of early 2025, single-junction perovskite solar cells have reached certified 27% efficiency (figure 6.1(b)) [3], while tandem cells have reached up to 30.1%; with continued advancements, these efficiencies are expected to push these figures even higher [3].

Despite the significant progress, PSCs still face challenges including instability, ion migration, and structural complexity. The stability of PSCs can be assessed through performance studies such as current–voltage (I–V) measurements in dark and illumination conditions. In addition, advanced simulations offer fast, cost-effective insights into material and device behavior, aiding experimental efforts through material selection and design optimization.

The performance parameters such as I_{sc}, V_{oc}, FF, and η- are extracted from the illuminated I–V curve. These parameters are influenced by other diode parameters, such as I_{ph}, n, I_0, R_s, and R_{sh}, all of which are affected by the environmental conditions such as light intensity (P_{in}) and the cell temperature (T).

In order to accurately evaluate the behavior and efficiency of solar cells, these parameters must be determined precisely. The single-diode model (SDM) and double-diode model (DDM) are commonly employed to represent the I–V characteristics. Both of these models are based on the Shockley equation that includes R_s and R_{sh}, considering current-dependent and voltage-dependent losses [4, 5]. While SDM offers a simplified approach, it ignores certain recombination mechanisms, making it less effective for some PV technologies [6, 7]. In such cases, the DDM provides a more accurate and comprehensive representation of the device physics [8, 9]. Although more complex models such as the three-diode model exist, their practical use is often limited due to increased computational demands [5].

Both SDM and DDM involve nonlinear, implicit, and multivariable equations, which makes direct extraction of the device parameters from I–V data very challenging. Consequently, various analytical and numerical techniques have been developed in the literature to estimate the five diode parameters in SDM (I_{ph}, n, I_0, R_s, R_{sh}) [10–12], and the seven parameters in DDM (I_{ph}, n_1, n_2, I_{01}, I_{02}, R_s, R_{sh}). These approaches include methods that are based on equivalent electrical circuits [4, 10], semiconductor physics or partial differential equations [13], and empirical curve-fitting techniques [14].

Atomistic methods such as density functional theory (DFT) and molecular dynamics (MD) help identify defect pathways and material instabilities, while device-level drift-diffusion models capture ionic transport and hysteresis. Optical simulation techniques such as transfer matrix method (TMM) and finite-difference time-domain (FDTD) enable light management, and integrated optoelectronic simulations assess performance metrics. Tools such as SCAPS-1D, SETFOS, COMSOL, and Silvaco ATLAS support multiscale modeling, and help in the design of stable and efficient PSCs.

This chapter describes the analytical and numerical simulation-based methods for assessing the PSC performance, covering important metrics (I_{sc}, V_{oc}, FF, η), diode parameters (I_{ph}, n, I_0, R_s, R_{sh}), spectral response (SR), quantum efficiency (QE), and charge carrier lifetime (τ). It highlights the optical, electrical, and Multiphysics simulation tools, in the context of providing a detailed analysis and leading to the rapid advancement of efficient, stable perovskite technologies.

6.2 Fundamentals

While earlier chapters have explored PSC architecture in depth, a brief overview of the fabrication methods and structural configurations is provided for a fundamental understanding of the computation and simulation techniques discussed in this chapter. Given the focus on photovoltaic applications, the emphasis here will be on solar cell architectures and their relevance to computation and numerical simulation-based modeling.

6.2.1 Perovskite solar cell structure

The development of solid-state PSCs advanced significantly in 2012 with the integration of a solid thin film of hole transport material spiro-OMeTAD, replacing liquid electrolytes and significantly accelerating the progress in PSCs [15, 16]. Shortly thereafter, the mesoporous architecture (figure 6.2(a)) gained attention for its superior charge transport, although it involves more complex fabrication with sequential deposition of a compact layer and a mesoporous TiO_2 scaffold [17–19].

Compared to mesoporous PSCs, planar PSCs offer simpler fabrication and are broadly categorized into two configurations: conventional n–i–p and inverted p–i–n, distinguished by the sequence of charge transport layers relative to the intrinsic perovskite absorber (figure 6.2(b) and figure 6.2(c)). The n–i–p layout sequentially includes a transparent conducting oxide (ITO or FTO) electrode, electron transport layer (ETL), perovskite absorber, hole transport layer (HTL), and metal electrode. The p–i–n structure reverses this order, starting with the HTL on transparent electrode, followed by the perovskite layer, ETL, and metal back contact. These planar configurations are advantageous for their ease of fabrication, well-defined

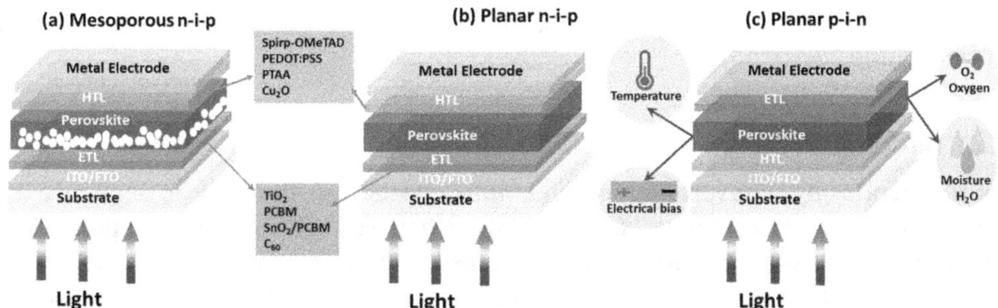

Figure 6.2. Schematic illustration of PSCs having (a) mesoporous n–i–p structure, (b) regular planar structure (n–i–p) and (c) inverted planar structure (p–i–n). The factors causing degradation in PSCs are also shown in (c).

interfaces, and compatibility with low-temperature processing, making them suitable for tandem solar cell integration and scalable manufacturing.

Efficient ETLs are vital in planar PSCs to ensure effective electron extraction, suppressed recombination, and enhanced device performance [20, 21]. In n–i–p perovskite solar cells, common ETLs include TiO_2, SnO_2, and ZnO, with SnO_2 preferred for its low-temperature processing and superior stability (figure 6.2(a) and (b)). In contrast, p–i–n architectures typically employ fullerene derivatives such as PCBM and C_{60} as ETLs due to their low-temperature solution processability. HTLs, in n–i–p devices, often include spiro-OMeTAD and PTAA, while PEDOT:PSS and NiO_x are standard in p–i–n cells. Recently, inorganic HTLs such as CuSCN and MoO_3 have also attracted interest due to their enhanced stability and optical transparency.

Currently, the n–i–p structure dominates in laboratory-scale research and high-efficiency devices due to the maturity as can be assessed by the history of its development and well-established material systems. However, the p–i–n configuration is rapidly gaining interest for its flexibility, low-temperature processing, and the potential for large-scale manufacturing, along with reduced hysteresis and improved interface stability. Thus, while n–i–p structures continue to lead in record efficiencies, p–i–n architectures are emerging as promising alternatives for the next-generation scalable perovskite photovoltaics [22].

As illustrated in figure 6.2(c), the environmental conditions significantly influence PSC performance and accelerate degradation. The degradation pathways include moisture ingress, thermal stress, UV irradiation, and ion migration, each compromising device stability. Moisture triggers hydrolysis in perovskite films such as $MAPbI_3$, leading to the formation of PbI_2 and other volatile organic species, thereby compromising film integrity and device performance [23]. Elevated temperatures (>85 °C) induce phase segregation and degradation in perovskite and charge transport layers [24], while UV exposure accelerates decomposition through photochemical reactions, especially at the interfacial layers, such as TiO_2 [25]. Ion migration (e.g. I^-, MA^+, FA^+) under electric field and thermal gradients causes hysteresis, interface degradation, and ultimately, long-term performance losses [26].

Degradation in ETLs and HTLs (e.g. spiro-OMeTAD, PTAA, TiO_2) further reduce efficiency and lifespan [27].

6.2.2 Basic equations

PSCs operate by converting sunlight into electricity through a sequence of fundamental processes (figure 6.3(a)) [28]. When sunlight strikes the perovskite active layer, it absorbs photons and generates bound electron–hole pairs known as excitons. Due to the low exciton binding energy in perovskite materials, these excitons quickly dissociate into free electrons and holes. The electrons transport toward ETL, while the holes move toward HTL, driven by the internal electric fields and carrier concentration gradients within the device. The separated charges are collected at the respective electrodes, electrons at the cathode and holes at the anode. Among these charge carriers, which flow across the junction to complete the circuit generate a photocurrent and those, which accumulate at the junction, generate a photovoltage. Dynamical behavior of electrons and holes under illumination can be described with the basic equations in p–n junctions. It is essential to address the fundamental equations governing p–n junction behavior, as they form the foundation of simulation techniques used to model charge transport, electrostatic potential, and carrier dynamics in PSCs. The Poisson's equation relates the gradient of the electric field (E) to the space charge density (ρ) and is used to determine the width of the space charge region in a p–n junction. In one-dimensional form, it is expressed as [29, 30]

$$\frac{\partial E}{\partial x} = \frac{\rho}{\varepsilon}, \tag{6.1}$$

where ε is the permittivity of the material.

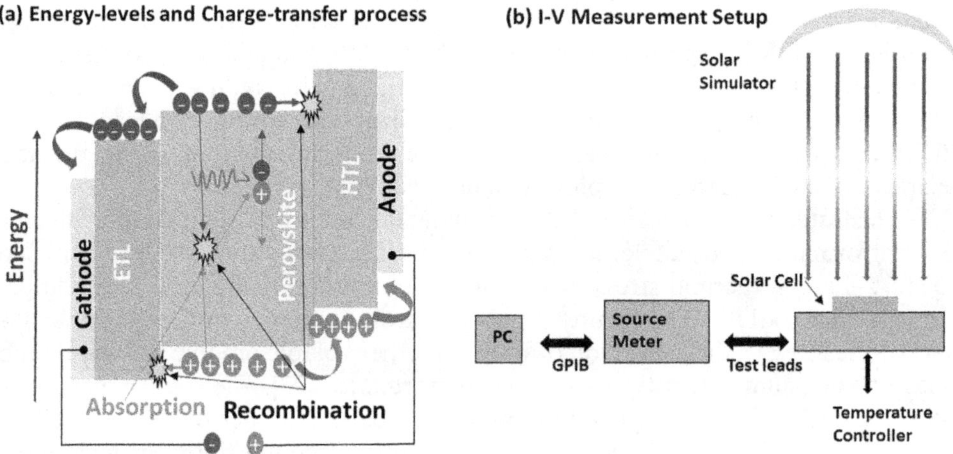

Figure 6.3. (a) Schematic illustration of energy band diagram of a p–i–n PSC under illumination. Incident photons are absorbed in the perovskite layer and generate electron–hole pairs that are separated into free electrons and holes. The electrons transport through ETL and are collected at the cathode whereas the holes transport through HTL and are collected at the anode. (b) Schematic of the current–voltage (I–V) characterization set-up. ((b) Reproduced with permission from [33]. Copyright 2025 IOP Publishing.)

In simulations, the general form of Poisson's equation is expressed as [31, 32]

$$\frac{\partial^2\psi}{\partial^2 x} = -\frac{\partial E}{\partial x} = -\frac{\rho}{\varepsilon} = -\frac{q}{\varepsilon}\left[p - n + N_D^+(x) - N_A^-(x) \pm N_{\text{def}}(x)\right], \tag{6.2}$$

where, ψ is the electrostatic potential, q is elementary charge, n is the electron density, p is the hole density, $N_D{}^+(x)$ is the density of ionized donors, $N_A{}^-(x)$ is the density of ionized acceptors, and N_{def} is the possible defect acceptor and donor density [31, 32].

The gradient of the current density is related to the recombination and generation rates of charge carriers by the continuity equation. The electron and hole continuity equations are given by

$$\frac{\partial n}{\partial t} = G(x) - R(x) + \frac{1}{q}\frac{\partial}{\partial x}J_n(x) \tag{6.3}$$

$$-\frac{\partial p}{\partial t} = G(x) - R(x) - \frac{1}{q}\frac{\partial}{\partial x}J_p(x), \tag{6.4}$$

where, J_n and J_p are the current densities due to electrons and holes, R and G refer to the recombination and generation rates, respectively, and t is time.

Minority carrier current in a PV device is a result of the flow of electrons and holes, which is caused by the electric field, and the gradient of the carrier density. The drift-diffusion current relations are described by

$$J_n = qn\mu_n E + qD_n\frac{\partial n}{\partial x} \tag{6.5}$$

$$J_p = qp\mu_p E - qD_p\frac{\partial p}{\partial x}, \tag{6.6}$$

where μ_n and μ_p are the mobilities of electrons and holes, D_n and D_p are their respective diffusion coefficients, and E is the electric field. The first term in equations (6.5) and (6.6) describes the drift current driven by the electric field E and the second term is the diffusion current driven by a concentration gradient. The mobilities $\mu_n(\mu_p)$ and diffusion constants $D_n(D_p)$ are related through the Einstein relationships $(D_n = \mu_n(kT/q), D_p = \mu_p\,(kT/q))$ [29, 30].

6.3 Characterization techniques

The characterization of PSCs primarily involves current–voltage (I–V), spectral response (SR), quantum efficiency (QE) and lifetime (τ) measurements.

6.3.1 Illuminated current–voltage characteristics

I–V measurements are essential for the characterization of solar cells, as they enable the calculation of the performance parameters such as I_{sc}, V_{oc}, FF and η, and the diode parameters I_{ph}, n, I_0, R_s, and R_{sh}. These parameters provide crucial insights into the overall behavior and efficiency of the devices. Although dark I–V curves are more sensitive for evaluating the diode-related characteristics [34], the extraction of solar cell parameters is generally carried out under illuminated conditions to better reflect the real-world operating performance [6, 35]. The schematic diagram of the I–V measurement set-up is shown in figure 6.3(b). In this set-up, the solar cell is illuminated with a simulated AM 1.5 spectrum at 100 mW cm^{-2}, while its temperature is maintained at 25 °C (298 K) using a temperature controller, as shown in figure 6.3 (b). A source meter records the I–V characteristics. The light intensity and temperature can be kept constant or varied based on the needs of measurements. The light source is calibrated using a standard reference solar cell with a known I_{sc} under standard test conditions (STC-AM 1.5, 100 mW cm^{-2}, 25 °C).

Both SDM and DDM have been widely used to analyse the electrical characteristics of solar cells and modules [36]. Numerous studies [4, 10, 37, 38] have explored the advantages and limitations of each model. Comparative analysis have highlighted a trade-off between the computational efficiency and modeling accuracy [39]. Despite its simplified representation, the SDM remains the preferred choice for many PV technologies due to its reduced complexity and ease of implementation [40, 41], particularly under varying environmental conditions [42, 43]. Our recent work [33] reviews the analytical and numerical simulation techniques for extracting diode parameters from single- and double-diode models of solar cells. The following section outlines the analytical methods that are used to derive solar cell performance metrics from these models, as presented in the literature [33].

6.3.1.1 Single-diode model

The SDM of a solar cell, incorporating R_s and R_{sh} based on the Shockley equation, is illustrated in figure 6.4(a) [44, 45]. Under illumination, the I–V characteristics of PSC can be theoretically described using the following equations [46, 47]:

$$I = -I_{ph} + I_d + \frac{V_j}{R_{sh}}, \tag{6.7}$$

where I_{ph} is the photogenerated current and I_d is the current flowing through the diode due to the applied voltage. V_j is the voltage developed across the junction and the third term on the right-hand side accounts for the leakage current through the shunt resistance R_{sh}. The expressions for I_d and V_j are defined as follows:

$$I_d = I_0\left\{\exp\left(\frac{V_j}{nV_t}\right) - 1\right\} \tag{6.8}$$

$$V_j = V - IR_s. \tag{6.9}$$

Substituting the values of I_d and V_j from equation (6.8) and equation (6.9), respectively, in equation (6.7),

$$I = -I_{ph} + I_0\left\{\exp\left(\frac{V - IR_s}{nV_t}\right) - 1\right\} + \frac{V - IR_s}{R_{sh}}. \tag{6.10}$$

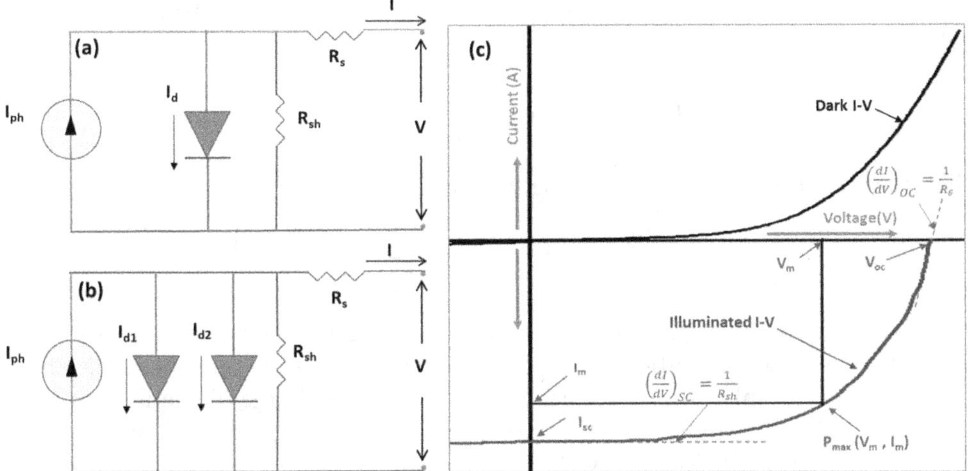

Figure 6.4. The equivalent circuit diagram of a solar cell described by (a) SDM and (b) DDM having R_s and R_{sh}, (c) dark and illuminated I–V curves of a solar cell to denote the parameters V_{oc}, I_{sc}, V_m and I_m at maximum power output (P_{max}) and slope dI/dV at open circuit and short circuit conditions. (Adapted with permission from [33]. Copyright 2025 IOP Publishing.)

In equations (6.9) and (6.10), I and V are output current and applied voltage, respectively as illustrated in figure 6.4(a) and in figure 6.4(b). Here $V_t = kT/q$ is thermal voltage, $q = 1.602 \times 10^{-19}$ C is the electron charge, $k = 1.38 \times 10^{-23}$ J K^{-1} is Boltzmann's constant, and I_0 is the reverse saturation current. SDM comprises five unknown diode parameters (I_{ph}, n, I_0, R_s and R_{sh}), each dependent on the solar irradiance and temperature conditions [47]. Various analytical models have been developed to estimate these parameters, generally applying some assumptions to ease the computational process [48, 49]. For instance, certain models consider R_{sh} to be infinite [46], while others simplify the I_{ph} by equating it to the I_{sc}, effectively ignoring the diode current [50]. Several review papers explore strategies to streamline the model complexity, often by approximating or omitting parameters, potentially reducing the parameter set from five to just one [4, 37].

6.3.1.2 Double diode model

The SDM commonly assumes that the recombination losses within the space charge region (depletion region) are negligible. Additionally, for most silicon (Si) solar cells, it is generally valid to assume that $R_{sh} >> R_s$. However, this assumption does not apply to thin-film and organic solar cells, which exhibit more complex electrical behavior. In such cases, the DDM provides a more accurate representation. As illustrated in figure 6.4(b) [51, 52], DDM incorporates two additional parameters; reverse saturation current (I_{02}) and ideality factor (n_2) corresponding to another diode that accounts for recombination processes occurring within the space charge region. This recombination current is incorporated into the extended diode equation, making the DDM suitable for more accurate modeling of the I–V

characteristics of photovoltaic cells. The refined model is described by equations (6.11) and (6.12):

$$I = -I_{\text{ph}} + I_{d1} + I_{d2} + \frac{V_j}{R_{\text{sh}}}. \tag{6.11}$$

The above equation (6.11) can be written as

$$I = -I_{\text{ph}} + I_{01}\left\{\exp\left(\frac{V - IR_s}{n_1 V_t}\right) - 1\right\} + I_{02}\left\{\exp\left(\frac{V - IR_s}{n_2 V_t}\right) - 1\right\} + \frac{V - IR_s}{R_{\text{sh}}}. \tag{6.12}$$

In DDM, I_{01} and n_1 represent the reverse saturation current and ideality factor associated with recombination in the quasi-neutral (bulk) regions, while I_{02} and n_2 correspond to recombination occurring within the space charge (depletion) region of the solar cell. Although DDM offers enhanced accuracy in the modeling of photovoltaic cell behavior, it involves the extraction of seven parameters, I_{ph}, n_1, n_2, I_{01}, I_{02}, R_s and R_{sh} which requires a computationally intensive process [53]. In order to address this challenge, researchers have proposed methods to simplify parameter extraction using reduced parameter sets, such as seven [54], six [55], five [56], and even four parameters [57]. These approaches often employ numerical algorithms, including the Newton–Raphson method and particle swarm optimization, which rely on extensive data inputs and involve numerous mathematical iterations to accurately determine the unknown values [58].

6.3.2 Dark current–voltage characteristics

The dark I–V characteristics of a PV cell, considering finite R_s and R_{sh}, can be characterized by setting the photogenerated current $I_{\text{ph}} = 0$ in equation (6.10) for SDM and in equation (6.12) for the DDM. Under these conditions, the corresponding expressions simplify to represent the behavior of the cell in the absence of illumination and are given by

$$I = I_0\left\{\exp\left(\frac{V - IR_s}{nV_t}\right) - 1\right\} + \frac{V - IR_s}{R_{\text{sh}}} \tag{6.13}$$

$$I = I_{01}\left\{\exp\left(\frac{V - IR_s}{n_1 V_t}\right) - 1\right\} + I_{01}\left\{\exp\left(\frac{V - IR_s}{n_2 V_t}\right) - 1\right\} + \frac{V - IR_s}{R_{\text{sh}}}. \tag{6.14}$$

6.3.3 Solar cell performance parameters

In the open circuit condition, $I = 0$ and $V = V_{\text{oc}}$, and from equation (6.10), the relationship between V_{oc}, I_0 and I_{ph} for single exponential model is defined as

$$V_{\text{oc}} = \frac{kT}{q}\ln\left\{\frac{I_{\text{ph}}}{I_0} + 1\right\}. \tag{6.15}$$

V_{oc} is related to I_{ph} and I_0. For a high V_{oc}, a low value of I_0 is absolutely necessary.

The fill factor is defined as the ratio of maximum power (P_{max}) at maximum power point (MPP) to the product of V_{oc} and I_{sc} ($\approx I_{ph}$) and its value is always less than one and can be expressed as

$$\text{FF} = \frac{P_{max}}{V_{oc}I_{sc}}. \tag{6.16}$$

The condition for MPP and expression for P_{max} is given by

$$\frac{dP}{dV} = 0 \Rightarrow \frac{d(IV)}{dV} = 0 \tag{6.17}$$

$$P_{max} = I_m V_m, \tag{6.18}$$

where I_m, V_m represent the current and voltage at MPP (figure 6.4(c)). The output power per unit area of the cell can be determined from the illuminated I–V characteristics by evaluating the area of the rectangle formed between the MPP and the two axes (as shown in figure 6.4(c)). The operating point depends on the load resistance.

The power conversion efficiency (η) of a solar cell is defined as the ratio of P_{max} per unit area to P_{in} and is represented as

$$\eta = \frac{P_{max}}{P_{in}.\,\text{Area}}, \tag{6.19}$$

where P_{in} is the intensity of the incident radiation.

In terms of V_{oc}, J_{sc} and FF, the efficiency of the cell can be represented as

$$\eta = \frac{V_{oc}.\,J_{sc}.\,\text{FF}}{P_{in}}, \tag{6.20}$$

where J_{sc} is the short circuit current density. The J_{sc} of the cell depends on the solar spectral irradiance and is given by

$$J_{sc} = q\int_{h\nu=E_g}^{\infty} \frac{dN_{ph}}{dh\nu} d(h\nu), \tag{6.21}$$

where N_{ph} is the incident photon flux. In a practical solar cell, the value of J_{sc} may be limited by reflection losses, ohmic losses (R_s and R_{sh}) and recombination losses.

6.3.4 Spectral response

The spectral response (SR) measurement characterizes the response of a solar cell to various light wavelengths, and help determine the spectral range where it operates efficiently. It provides insights into material properties and device design. SR is defined as the ratio of the J_{sc} produced under monochromatic illumination at a specific wavelength (λ) to P_{in} at that wavelength.

SR can be expressed mathematically as

$$\mathrm{SR}(\lambda) = \left(\frac{J_{sc}}{P_{in}}\right)_{\lambda}. \tag{6.22}$$

6.3.5 Quantum efficiency

The quantum efficiency (QE) of a solar cell is defined as the ratio of photo-generated electrons to the number of incident photons at a particular wavelength. It is categorized into two types: internal quantum efficiency (IQE) and external quantum efficiency (EQE). EQE is determined by considering all incident photons, whereas IQE accounts only for photons that are not reflected from the cell. EQE measurement permits to assess how efficiently a solar cell converts incident photons into electrical current across different wavelengths. This technique provides insights into SR of the solar cell and helps identify factors affecting light absorption and carrier generation. Conceptually, SR and QE are closely related.

Mathematically EQE and IQE can be defined as

$$\mathrm{EQE} = \frac{hc}{q\lambda}\mathrm{SR}(\lambda) \tag{6.23}$$

$$\mathrm{IQE} = \frac{\mathrm{EQE}}{(1 - R_{\lambda})}, \tag{6.24}$$

where h is the Planck's constant, c is the velocity of light, and R_{λ} is the reflection coefficient.

6.3.6 Minority carrier lifetime

Minority carrier lifetime (τ) is a critical parameter for evaluating the solar cell performance. It represents the average time a minority carrier persists before recombination. It can be measured using either transient or steady-state methods. Transient techniques track changes in carrier concentration over time after a light or voltage pulse, while steady-state methods monitor the response of cell to continuous excitation. In PSCs, due to their thin-film nature and ambipolar transport, several tailored techniques are employed [59]. Time-resolved photoluminescence (TRPL) is a widely used, non-contact technique for measuring the minority carrier lifetimes. Other important methods include transient photovoltage (TPV) and transient photocurrent (TPC), which probe carrier recombination and transport under open-circuit and short-circuit conditions, respectively, although TPV can be influenced by resistance–capacitance effects. Time-resolved terahertz spectroscopy (TRTS) uses ultrafast laser pulses to assess ultrafast mobility and carrier dynamics through terahertz conductivity, and is ideal for early-stage materials. Photoluminescence quantum yield (PLQY) [60] evaluates non-radiative losses and effective carrier lifetimes by measuring absolute photoluminescence across varying excitation intensities. Intensity-modulated photocurrent and photovoltage spectroscopy (IMPS/IMVS) use modulated light to reveal

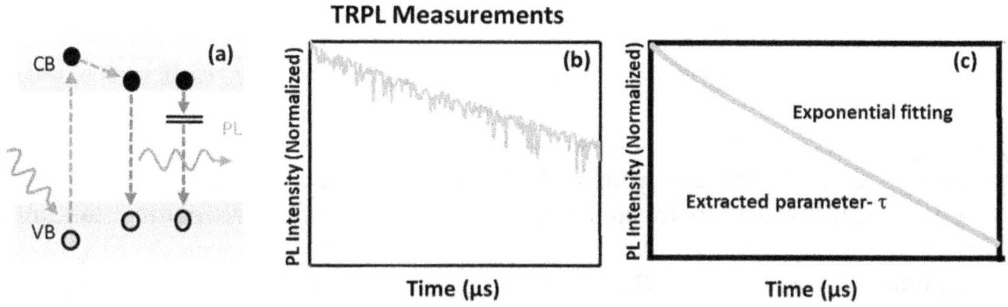

TRPL Measurements

Figure 6.5. (a) TRPL signals generated from recombination pathways, (b) measured PL signal as a function of time using a sensitive detector, (c) conventional fitting method of TRPL readings (logarithmic scale). PL curves are commonly fitted by exponential function to derive lifetime τ (for $\beta = 1$). (Adapted with permission from [61]. Copyright 2018 American Chemical Society.)

recombination lifetimes and charge transport behavior under operational conditions. Each technique complements the others, offering a comprehensive understanding of recombination and transport dynamics in PSCs. Among these methods, TRPL is the most widely used to measure carrier lifetimes by tracking PL decay after pulsed excitation. Figure 6.5 illustrates the TRPL set-up and standard curve-fitting approach that is used in lifetime measurements.

In TRPL technique, a semiconductor sample is stimulated by a short pulse of light, usually a laser [61]. This pulse produces an initial excess carrier density (electrons or holes) in the sample. After the excitation pulse, the sample emits PL as the excess carriers recombine with the majority carriers as shown in figure 6.5(a). This PL signal is typically measured (depicted in figure 6.5(b)) as a function of time using a sensitive detector. The PL signal decays exponentially over time as the excess carriers recombine and can be represented by the following equation (6.25):

$$I(t) = I_0 \exp [(-t/\tau)^{\beta}], \tag{6.25}$$

where $I(t)$ is the PL intensity at time t and I_0 is the initial PL intensity at $t = 0$. τ is a time constant and represents the minority carrier lifetime and β ($0 \leqslant \beta \leqslant 1$) is the stretching index. The stretched exponential function is well known to describe the PL decay and transport properties of disordered systems [62]. The stretching index is a measure of the degree of disorder in the material.

For $\beta = 1$, equation (6.25) reduces to

$$I(t) = I_0 \exp [(-t/\tau)]. \tag{6.26}$$

In logarithmic scale, equation (6.26) can be written as

$$\ln I(t) = \ln I_0 - t/\tau. \tag{6.27}$$

The slope of the curve, $\ln I(t)$ versus t, can be used to calculate τ, in accordance with the above equation (6.27). TRPL measurements can be used to investigate τ at different temperatures or excitation conditions. Generally, the decay of PL intensity is fitted by an exponential function according to equation (6.27) as shown in figure 6.5(c).

6.4 Analytical techniques

Analytical approaches typically utilize certain approximations to derive standard equations to describe solar cell characteristics from equation (6.10), based on the operating points of the I–V curve, namely, the short-circuit, open-circuit, and maximum power point conditions. These characteristic points are strategically applied to equation (6.10) to facilitate the evaluation of diode parameters under different operating scenarios. At short circuit condition ($I = -I_{sc}$ and $V = 0$), I_{ph} can be evaluated in terms of n, I_0, R_s, and R_{sh} using equation (6.10) and can be expressed as

$$I_{ph} = I_{sc}\left(1 + \frac{R_s}{R_{sh}}\right) + I_0\left\{\exp\left(\frac{I_{sc}R_s}{nV_t}\right) - 1\right\}. \tag{6.28}$$

Similarly, at the open circuit condition ($V = V_{oc}$, $I = 0$), I_{ph} in terms of V_{oc}, n, I_0, and R_{sh} from equation (6.10) can be written as

$$I_{ph} = \frac{V_{oc}}{R_{sh}} + I_0\left\{\exp\left(\frac{V_{oc}}{nV_t}\right) - 1\right\}. \tag{6.29}$$

Equation (6.10) at MPP ($I = -I_m$, $V = V_m$), I_{ph} in terms of V_m, I_m, n, I_0, R_s, and R_{sh} can be expressed as

$$I_{ph} = I_m + I_0\left\{\exp\left(\frac{V_m + I_mR_s}{nV_t}\right) - 1\right\} + \left(\frac{V_m + I_mR_s}{nV_t}\right). \tag{6.30}$$

Analytical techniques provide faster, empirical solutions without the need for iterative procedures [63–65], although their accuracy can depend on the specific PV technology being studied [66]. In contrast, numerical methods use iterative algorithms and solvers to achieve higher precision [6, 67].

6.4.1 Solar cell diode parameters

The diode parameters of a solar cell play a critical role in determining its performance characteristics [68, 69]. An increase in R_s and a decrease in R_{sh} significantly reduce FF. Specifically, high R_s leads to reduction in I_{sc}, while low R_{sh} contributes to decrease in V_{oc}. A lower R_{sh} allows more leakage current, increasing I_0, which in turn negatively affects V_{oc}, FF, and ultimately the power conversion efficiency [68, 70]. Moreover, n and I_0 provide insights into the recombination mechanisms within the surface, bulk, and space charge regions of the cell. Due to the inverse relationships of I_0 with V_{oc} (as seen in equation 6.15) and that of n with FF (as in equation 6.16), increase in either I_0 or n can result in inferior performance of the solar cell. Hence, accurate determination of these parameters is essential for optimizing the cell efficiency. Various parameter extraction strategies, including analytical, numerical, hybrid, and optimization-based methods, have been explored in the literature [71–73], with most approaches classified as either analytical or numerical [6, 10, 38, 63].

Diode parameter extraction is generally classified into three distinct models, as commonly discussed in the literature. Model I represents the most simplified scenario, where only three parameters (I_{ph}, n, and I_0) are extracted, assuming both R_s and R_{sh} to be negligible [74]. Model II simplifies by considering four parameters; namely I_{ph}, n, I_0, and R_s while neglecting the influence of R_{sh}. Model III involves the extraction of all five key parameters: I_{ph}, n, I_0, R_s and R_{sh}. The governing equation (6.28), (6.29), and (6.30) are employed to compute the diode parameters.

In ideal simulations or simplified analytical models, model I is commonly adopted to perform the analysis with ease and may be appropriate during initial curve fitting or parameter extraction, in particular when the solar cell demonstrates near-ideal behavior with very high R_{sh} and negligible R_s. However, for accurate and realistic modeling of solar cell performance, it is crucial to account for both R_s and R_{sh}, as they play a significant role in defining the practical limitations and overall efficiency. Therefore, in the following section for simplicity and clarity, the equations are first derived for the general case (model III), followed by a focused discussion on model I. The equations for model II can be obtained from model III by applying the condition $R_{sh} \to \infty$.

6.4.1.1 Photocurrent

I_{ph} in a solar cell is directly proportional to the incident light intensity and varies with temperature. When all the solar cell parameters (I_{sc}, V_{oc}, V_m, I_m, I_0, n, R_s, and R_{sh}) are known, I_{ph} can be determined from any of the equations (6.28), (6.29), and (6.30). The selection of a specific equation depends on the availability of known parameters in a given scenario, allowing model I, model II, or model III to be applied as appropriate.

Model I illustrates SDM under ideal conditions, where no resistive losses are considered [75–77]. Specifically, it assumes $R_s = 0$ and $R_{sh} \to \infty$, simplifying the behavior of the solar cell. Under these assumptions, equation (6.28), (6.29) and (6.30) reduce accordingly. The resulting expression for model I is given as follows [76]:

$$I_{ph} = I_{sc}. \tag{6.31}$$

6.4.1.2 Ideality factor and reverse saturation current

Equations (6.29) and (6.30) can be solved for n and this is approximated for model III as [33, 74]

$$n = \frac{V_m + I_m R_s - V_{oc}}{V_t\left[\ln\left(I_{sc} - \frac{V_m}{R_{sh}} - I_m\right) - \ln\left(I_{sc} - \frac{V_{oc}}{R_{sh}}\right) + \frac{I_m}{I_{sc} - (V_{oc}/R_{sh})}\right]}, \tag{6.32}$$

which can be simplified for model I as

$$n = \frac{V_m - V_{oc}}{V_t\left[\ln\left(1 - \frac{I_m}{I_{sc}}\right)\right]}. \tag{6.33}$$

The ideality factor n is the most frequently utilized parameter for various solar cell technologies and generally ranges from 1 to 2. The value of n close to 1 represents ideal junctions, while n approaching to 2 is associated with solar cell deterioration, non-uniformities or recombination centers [78]. This factor is more than 2 in organic solar cells [44] whereas in dye-sensitized solar cells, it is between 2 and 3 [79]. Ideality factor in the vicinity of 2 has been reported in PSCs as a result of carrier recombination and trap-assisted recombination in dark conditions [80, 81].

Three equations for computation of I_0 for model III can be obtained by eliminating I_{ph} from equation (6.28) (6.29) and (6.30). First equation for I_0 can be obtained from the short circuit equation (6.28) and open circuit equation (6.29) conditions as in [33, 74]

$$(I_0)_{SC,\,OC} = \frac{I_{sc} + \frac{I_{SC}R_S}{R_{sh}} - \frac{V_{OC}}{R_{sh}}}{\left\{\exp\left(\frac{V_{OC}}{nV_t}\right) - \exp\left(\frac{I_{SC}R_S}{nV_t}\right)\right\}}. \tag{6.34}$$

Similarly, eliminating I_{ph} from equations (6.29) and (6.30), we obtain I_0 at open circuit and MPP conditions:

$$(I_0)_{OC,\,MPP} = \frac{I_m + \frac{I_mR_S}{R_{sh}} + \frac{V_m}{R_{sh}} - \frac{V_{OC}}{R_{sh}}}{\left\{\exp\left(\frac{V_{OC}}{nV_t}\right) - \exp\left(\frac{V_m + I_mR_S}{nV_t}\right)\right\}}. \tag{6.35}$$

Similarly, calculating I_0 from equations (6.28) and (6.30) by eliminating I_{ph} at short circuit and MPP conditions we obtain

$$(I_0)_{SC,\,MPP} = \frac{I_m + \frac{I_mR_S}{R_{sh}} + \frac{V_m}{R_{sh}} - \frac{I_{SC}R_S}{R_{sh}} - I_{sc}}{\left\{\exp\left(\frac{I_{SC}R_S}{nV_t}\right) - \exp\left(\frac{V_m + I_mR_S}{nV_t}\right)\right\}}. \tag{6.36}$$

Equations (6.34), (6.35) and (6.36) can be used to compute I_0 at a given temperature T. However, these equations are dependent on three unknown diode parameters: R_s, R_{sh}, and n (model III). Depending on which parameters are estimated first, different approaches yield different values of I_0. For instance, if n, R_s, and R_{sh} are computed first, I_0 can be derived using equations (6.34) and (6.36). It has been discovered that equation (6.35) is inappropriate for the determination of I_0 as it produces negative values for I_0 [82]. Various alternative methods have been explored to determine n and I_0 [14, 83]. One such approach is the simple variable intensity method [84, 85], which enables the simultaneous determination of n and I_0 for SDM and DDM models. In this method, the incident radiation intensity on the PV cell is varied, and the corresponding V_{oc} and I_{sc} are measured.

Equations (6.34), (6.35) and (6.36) reduce for model I as

$$(I_{0I})_{SC,\,OC} = \frac{I_{sc}}{\left\{\exp\left(\frac{V_{OC}}{nV_t}\right) - 1\right\}} \tag{6.37}$$

$$(I_{01})_{\text{OC, MPP}} = \frac{I_m}{\left\{\exp\left(\frac{V_{\text{OC}}}{nV_t}\right) - \exp\left(\frac{V_m}{nV_t}\right)\right\}} \tag{6.38}$$

$$(I_{01})_{\text{SC, MPP}} = \frac{I_{\text{sc}} - I_m}{\left\{\exp\left(\frac{V_m}{nV_t}\right) - 1\right\}}. \tag{6.39}$$

Nonetheless, the simplicity of the Model I is achieved at the expense of decreased accuracy, particularly at low irradiance due to the effect of the PV resistances. A recent study [74] evaluated the accuracy and complexity of these analytical techniques (model I, model II and model III) for parameter extraction. The findings suggest that analytical procedures can produce results that are comparable to those obtained by numerical techniques [74].

6.4.1.3 Series and shunt resistances

The series and shunt resistances can be evaluated by analysing the slopes of the *I–V* curve under three conditions: open circuit, short circuit, and MPP. The equations for R_s and R_{sh} are derived from the *I–V* characteristics [86, 87] via differentiation of equation (6.10) with respect to V:

$$\left(\frac{dI}{dV}\right) = \frac{I_0}{nV_t}\left(1 - R_s\frac{dI}{dV}\right)\exp\left(\frac{V_j}{nV_t}\right) + \frac{1}{R_{\text{sh}}}\left(1 - R_s\frac{dI}{dV}\right). \tag{6.40}$$

As demonstrated in [71], the slope at short circuit condition, using equation (6.40), can be expressed as

$$\left(\frac{dI}{dV}\right)_{\text{SC}} = \frac{1}{R_{\text{sh}}}. \tag{6.41}$$

The slope at open circuit condition, using equation (6.40), can be approximated as

$$\left(\frac{dI}{dV}\right)_{\text{OC}} = \frac{1}{R_s}. \tag{6.42}$$

The slopes $(dI/dV)_{\text{sc}}$ and $(dI/dV)_{\text{oc}}$ are depicted in figure 6.4(c).

At MPP, the derivative of $P = IV$ with respect to V can be determined [47, 86, 88] by substituting $dP/dV = 0$ for MPP condition, yielding R_{sh} in terms of R_s, V_m, I_m, n and I_0, for model III as

$$(R_{\text{sh}})_{\text{III}} = \left[\frac{V_m - I_m R_s}{I_m + \frac{I_0}{nV_t}(V_m - I_m R_s)\exp\left(\frac{V_m + I_m R_s}{nV_t}\right)}\right]. \tag{6.43}$$

Equations (6.41) and (6.42) can be used to determine R_{sh} and R_s. Equation (6.43) represents a relationship between R_{sh} and R_s. In general, R_s includes the material resistance, the resistances of the front and rear metallic contacts, and the contact resistances of the metallic contacts to the front and back surfaces [89]. R_{sh} is a parallel high conductive path across the p–n junction or at the cell edges, and it is related to the leakage current across the surfaces involving pin-holes, grain boundaries, and charge recombination processes [90]. Several approaches are available in the literature for measuring R_s and R_{sh} of a PV cell [69, 91–94]. It is important to mention that in both laboratory and commercial solar cells, FF is limited not only by high values of R_s but also by low R_{sh} [95] values. Furthermore, resistive losses in commercial solar cells become larger as substrate size increases [69].

In a related study [96], in order to extract accurate SDM parameters, 11 p–i–n planar PSCs (~1 cm^2, 4.6–12.2% efficiency) were fabricated across multiple batches using a reproducible protocol [96–99]. 11 cells (cell 1 to cell 11) with varying *I–V* characteristics were selected, as shown in figure 6.6(a). The *I–V* characteristics were fitted using a genetic algorithm combined with the Nelder–Mead method, minimizing mean square errors below 9.1×10^{-9}. This hybrid optimization approach outperformed five conventional methods [11], which often yielded non-physical parameters such as negative resistances or abnormally high ideality factors. The results validated the applicability of SDM for PSCs, with n averaging ~2 for devices with FF > 0.5, consistent with electroluminescence [100] and electrochemical impedance spectroscopy (EIS) data [83]. In contrast, cells with FF < 0.5 showed higher n (> 3), low R_{sh}, and increased losses. The best-performing cell had an ideality factor of 1.4. Table 6.1 summarizes the performance and diode parameters for these 11 cells [96].

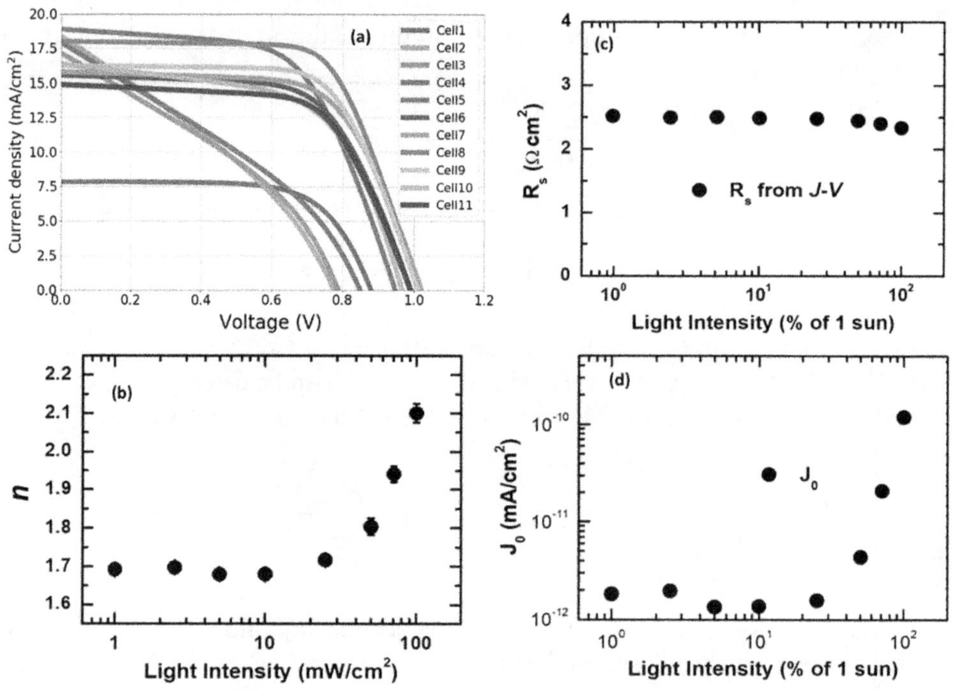

Figure 6.6. (a) *I–V* curve of perovskite solar cells. Variations of (a) ideality factor (n), (b) series resistance (R_s) and (c) reverse saturation current density (J_0). ((a) Reproduced from [96]. CC BY 4.0. (b) and (c) Adapted from [101]. CC BY 4.0.)

Table 6.1. Performance parameters of perovskite solar cells and diode parameters extracted using the single diode model (SDM) [96].

		Performance parameters				Error	Diode parameters				
Cell	Area (cm^2)	PCE (%)	FF	V_{oc} (V)	J_{sc} (mA cm^{-2})	MSE ($\times 10^{-9}$)	I_{ph} (mA cm^{-2})	I_0 (nA)	n	R_s ($\Omega.cm^2$)	R_{sh} ($\Omega.cm^2$)
1	1.000	4.674	0.674	0.878	7.889	0.480	7.899	0.023	2.697	6.443	3392.119
2	1.000	4.683	0.328	0.776	18.376	0.721	18.509	12 943.355	5.014	0.460	59.020
3	1.000	4.622	0.344	0.789	17.044	7.833	17.160	3088.873	4.100	0.200	67.629
4	1.000	5.024	0.331	0.850	17.858	1.102	17.933	1401.235	4.057	0.200	64.984
5	1.200	10.684	0.598	0.947	18.866	1.784	0.019	0.331	2.074	9.648	586.058
6	1.500	9.813	0.638	0.992	15.508	1.176	15.662	0.238	2.151	9.981	1195.976
7	1.350	10.458	0.652	1.021	15.722	9.913	15.852	0.240	2.216	9.661	1491.744
8	1.000	12.188	0.671	1.010	17.978	5.195	18.051	5.99×10^{-5}	1.486	9.747	3184.295
9	1.000	10.948	0.668	1.011	16.215	2.060	16.263	3.976×10^{-3}	1.780	9.979	3951.715
10	1.000	9.349	0.592	0.968	16.306	0.420	16.581	39.718	2.949	5.514	352.362
11	1.000	9.408	0.642	0.990	14.811	3.974	15.061	0.002	1.705	11.318	828.667

In another study [101] the performance of planar $MAPbI_3$ PSC with the structure FTO/Cu-doped NiO_x/$MAPbI_3$/PCBM/LiF/Al was investigated under 0.01–1 sun illumination. *J–V* characteristics revealed space-charge-limited current (SCLC) behavior dominated by drift at low voltages and diode-like diffusion currents at higher voltages. Increasing light intensity reduced defect-induced charge trapping, resulting in quasi-ohmic SCLC behavior. Ideality factor and efficiency were extracted by fitting $(J + J_{sc})$ versus $(V–JR_s)$ under high-bias conditions [101]. As shown in figure 6.6(b) n showed two regimes: it remained stable between 1.62 and 1.65 at low intensities ($\leqslant$0.1 sun), reflecting mixed monomolecular and bimolecular recombination, but increased steadily beyond 0.25 sun, approaching $n \approx 2$, signaling dominant monomolecular recombination [102–104]. It has been found that as intensity increases, efficiency decreases from 1.36 at 0.01 sun to 1.16 at 0.8 sun, indicating light-induced redistribution of trapped charges and suppressed SCLC behavior [102, 105]. Additionally, R_s and J_0 were extracted from the *J–V* fitting. R_s remained nearly constant across light intensities (figure 6.6(c)) and aligned well with the values obtained via EIS, validating the model. In contrast, J_0 exhibited an exponential increase with increment in illumination intensity (figure 6.6(d)).

6.5 Simulation techniques

While analytical models offer valuable insights into the functioning of PV cells and are relatively straightforward to solve manually, they become increasingly complex when additional operational conditions are introduced. To address this, a wide range of analytical and numerical simulation models have been developed worldwide to accurately represent the behavior of solar cells [106].

In the context of PSCs, simulation encompasses a broad spectrum of physical phenomena spanning multiple spatial and temporal scales. It can be broadly categorized into electrical, optical and multiphysics domains, each focusing on specific aspects of the device operation [107–110]. A diverse range of specialized tools has been developed to support these simulations, each tailored to a particular level of complexity and physical modeling. These simulation tools allow to investigate the performance, stability, and degradation mechanisms of PSCs.

6.5.1 Electrical simulation

Electrical simulations are essential for evaluating the performance of PSCs by modeling charge carrier generation, transport, recombination, and extraction using the drift-diffusion equations. The simulation accurately predicts performance parameters such as *I–V* characteristics, V_{oc}, J_{sc}, FF, and η by incorporating various recombination mechanisms, including Shockley–Read–Hall (SRH) recombination, radiative, and Auger processes, as well as trap-assisted and interfacial effects. Time-dependent simulations are used to study the dynamic behaviors such as hysteresis, ion migration, and interface quality, supporting analysis via techniques such as impedance spectroscopy and capacitance–voltage (*C–V*) profiling method.

SCAPS-1D, developed by Professor M Burgelman [111, 112] at the University of Ghent, is a widely used tool for simulating one-dimensional multi-layered solar cell

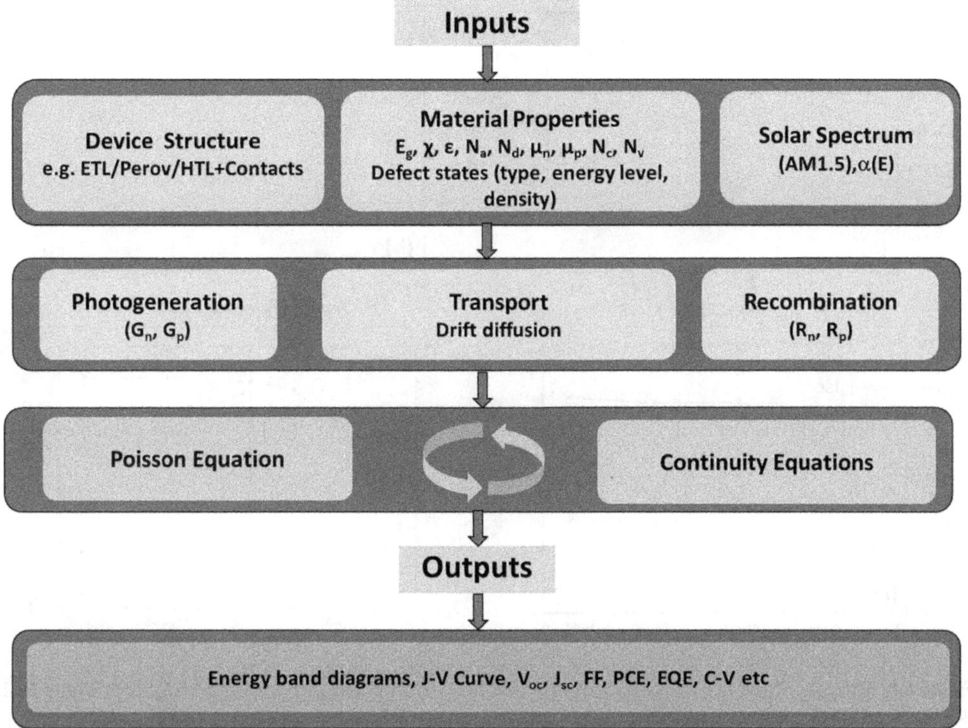

Figure 6.7. Simplified flowchart illustrating the simulation process, including key inputs and outputs of the SCAPS-1D simulator. (Adapted from [114]. CC BY 4.0.)

structures. It employs a self-consistent iterative approach combining the Gummel method with Newton–Raphson iteration to solve semiconductor equations. Using the drift-diffusion model, SCAPS-1D accurately simulates photogenerated carrier transport by accounting for electric fields, concentration gradients, and boundary conditions at interfaces and contacts.

wxAMPS, an extension of the AMPS [113] family, is a flexible simulation tool well-suited for modeling multijunction, tunnel-junction devices, and also has advanced options for defining recombination mechanisms. While powerful, it is less intuitive and has a steeper learning curve compared to SCAPS-1D. Together, tools such as SCAPS-1D and wxAMPS enable comprehensive modeling and optimization of both lead-based and lead-free PSCs for high-efficiency and stability through simulation-driven design.

A simplified flowchart in figure 6.7 [114] illustrates the important equations, inputs, and outputs of the simulator. The basic equations involved in the simulation are presented in section 6.3.2. The simulation uses the standard AM 1.5 G solar spectrum (1000 W m^{-2}) and incorporates the absorption coefficient, α, which is defined as a function of photon energy [112]. For each layer in the device, essential parameters are provided, including thickness, dielectric permittivity (ε), bandgap (E_g), electron affinity (χ), effective density of states (DOS) in the conduction (N_C) and valence bands (N_V), as well as charge carrier mobilities. Doping concentrations, either acceptor (N_A) or

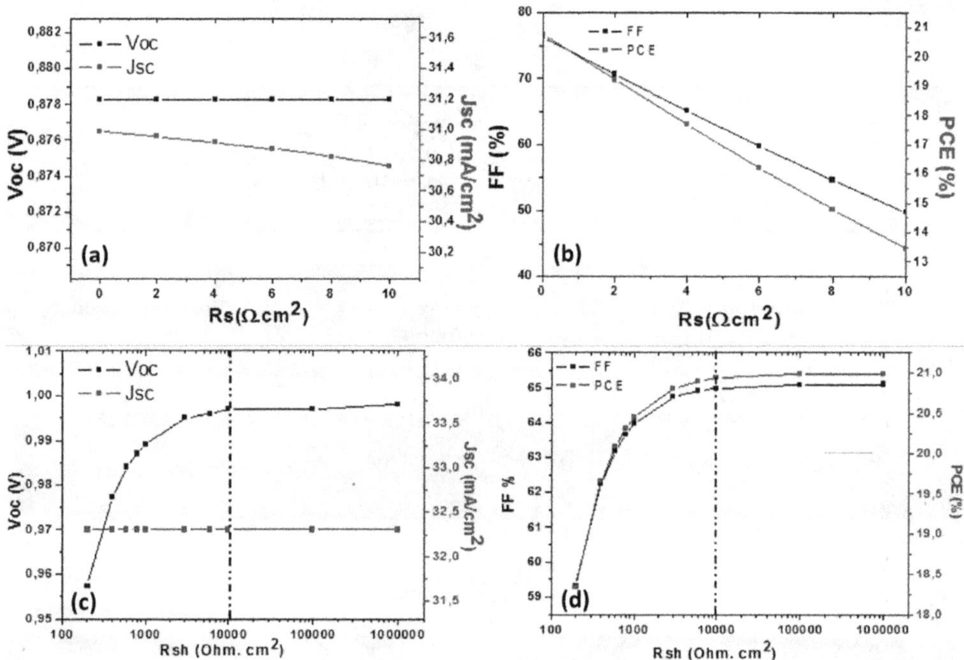

Figure 6.8. Effect of R_s on (a) V_{oc} and J_{sc}, (b) FF and PCE. Effect of R_{sh} on (c) V_{oc} and J_{sc}, (d) FF and PCE. (Reproduced from [115]. CC BY 4.0.)

donor (N_D), are also specified. The simulator accounts for both bulk and interface defects. As depicted in figure 6.7, the outputs include physical parameters such as energy band diagrams, electric field distribution, and recombination rates, along with terminal characteristics such as I–V, C–V, and EQE curves [112].

SCAPS-1D simulations were carried out on PSCs structure, FTO/ZnO/$MASnI_3$/NiO_x/Au to optimize the device performance parameters [115]. The optimal absorber thickness was identified as 0.6 μm, for achieving maximum PCE. As R_s values increased from 0 to 10 $\Omega \cdot cm^2$, FF and PCE declined sharply due to increased leakage currents, while V_{oc} remained relatively stable and J_{sc} notably decreased (figures 6.8(a) and (b)), consistent with observations reported in [116]. Conversely to the impact of R_s, J_{sc} remained largely unaffected by changes in R_{sh}, as shown in figure 6.8(c). Instead, V_{oc} significantly increased with R_{sh} values rising from 200 to 1000 $\Omega \cdot cm^2$, plateauing at higher values. This improvement in V_{oc} led to enhanced FF and overall PCE, as seen in figure 6.8(c). These results align with earlier reports [117, 118], confirming that higher R_{sh} effectively reduces power losses. R_{sh} in the range of 10^4 to 10^6 $\Omega.cm^2$ is recommended to achieve high and stable solar cell efficiency.

6.5.2 Optical simulation

Optical simulations are essential for optimizing the light management in PSCs. They are performed by analysing the interactions of light with the various layers in the device. Analytical approaches such as the transfer matrix method (TMM) [119] are particularly effective for modeling the planar PSC structures, enabling quick and accurate calculations of reflectance, transmittance, and absorptance by solving Maxwell's equations. Tools such as GenPro4 [120], OpenFilters [121], and TMM-Sim [122] utilize this method to optimize coatings and layer thicknesses, with TMM-Sim offering intuitive visualizations.

For more advanced architectures with textures or nanostructures, finite-difference time-domain (FDTD) methods [123] are preferred, as they solve Maxwell's equations in space and time, accurately modeling non-planar and sub-wavelength features. Commercial tools such as Lumerical FDTD [124] are adopted widely in the field for designing nanophotonic structures and antireflection coatings. Additionally, the finite-element-based solver JCMsuite [125] is gaining prominence for its high spatial resolution, making it well-suited for analysing optical fields in complex perovskite nanostructures.

A study [126] explores the impact of perovskite layer thickness and anti-reflective (AR) structures on optical absorption and PCE in planar PSCs under varying illumination. Thicker $MAPbI_3$ layers improve light harvesting in low-light conditions (0.1–0.5 sun), while thinner layers are advantageous under strong illumination (> 1 sun) due to reduced recombination [127]. A 3D FDTD simulation [126] using Lumerical analyses, an ITO/TiO_2/$MAPbI_3$/Spiro-OMeTAD/Ag structure, incorporating a corrugated void-like AR layer to minimize reflectance and improve light coupling has been discussed in the literature [128, 129].

In this simulation, Maxwell's equations were employed to simulate light propagation and absorption, while Poisson's equation was used to account for photo-induced electric fields and charge transport [31]. The absorbed optical power per unit volume P_{Abs} was calculated using the following equations:

$$P_{Abs} = 1/2\, \omega\varepsilon''|\vec{E}|^2 \tag{6.44}$$

$$\text{Abs}(\lambda) = \int P_{Abs}(\lambda)dV. \tag{6.45}$$

Here, ω is the angular frequency, (ε'') is the imaginary component of the dielectric function, and ($|E|^2$ in $V^2{\cdot}m^{-2}{\cdot}Hz^{-1}$) is the square of the electric field intensity.

Spatial absorption spectra and electric field maps showed that the absorption below 2.5 eV is highly thickness-dependent, with a 750 nm $MAPbI_3$ layer delivering the highest absorption due to extended optical paths (figure 6.9(a)). At photon energies above 2.5 eV, absorption becomes thickness independent. Reflectance data (figure 6.9(b)) revealed that the thinner perovskite layers reflect more light, confirming the benefit of using thicker layers in terms of light trapping. The corrugated AR design further enhances light trapping, confirming that combining absorber thickness optimization with AR structures significantly improves optical performance and the overall device efficiency.

SCAPS-1D simulation study [130] evaluated $NH_3(CH_2)_2NH_3MnCl_4$ as a lead-free absorber in PSC structure FTO/TiO_2/$NH_3(CH_2)_2NH_3MnCl_4$/Spiro-OMeTAD/Au. As shown in figure 6.9(c), increment in the absorber thickness from 50 to 1000 nm enhances J_{sc} due to improved photon absorption, but leads to reduced V_{oc} and FF due to greater carrier recombination. An optimal thickness of ~600 nm achieves the highest PCE by balancing absorption and recombination losses.

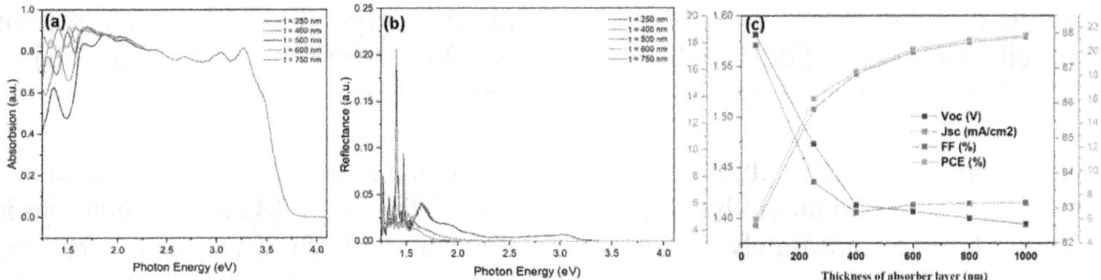

Figure 6.9. (a) The absorption and (b) reflectance spectra of PSC with different thicknesses of $MAPbI_3$ perovskite layer. (c) Photovoltaic performance parameters of the device structure: FTO/TiO_2/$NH_3(CH_2)_2NH_3MnCl_4$/Spiro-OMeTAD/Au for varying perovskite absorbing layer. ((a) and (b) Reproduced with permission from [126]. Copyright 2023 The Author(s). Published by IOP Publishing Ltd. CC BY 4.0. (c) Reproduced from [130]. CC BY 4.0.)

Combining optical simulations with electrical modeling enables a multiscale approach that links structural and material properties to overall device performance. This integrated approach links material and structural design to device-level performance, enabling more efficient light trapping and reduced optical losses in high-efficiency perovskite photovoltaics.

6.5.3 Multiphysics simulation

Multiphysics simulations integrate electrical, thermal, mechanical, and ionic transport phenomena into a single modeling, enabling a comprehensive understanding of the behavior of PSCs. These simulations are crucial for capturing the complex interactions between performance-limiting processes and degradation pathways, offering more accurate and predictive analysis compared to conventional single-physics models. An important aspect is ion migration, where mobile ions such as I^-, MA^+, and FA^+ redistribute under electric fields and thermal gradients, impacting internal fields, inducing hysteresis, and contributing to long-term instability. Ion migration modeling generally couples Poisson's equation, drift-diffusion, and Nernst–Planck formulations. Thermal simulations further contribute to understanding PSC stability, as elevated temperatures—caused by light absorption and resistive heating—accelerate degradation through enhanced ion movement, phase transitions, and chemical decomposition.

Finite-element modeling (FEM) [131] tools such as COMSOL Multiphysics [108], ANSYS [132], and Sentaurus TCAD [133] have capabilities for multi-domain simulations. COMSOL stands out as a versatile commercial platform that enables the coupling of multiple physical phenomena, making it especially valuable for studying stability, degradation, and ion migration. It also facilitates mechanical simulations to assess stress and deformation, informing strategies such as flexible substrates and stress-buffering layers. While Sentaurus TCAD and Silvaco ATLAS [134] provide advanced semiconductor modeling, their complexity and cost can limit academic use. Nevertheless, such tools are critical for high-fidelity modeling of PSC operation and reliability under real-world conditions.

TMM and SCAPS-1D simulations [135] were used to evaluate the J–V characteristics and quantum efficiency of $CH_3NH_3PbI_3$-based perovskite solar cells with various ETL and HTL combinations. Figure 6.10(a) and (b) show the effects of various ETLs on J–V and QE of PSCs having P3HT as HTL. Among the materials studied, ITO exhibits the lowest J_{sc} of 22.728 mA cm^{-2} and poor QE (figure 6.10(b)), attributed to its low carrier mobilities (10 cm^2 V^{-1} s^{-1}), which hinder effective charge collection. In contrast, TiO_2 achieves the highest V_{oc} of 1.266 V and FF of 71.49%, resulting in PCE of 20.71%. SnO_2 and ZnO also demonstrate strong performance, with improved QE and comparable PCEs of 20.88% and 20.95%, respectively, supported by similar device parameters ($V_{oc} \approx 1.27$ V, $J_{sc} \approx 23.08$ mA cm^{-2}, FF $\approx$ 71.5%).

Figures 6.10(c) and (d) display the J–V characteristics and QE of PSCs for various HTL combinations. The results indicate that the choice of HTLs has minimal impact on solar cell performance, primarily due to their placement at the rear of the device. However, the J–V curve reveals a notable improvement in performance, particularly in terms of J_{sc} and FF.

The results highlight the superiority of the optimized ZnO/NiO-based PSC over the conventional ITO/P3HT structure, showing significant improvements of

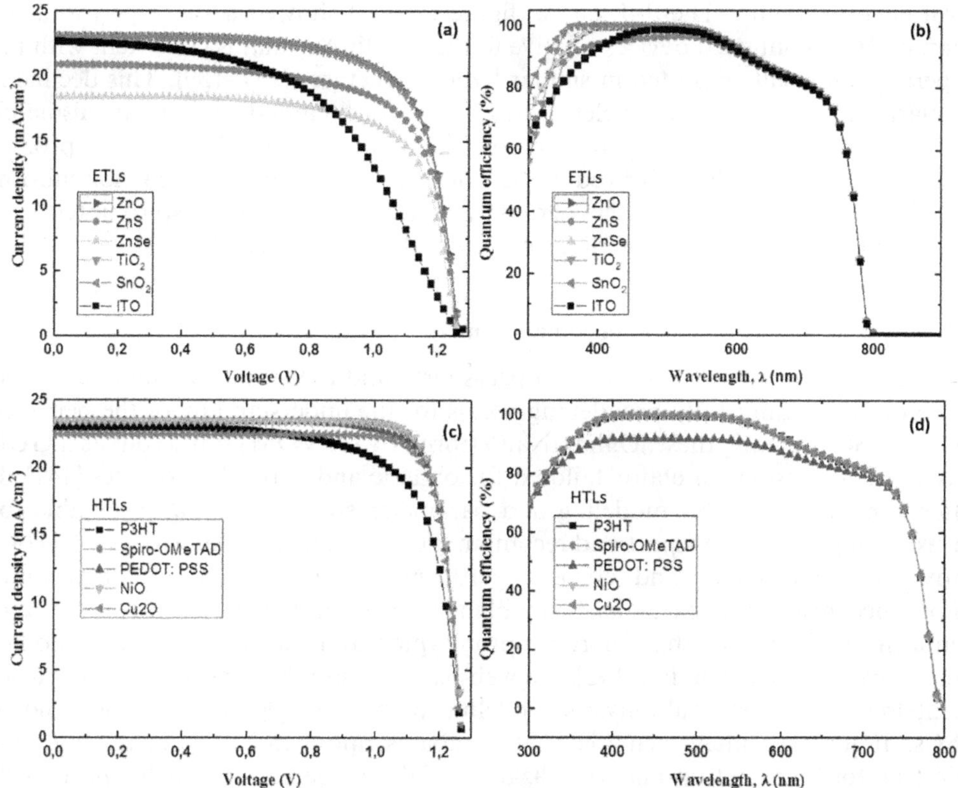

Figure 6.10. Effect of various ETLs on the performance of $CH_3NH_3PbI_3$-based PSCs using P3HT as HTL: (a) current–voltage (J–V) characteristics; (b) quantum efficiency (QE) spectra. Effect of HTLs on the performance of $CH_3NH_3PbI_3$-based PSCs employing ZnO as ETL: (c) J–V characteristics; (d) QE spectra. (Reproduced with permission from [135]. Copyright 2020 Elsevier.)

approximately 9.81% in PCE and 31.25% in FF. This integrated optical and electrical simulation approach effectively optimizes the device performance by evaluating both light absorption and charge transport, providing valuable insights into the material selection for high-efficiency PSC design.

Another recent study [136] shows that Multiphysics simulations streamline the design of semitransparent (ST) PSCs by optimizing the materials and minimizing prototyping. $MAPbI_3$ is identified as an efficient absorber, achieving 4.15% light utilization at 100 nm thickness. A MoO_3/Ag/WO_3 rear electrode further enhances both PCE and average visible transmittance, supporting the potential of ST-PSCs for applications such as in transparent windows and building-integrated photovoltaics [136].

A 3D Multiphysics simulation using COMSOL was conducted to analyse the optical generation, electrical behavior, and heat distribution in PSCs [108]. The study examined the effect of replacing the conventional metallic bottom contact with reduced graphene oxide (RGO) to improve thermal stability. Although RGO had minimal influence on the optical and electrical performance, its high thermal conductivity significantly enhanced heat dissipation, reducing heat accumulation and thermal gradients at the RGO/Spiro interface when compared to Au contacts [108]. This suggests improved thermal management and potential for enhanced long-term device stability. The J–V curve (figure 6.11(a)) showed a slight performance drop in RGO-contacted devices relative to those with Au; that is consistent with the experimental results reported in similar hybrid structures [137–139]. This decline is primarily attributed to incomplete carrier extraction caused by orbital mismatch between graphene and the planar perovskite lattice [140]. Electric field profiles (figures 6.11(b) and (c)), derived from Poisson's equation, revealed nonuniform distributions with peaks at the perovskite/TiO_2 and perovskite/spiro interfaces, resulting from charge accumulation in heavily doped regions [108].

6.5.4 Emerging and open-source simulation

In recent years, a new generation of open-source and user-friendly simulation tools has emerged, significantly broadening access for the understanding of the performance of PSCs. Among these, OghmaNano (formerly GPVDM) stands out as a GUI-based drift-diffusion simulator tailored for organic and perovskite devices [141]. It offers support for defect modeling and parameter sweeps, making it suitable for investigating charge transport and recombination mechanisms. IonMonger 2.0 [142] provides a lightweight and efficient platform for simulating J–V curves and impedance responses, particularly in planar device architectures. Its speed and simplicity make it valuable for rapid prototyping and screening studies. Another innovative tool, SolarDesign [143], is a web-based, customizable simulation environment that facilitates Multiphysics modeling of both single-junction and tandem PSCs. It includes integrated material databases and band alignment diagrams, allowing for quick set-up and visualization of device behavior. Furthermore, dPV [144] introduces a differentiable drift-diffusion simulation, which incorporates automatic differentiation to enable machine-learning-based optimization of device parameters. Collectively, these emerging tools provide flexible, accessible options

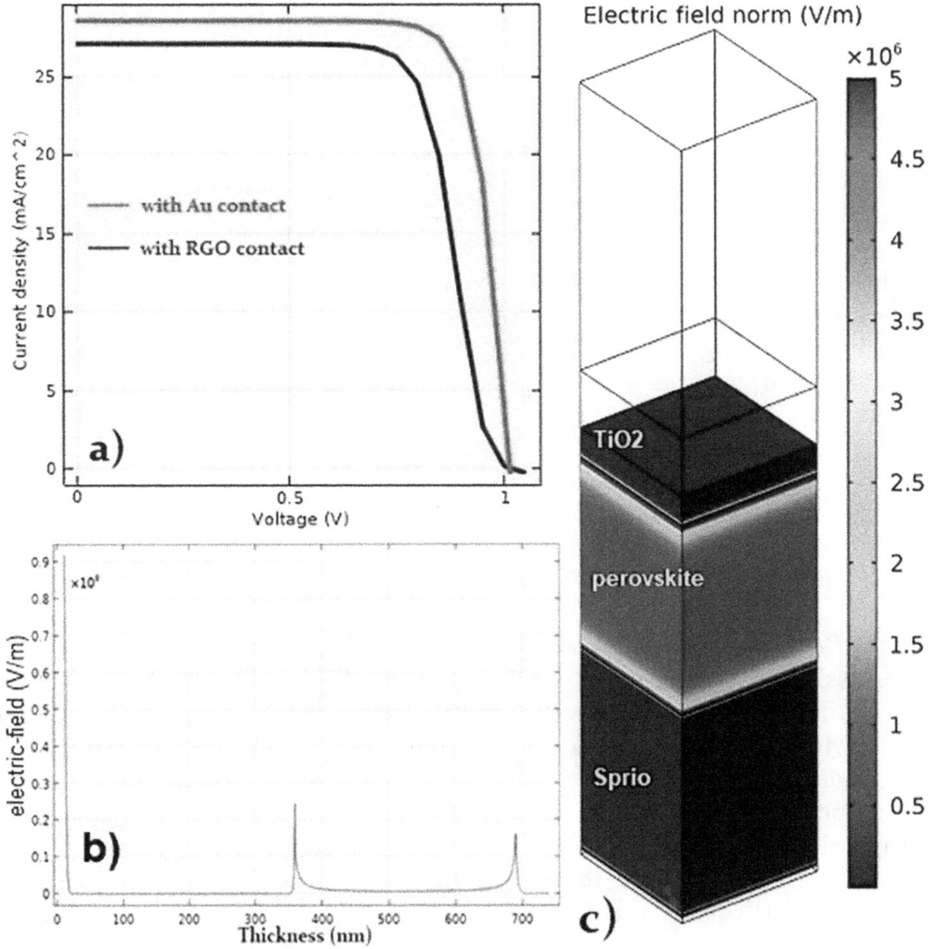

Figure 6.11. (a) *J–V* characteristics of the PSC with RGO and conventional Au electrodes, (b) 1D electric field distribution across the device structure, and (c) 3D electric field profile highlighting peak intensities at the spiro-MeOTAD/perovskite and perovskite/TiO_2 interfaces. (Reproduced with permission from [108]. Copyright 2020 Elsevier.)

that complement traditional simulation platforms and support accelerated development and understanding of the next-generation PSCs.

In order to capture long-term behavior and ion-related phenomena in PSCs, Driftfusion [145] serves as a powerful open-source simulation tool. It is specifically designed to model mobile ion dynamics, charge trapping, and interfacial recombination-critical processes that influence hysteresis, stability, and degradation. While the tool requires a scripting-based approach, it offers high flexibility and precision for investigating the transient and degradation effects. Although not semiconductor-specific, ANSYS [132] contributes significantly to mechanical reliability assessments and optimization of device packaging strategies. These tools enable deeper insights and more effective design of stable, high-performance PSCs through comprehensive modeling at multiple scales.

6.5.5 Density functional theory

At the atomistic scale, density functional theory (DFT) tools such as VASP [146] and Quantum ESPRESSO [147] are used to investigate the electronic structure, defect formation, and ion migration in perovskite materials, aiding the development of lead-free and hybrid compositions. Complementary molecular dynamics (MD) simulation tools such as GROMACS [148] and LAMMPS [149] explore the structural dynamics, ion diffusion, and thermal properties of perovskite layers, revealing degradation mechanisms and phase transitions.

MD and DFT together offer a robust framework for predicting the material properties, evaluating device performance, and understanding long-term stability. Their combined use enables detailed insights into the thermal effects, phase transitions, and ion migration-key factors in assessing the degradation of perovskites under real-world conditions. DFT, particularly with hybrid functionals (e.g. HSE06), is crucial for analysing the electronic and optoelectronic properties of perovskites, including the band structure, carrier mobility, and light absorption [150]. Advanced DFT methods are essential for accurate predictions, especially for materials such as $MAPbI_3$ and $FAPbBr_3$.

6.5.5.1 Effect of composition on electronic properties

The perovskite crystal structure—ABX_3 (figure 6.1(a)) offers significant compositional flexibility, allowing for tailored bandgap tuning and improved structural stability. Modifying the A-site cation—such as replacing methylammonium (MA^+) with formamidinium (FA^+) or cesium (Cs^+)—alters lattice symmetry and stabilizes different phases, which can narrow the bandgap and enhance light absorption. Adjusting the halide at the X-site, for instance substituting iodide (I^-) with bromide (Br^-) or chloride (Cl^-), shifts the absorption edge toward shorter wavelengths; this approach is particularly beneficial for tandem and color-specific optoelectronic applications.

DFT simulations have played a crucial role in quantifying these effects, offering atomistic insight into how chemical substitutions across the A, B, and X sites influence key electronic properties, including bandgap nature (direct versus indirect), defect states, and overall material stability. These insights support rational design strategies for high-performance and application-specific perovskite materials [151].

6.5.5.2 Defect and interface modeling

Defects significantly impact the performance and stability of perovskite materials. DFT enables the calculation of defect formation energies and charge transition levels, helping to identify deep versus shallow traps that affect recombination and carrier lifetimes. For example, iodine vacancies are commonly found to introduce shallow donor levels, whereas Pb-related defects often lead to deep traps. Interface modeling between perovskites and charge transport layers can also be performed to investigate band offsets, dipole formation, and potential barriers. These insights are crucial for engineering high-efficiency, low-loss interfaces in solar cells and LEDs [152].

6.5.5.3 Strain and phase engineering

Mechanical strain and phase transitions also play a role in tuning the electronic properties of perovskites. Using DFT, the effect of tensile or compressive strain on band alignment and mobility have been explored. Strain-induced changes in orbital overlap can lead to bandgap narrowing or widening. Additionally, perovskites undergo phase transitions (e.g. cubic to tetragonal) depending on the temperature and composition. These transitions alter the symmetry and bond angles in the crystal lattice, influencing the electronic behavior of perovskites. Simulations can predict these transitions and guide stability improvements under operational conditions [153]. Emerging applications of DFT in perovskite research focus on strain engineering, dimensional confinement, and surface termination effects to optimize the bandgap and charge carrier dynamics. Additionally, DFT supports the design of defect-tolerant materials and passivation strategies aimed at enhancing both performance and longevity.

A combined DFT and SCAPS-1D study [154] investigates $Cs_2SnI_{6-x}Br_x$ alloys as lead-free absorber materials for PSCs. Bromine substitution at iodine sites increases the bandgap from 1.33 to 2.24 eV and enhances visible light absorption. Device simulations using device structure FTO/TiO_2/Cs_2SnI_6/P_3HT/Ag showed a PCE improvement from 0.47% to 3.07% with optimized design. Band structure and DOS calculations (figure 6.12) for $x = 1–5$ using the TB-mBJ potential confirm a rising intermediate band energy and a direct bandgap that closely matches the experimental values, validating the accuracy of the TB-mBJ approach.

The simulation tools for PSCs, including their types, applications and limitations are summarized in table 6.2.

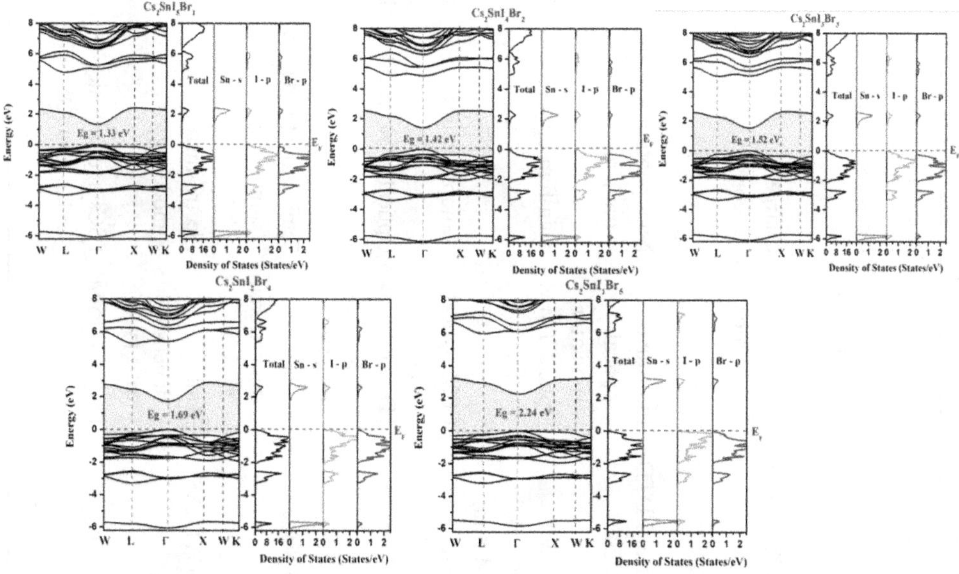

Figure 6.12. Electronic band structures and density of states (DOS) profiles for $Cs_2SnI_{6-x}Br_x$ alloys with bromine concentrations of $x = 1, 2, 3, 4$, and 5. (Reproduced from [154]. CC BY 4.0.)

Table 6.2. Summary of simulation tools for PSCs.

Tool name	Type	Application area	Key features/strengths	Limitations	Developer/source	References
SCAPS-1D	1D electrical simulator	Device-level modeling	User-friendly GUI, JV/EQE/CV simulation, drift-diffusion model	Limited to 1D, no complex geometry	University of Ghent	[111]
wxAMPS /AMPS-1D	1D electrical simulator	Charge transport, recombination	Trap modeling, tunnel/multijunction support	Older interface, less intuitive	Penn State University (original AMPS)	[113]
Lumerical FDTD / MODE	Full-wave optical simulator	Nanophotonics, light trapping	Solves Maxwell's equations, accurate field visualization	High computa tional demand	ANSYS (formerly Lumerical)	[124]
TMM Tools (GenPro4, TMM-Sim)	Optical TMM solver	Thin-film optics	Efficient for planar layers, generation profiles, field intensity	Limited to planar stacks	Open-source/academic	[119], [122]
JCMsuite	FEM optical simulator	Nanophotonics	Finite-element EM solver, field visualization	Commercial, set-up complexity	JCMwave	[125]
Setfos (Fluxim)	1D opto electronic simulator	Optical + electrical	Drift-diffusion + optical integration, commercial support	Paid license required	Fluxim AG	[155]
TCAD Sentaurus	Multi-dimensional TCAD	Semiconductor device simulation	Advanced physics (quantum, traps, doping), 2D/3D modeling	Commercial, complex set-up	Synopsys	[133]
Silvaco ATLAS	Multi-dimensional TCAD	Electro-thermal modeling	Field-driven transport, photo-generation, recombination	Steep learning curve	Silvaco	[134]
OghmaNano	Drift-diffusion electrical tool	PSCs and organic solar cells (OSCs)	Open-source, GUI-based, good for parameter sweeps	Limited documentation/support	Open-source (formerly GPVDM)	[141]

(*Continued*)

Table 6.2. (*Continued*)

Tool name	Type	Application area	Key features/strengths	Limitations	Developer/source	References
IonMonger 2.0	Compact electrical model	Planar PSCs	Fast JV/impedance simulations, ionic transport modeling	Simplified physics versus full simulators		[142]
SolarDesign	Web-based multi-physics tool	Tandem and single-junction PSCs	Online interface, editable material libraries, drift-diffusion + optics	Web-based, evolving feature set	SolarDesign	[143]
dPV	Differentiable drift-diffusion	Optimization, ML integration	Auto-differentiation, Python-based optimization	Scripting required	Open-source	[144]
COMSOL Multiphysics	FEM multi-physics tool	Coupled optical, thermal, ionic modeling	Strong multi-physics capabilities: thermal, electrical, structural	Commercial, steep learning curve	COMSOL Inc.	[108]
Driftfusion	Drift-diffusion solver	Ionic-electronic PSCs	Open-source, simulates mobile ions, recombination	Requires scripting	Open-source	[145]
VASP / QE	DFT atomistic simulators	Electronic structure, defect modeling	Band structure, defect formation energy, migration paths	Very high computational cost	VASP GmbH / Quantum ESPRESSO Consortium	[146, 147]
GROMACS / LAMMPS	Molecular dynamics	Ion dynamics, grain boundaries	Atomic-level simulations, perovskite stability under thermal/mechanical stress	Not for direct solar cell simulation	Open-source	[148, 149]
ANSYS	Thermal and mechanical FEM	Packaging and reliability	Simulates delamination, cracking, heat dissipation	Not solar-specific	ANSYS Inc.	[132]

6.6 Conclusions

This chapter provides a comprehensive overview of the analytical and numerical simulation techniques for evaluating PSC performance parameters and diode parameters using single- and double-diode models. It discusses the impact of these parameters on device performance and provides insights into spectral response, quantum efficiency, and carrier lifetime, while emphasizing multiscale simulation approaches such as DFT, for developing efficient and stable perovskites and PSCs.

References

[1] Parida B, Iniyan S and Goic R 2011 A review of solar photovoltaic technologies *Renew. Sustain. Energy Rev.* **15** 1625–36

[2] Green M A, Ho-Baillie A and Snaith H J 2014 The emergence of perovskite solar cells *Nat. Photonics* **8** 506–14

[3] NREL 2025 Best Research-Cell Efficiency Chart *US Department of Energy, Office of Energy Efficiency and Renewable Energy* https://www2.nrel.gov/pv/cell-efficiency

[4] Humada A M *et al* 2016 Solar cell parameters extraction based on single and double-diode models: a review *Renew. Sustain. Energy Rev.* **56** 494–509

[5] Sarkar M N I 2016 Effect of various model parameters on solar photovoltaic cell simulation: a SPICE analysis *Renew.: Wind, Water, Solar* **3** 13

[6] Villalva M G, Gazoli J R and Ruppert Filho E 2009 Comprehensive approach to modeling and simulation of photovoltaic arrays *IEEE Trans. Power Electron.* **24** 1198–208

[7] Gontean A *et al* 2017 A novel high accuracy PV cell model including self heating and parameter variation *Energies* **11** 36

[8] Yahya-Khotbehsara A and Shahhoseini A 2018 A fast modeling of the double-diode model for PV modules using combined analytical and numerical approach *Solar Energy* **162** 403–9

[9] Chin V J, Salam Z and Ishaque K 2016 An accurate modelling of the two-diode model of PV module using a hybrid solution based on differential evolution *Energy Convers. Manag* **124** 42–50

[10] Cotfas D T, Cotfas P A and Kaplanis S 2013 Methods to determine the dc parameters of solar cells: a critical review *Renew. Sustain. Energy Rev.* **28** 588–96

[11] Bashahu M and Nkundabakura P 2007 Review and tests of methods for the determination of the solar cell junction ideality factors *Solar Energy* **81** 856–63

[12] Wang G *et al* 2017 An iterative approach for modeling photovoltaic modules without implicit equations *Appl. Energy* **202** 189–98

[13] Wong J 2013 Griddler: intelligent computer aided design of complex solar cell metallization patterns *2013 IEEE 39th Photovoltaic Specialists Conf. (PVSC)* (Piscataway, NJ: IEEE)

[14] Liao P *et al* 2018 A new method for fitting current–voltage curves of planar heterojunction perovskite solar cells *Nanomicro Lett.* **10** 5

[15] Im J-H *et al* 2011 6.5% efficient perovskite quantum-dot-sensitized solar cell *Nanoscale* **3** 4088–93

[16] Kim H-S *et al* 2012 Lead iodide perovskite sensitized all-solid-state submicron thin film mesoscopic solar cell with efficiency exceeding 9% *Sci. Rep.* **2** 591

[17] Burschka J *et al* 2013 Sequential deposition as a route to high-performance perovskite-sensitized solar cells *Nature* **499** 316–9

[18] Shin S S, Lee S J and Seok S I 2019 Metal oxide charge transport layers for efficient and stable perovskite solar cells *Adv. Funct. Mater.* **29** 1900455

[19] Shaikh S F *et al* 2018 Performance enhancement of mesoporous TiO_2-based perovskite solar cells by ZnS ultrathin-interfacial modification layer *J. Alloys Compd* **738** 405–14

[20] Zhou Y, Li X and Lin H 2020 To be higher and stronger—metal oxide electron transport materials for perovskite solar cells *Small* **16** 1902579

[21] Song J-X *et al* 2021 Low-temperature-processed metal oxide electron transport layers for efficient planar perovskite solar cells *Rare Metals* **40** 2730–46

[22] Warby J *et al* 2022 Understanding performance limiting interfacial recombination in *pin* perovskite solar cells *Adv. Energy Mater.* **12** 2103567

[23] Niu G, Guo X and Wang L 2015 Review of recent progress in chemical stability of perovskite solar cells *J. Mater. Chem.* A **3** 8970–80

[24] Conings B *et al* 2015 Intrinsic thermal instability of methylammonium lead trihalide perovskite *Adv. Energy Mater.* **5** 1500477

[25] Liang L *et al* 2018 All that glitters is not gold: recent progress of alternative counter electrodes for perovskite solar cells *Nano Energy* **52** 211–38

[26] Azpiroz J M *et al* 2015 Defect migration in methylammonium lead iodide and its role in perovskite solar cell operation *Energy Environ. Sci.* **8** 2118–27

[27] Kundu S and Kelly T L 2020 *In situ* studies of the degradation mechanisms of perovskite solar cells *EcoMat* **2** e12025

[28] Jung H S and Park N G 2015 Perovskite solar cells: from materials to devices *Small* **11** 10–25

[29] Kittel C and McEuen P 2018 *Introduction to Solid State Physics* (John Wiley and Sons)

[30] Sze S M, Li Y and Ng K K 2021 *Physics of Semiconductor Devices* (John Wiley and Sons)

[31] Shi J *et al* 2016 Interplays between charge and electric field in perovskite solar cells: charge transport, recombination and hysteresis arXiv:1604.02819

[32] Minemoto T and Murata M 2014 Impact of work function of back contact of perovskite solar cells without hole transport material analyzed by device simulation *Curr. Appl. Phys.* **14** 1428–33

[33] Ravindra N M, Lin L and Singh P 2025 *Recent Advances in Solar Cells* (IOP Publishing.)

[34] King D *et al* 1997 Dark current–voltage measurements on photovoltaic modules as a diagnostic or manufacturing tool *Conf. Record of the Twenty Sixth IEEE Photovoltaic Specialists Conf.* (Piscataway, NJ: IEEE)

[35] Ishaque K and Salam Z 2011 A comprehensive MATLAB Simulink PV system simulator with partial shading capability based on two-diode model *Solar Energy* **85** 2217–27

[36] Franzitta V, Orioli A and Gangi A D 2017 Assessment of the usability and accuracy of two-diode models for photovoltaic modules *Energies* **10** 564

[37] Chin V J, Salam Z and Ishaque K 2015 Cell modelling and model parameters estimation techniques for photovoltaic simulator application: a review *Appl. Energy* **154** 500–19

[38] Abbassi R *et al* 2018 Identification of unknown parameters of solar cell models: a comprehensive overview of available approaches *Renew. Sustain. Energy Rev.* **90** 453–74

[39] Shannan N M A A, Yahaya N Z and Singh B 2013 Single-diode model and two-diode model of PV modules: a comparison *2013 IEEE International Conference on Control System, Computing and Engineering* (Piscataway, NJ: IEEE)

[40] Chegaar M, Nehaoua N and Bouhemadou A 2008 Organic and inorganic solar cells parameters evaluation from single *I–V* plot *Energy Convers. Manag.* **49** 1376–9

[41] Easwarakhanthan T *et al* 1986 Nonlinear minimization algorithm for determining the solar cell parameters with microcomputers *Int. J. Solar Energy* **4** 1–12

[42] Boutana N *et al* 2017 An explicit *IV* model for photovoltaic module technologies *Energy Convers. Manag.* **138** 400–12

[43] Lim L H I *et al* 2015 A linear method to extract diode model parameters of solar panels from a single *I–V* curve *Renew. Energy* **76** 135–42

[44] Chegaar M, Azzouzi G and Mialhe P 2006 Simple parameter extraction method for illuminated solar cells *Solid-State Electron.* **50** 1234–7

[45] Ruschel C S, Gasparin F P and Krenzinger A 2021 Experimental analysis of the single diode model parameters dependence on irradiance and temperature *Solar Energy* **217** 134–44

[46] Celik A N and Acikgoz N 2007 Modelling and experimental verification of the operating current of mono-crystalline photovoltaic modules using four- and five-parameter models *Appl. Energy.* **84** 1–15

[47] Kennerud K L 1969 Analysis of performance degradation in CdS solar cells *IEEE Trans. Aerosp. Electron. Syst.* **AES-5** 912–7

[48] Shongwe S and Hanif M 2015 Comparative analysis of different single-diode PV modeling methods *IEEE J. Photovolt.* **5** 938–46

[49] Ortiz-Conde A *et al* 2014 A review of diode and solar cell equivalent circuit model lumped parameter extraction procedures *Facta Univ. Ser.: Electron. Energ.* **27** 57–102

[50] Tivanov M *et al* 2005 Determination of solar cell parameters from its current–voltage and spectral characteristics *Sol. Energy Mat. Sol. Cells.* **87** 457–65

[51] Sulyok G and Summhammer J 2018 Extraction of a photovoltaic cell's double-diode model parameters from data sheet values *Energy Sci. Eng.* **6** 424–36

[52] Dehghanzadeh A, Farahani G and Maboodi M 2017 A novel approximate explicit double-diode model of solar cells for use in simulation studies *Renew. Energy* **103** 468–77

[53] Hovinen A 1994 Fitting of the solar cell IV-curve to the two diode model *Phys. Scr.* **1994** 175

[54] Sandrolini L, Artioli M and Reggiani U 2010 Numerical method for the extraction of photovoltaic module double-diode model parameters through cluster analysis *Appl. Energy* **87** 442–51

[55] Garrido-Alzar C 1997 Algorithm for extraction of solar cell parameters from *I–V* curve using double exponential model *Renew. Energy* **10** 125–8

[56] Hejri M *et al* 2014 On the parameter extraction of a five-parameter double-diode model of photovoltaic cells and modules *IEEE J. Photovolt.* **4** 915–23

[57] Ishaque K, Salam Z and Taheri H 2011 Simple, fast and accurate two-diode model for photovoltaic modules *Sol. Energy Mat. Sol. Cells.* **95** 586–94

[58] Gow J and Manning C 1996 Development of a model for photovoltaic arrays suitable for use in simulation studies of solar energy conversion systems *Sixth International Conference on Power Electronics and Variable Speed Drives* (Piscataway, NJ: IEEE)

[59] Bisquert J 2022 Interpretation of the recombination lifetime in halide perovskite devices by correlated techniques *J. Phys. Chem. Lett.* **13** 7320–35

[60] Vonk S *et al* 2020 Trapping and de-trapping in colloidal perovskite nanoplatelets: elucidation and prevention of nonradiative processes through chemical treatment *J. Phys. Chem.* C **124** 8047–54

[61] Baloch A A B *et al* 2018 Analysis of photocarrier dynamics at interfaces in perovskite solar cells by time-resolved photoluminescence *J. Phys. Chem.* C **122** 26805–15

[62] Sturman B, Podivilov E and Gorkunov M 2003 Origin of stretched exponential relaxation for hopping-transport models *Phys. Rev. Lett.* **91** 176602

[63] Batzelis E 2019 Non-iterative methods for the extraction of the single-diode model parameters of photovoltaic modules: a review and comparative assessment *Energies* **12** 358

[64] Jain A, Sharma S and Kapoor A 2006 Solar cell array parameters using Lambert *W*-function *Sol. Energy Mat. Sol. Cells.* **90** 25–31

[65] Song Z *et al* 2021 An effective method to accurately extract the parameters of single diode model of solar cells *Nanomaterials* **11** 2615

[66] Pindado S *et al* 2018 Assessment of explicit models for different photovoltaic technologies *Energies* **11** 1353

[67] De Soto W, Klein S A and Beckman W A 2006 Improvement and validation of a model for photovoltaic array performance *Solar Energy* **80** 78–88

[68] McIntosh K and Honsberg C 2000 The influence of edge recombination on a solar cell's IV curve *16th European Photovoltaic Solar Energy Conf.*

[69] Wolf M and Rauschenbach H 1963 Series resistance effects on solar cell measurements *Adv. Energy Convers.* **3** 455–79

[70] Singh P and Ravindra N 2012 Analysis of series and shunt resistance in silicon solar cells using single and double exponential models *Emerg. Mater. Res.* **1** 33–8

[71] Phang J, Chan D and Phillips J 1984 Accurate analytical method for the extraction of solar cell model parameters *Electron. Lett.* **20** 406–8

[72] Khan F *et al* 2013 Extraction of diode parameters of silicon solar cells under high illumination conditions *Energy Convers. Manage.* **76** 421–9

[73] Zidan M N *et al* 2021 Organic solar cells parameters extraction and characterization techniques *Polymers* **13** 3224

[74] Ibrahim H and Anani N 2017 Evaluation of analytical methods for parameter extraction of PV modules *Energy Procedia* **134** 69–78

[75] Xiao W, Dunford W G and Capel A 2004 A novel modeling method for photovoltaic cells *2004 IEEE 35th Annual Power Electronics Specialists Conf.* (Piscataway, NJ: IEEE)

[76] Saloux E, Teyssedou A and Sorin M 2011 Explicit model of photovoltaic panels to determine voltages and currents at the maximum power point *Sol. Energy.* **85** 713–22

[77] Mahmoud Y, Xiao W and Zeineldin H 2011 A simple approach to modeling and simulation of photovoltaic modules *IEEE Trans. Sustain. Energy* **3** 185–6

[78] Jain A and Kapoor A 2005 A new method to determine the diode ideality factor of real solar cell using Lambert W-function *Sol. Energy Mater. Sol. Cells* **85** 391–6

[79] Murayama M and Mori T 2006 Equivalent circuit analysis of dye-sensitized solar cell by using one-diode model: effect of carboxylic acid treatment of TiO_2 electrode *Jpn. J. Appl. Phys.* **45** 542

[80] Wetzelaer G-J A *et al* 2015 Trap-assisted non-radiative recombination in organic–inorganic perovskite solar cells *Adv. Mater.* **27** 1837–41

[81] Agarwal S *et al* 2014 On the uniqueness of ideality factor and voltage exponent of perovskite-based solar cells *J. Phys. Chem. Lett.* **5** 4115–21

[82] Ndegwa R *et al* 2020 A fast and accurate analytical method for parameter determination of a photovoltaic system based on manufacturer's data *J. Renew. Energy* **2020** 1 7580279

[83] Almora O *et al* 2018 Discerning recombination mechanisms and ideality factors through impedance analysis of high-efficiency perovskite solar cells *Nano Energy* **48** 63–72

[84] Arora N 1982 Studies on Solar Grade Polycrystalline Silicon Solar Cells *Thesis* University of Delhi

[85] Singh P 2009 Fabrication, characterization and other related studies for performance improvement of crystalline silicon solar cells *PhD Thesis* Jamia Millia Islamia University and National Physical Laboratory

[86] Sera D, Teodorescu R and Rodriguez P 2007 PV panel model based on datasheet values *2007 IEEE Int. Symp. on Industrial Electronics* (Piscataway, NJ: IEEE)

[87] Lo Brano V and Ciulla G 2013 An efficient analytical approach for obtaining a five parameters model of photovoltaic modules using only reference data *Appl. Energy* **111** 894–903

[88] El Achouby H *et al* 2018 New analytical approach for modelling effects of temperature and irradiance on physical parameters of photovoltaic solar module *Energy Convers. Manage.* **177** 258–71

[89] Li Y *et al* 2013 Evaluation of methods to extract parameters from current–voltage characteristics of solar cells *Sol. Energy* **90** 51–7

[90] Mialhe P *et al* 1986 The diode quality factor of solar cells under illumination *J. Phys. D Appl. Phys.* **19** 483

[91] Agarwal S K *et al* 1981 A new method for the measurement of series resistance of solar cells *J. Phys. D Appl. Phys.* **14** 1643

[92] Singh V N and Singh R P 1983 A method for the measurement of solar cell series resistance *J. Phys. D Appl. Phys.* **16** 1823

[93] Sharma S *et al* 2000 Overcoming the problems in determination of solar cell series resistance and diode factor *J. Phys. D Appl. Phys.* **23** 1256

[94] Priyanka M, Lal S N and Singh 2007 A new method of determination of series and shunt resistances of silicon solar cells *Sol. Energy Mat. Sol. Cells* **91** 137–42

[95] Bowden S and Rohatgi A 2001 Rapid and accurate determination of series resistance and fill factor losses in industrial silicon solar cells *17th European Photovoltaic Solar Energy Conf. (Munich, Germany)*

[96] Velilla E *et al* 2018 Numerical analysis to determine reliable one-diode model parameters for perovskite solar cells *Energies* **11** 1963

[97] Fabregat-Santiago F *et al* 2017 Deleterious effect of negative capacitance on the performance of halide perovskite solar cells *ACS Energy Lett.* **2** 2007–13

[98] Ono L K *et al* 2017 Perovskite solar cells—towards commercialization *ACS Energy Lett.* **2** 1749–51

[99] Ciro J *et al* 2017 Optimization of the Ag/PCBM interface by a rhodamine interlayer to enhance the efficiency and stability of perovskite solar cells *Nanoscale* **9** 9440–6

[100] Tress W *et al* 2017 Interpretation and evolution of open-circuit voltage, recombination, ideality factor and subgap irreversible degradation of perovskite solar cells *Energy Environ. Sci.* **11** 151–65

[101] Ryu S *et al* 2019 Light intensity-dependent variation in defect contributions to charge transport and recombination in a planar $MAPbI_3$ perovskite solar cell *Sci. Rep.* **9** 19846

[102] Nelson J 2003 *The Physics of Solar Cells* (Imperial College Press)

[103] Streetman B G and Banerjee S 2015 *Solid State Electronic Devices* (Pearson)

[104] Cowan S R, Roy A and Heeger A J 2010 Recombination in polymer-fullerene bulk heterojunction solar cells *Phys. Rev.* B **82** 245207

[105] Kao K C, Hwang W and Choi S I 1983 Electrical transport in solids *Phys. Today* **36** 90–0

[106] Kowsar A *et al* 2025 An overview of solar cell simulation tools *Sol. Energy Adv.* **5** 100077

[107] Tesfancheal H Y *et al* 2024 Multi-physics device simulations of optimized semi-transparent perovskite solar cells: influence of material types and layer thicknesses on transmittance and electrical performance *Sol. Energy* **284** 113069

[108] Zandi S, Saxena P and Gorji N E 2020 Numerical simulation of heat distribution in RGO-contacted perovskite solar cells using COMSOL *Sol. Energy* **197** 105–10

[109] Devi R *et al* 2023 Efficient solar cell using COMSOL multiphysics *Int. Conf. on Power Engineering and Intelligent Systems (PEIS)* (Berlin: Springer)

[110] Jahantigh F and Bagher Ghorashi S 2019 Optical simulation and investigation of the effect of hysteresis on the perovskite solar cells *Nano* **14** 1950127

[111] Burgelman M, Nollet P and Degrave S 2000 Modelling polycrystalline semiconductor solar cells *Thin Solid Films* **361−362** 527–32

[112] Burgelman M *et al* 2016 *SCAPS Manual* (University of Ghent)

[113] Fonash S J *et al* 1997 A Manual for AMPS−1D *The Center for Nanotechnology Education and Utilization, Electronic Materials and Processing Research Laboratory* p 16802 https://academia.edu/42716428/AMPS_1D_A_One_Dimensional_Device_Simulation_Program_for_the

[114] Alanazi T I and Eid O I 2023 Simulation of triple-cation perovskite solar cells: key design factors for efficiency promotion *Energies* **16** 2717

[115] Mortadi A *et al* 2024 Analysis and optimization of lead-free perovskite solar cells: investigating performance and electrical characteristics *Mater. Renew. Sustain. Energy* **13** 219–32

[116] Şahin G and Alma M H 2019 Study of the static characteristic *IV* and the electrical parameters corresponding to the shunt resistance R_{sh} and series resistance R_s per unit area of a solar cell with grain size *Chin. J. Phys.* **62** 395–404

[117] Amiri O and Salavati-Niasari M 2015 High efficiency dye-sensitized solar cells (9.3%) by using a new compact layer: decrease series resistance and increase shunt resistance *Mater. Lett.* **160** 24–7

[118] Khalili S *et al* 2022 Synthesis characterization of SnO_2 nanofibers (NFs) and application of high-performing photodetectors based on SnO_2 NFs/n-Si heterostructure *Sens. Actuators* A **342** 113631

[119] Walder C *et al* 2012 Optical modeling of thin film silicon solar cells by combination of the transfer-matrix-method and the raytracer algorithm *Opt. Eng.* **51** 3801

[120] Santbergen R *et al* 2017 GenPro4 optical model for solar cell simulation and its application to multijunction solar cells *IEEE J. Photovolt.* **7** 919–26

[121] Larouche S and Martinu L 2008 OpenFilters: open-source software for the design, optimization, and synthesis of optical filters *Appl. Opt.* **47** C219–30

[122] Benatto L *et al* 2024 TMM−Sim: a versatile tool for optical simulation of thin-film solar cells *Comput. Phys. Commun.* **300** 109206

[123] McCoy D *et al* 2021 Finite-difference time-domain (FDTD) optical simulations: a primer for the life sciences and bio-inspired engineering *Micron* **151** 103160

[124] Kar S *et al* 2022 Optical simulations in perovskite devices: a critical analysis *ACS Photonics* **9** 3196–214

[125] Burger S *et al* 2008 JCMsuite: an adaptive FEM solver for precise simulations in nano-optics *Integrated Photonics and Nanophotonics Research and Applications* (Optica)
[126] Mulyanti B *et al* 2023 Light absorption enhancement of perovskite solar cells by a modified anti-reflection layer with corrugated void-like nanostructure using finite difference time domain methods *Phys. Scr.* **98** 065702
[127] Du T *et al* 2020 Light-intensity and thickness dependent efficiency of planar perovskite solar cells: charge recombination versus extraction *J. Mater. Chem.* C **8** 12648–55
[128] Yang Y and You J 2017 Make perovskite solar cells stable *Nature* **544** 155–6
[129] Chakraborty K, Choudhury M G and Paul S 2019 Numerical study of Cs_2TiX_6 ($X = Br^-$, I^-, F^- and Cl^-) based perovskite solar cell using SCAPS-1D device simulation *Sol. Energy* **194** 886–92
[130] Ahmad K *et al* 2022 Numerical simulation of $NH_3(CH_2)_2NH_3MnCl_4$ based Pb-free perovskite solar cells via SCAPS-1D *Nanomaterials* **12** 3407
[131] Zandi S and Razaghi M 2019 Finite element simulation of perovskite solar cell: a study on efficiency improvement based on structural and material modification *Sol. Energy* **179** 298–306
[132] Guo Q *et al* 2023 Numerical simulation on preparing uniform and stable perovskite wet film in slot-die coating process *ACS Omega* **8** 19547–55
[133] Ghosh R, Singh A and Agarwal P 2023 Study on effect of different HTL and ETL materials on the perovskite solar cell performance with TCAD simulator *Mater. Today Proc.* in press DOI 10.1016/j.matpr.2023.06.161
[134] Kumar A *et al* 2021 Simulation of perovskite solar cell employing ZnO as electron transport layer (ETL) for improved efficiency *Mater. Today Proc.* **46** 1684–7
[135] Bendib T *et al* 2020 Combined optical-electrical modeling of perovskite solar cell with an optimized design *Opt. Mater.* **109** 110259
[136] Yohanes Tesfancheal H *et al* 2024 Multi-physics device simulations of optimized semi-transparent perovskite solar cells: influence of material types and layer thicknesses on transmittance and electrical performance *Sol. Energy* **284** 113069
[137] Kang A and Zandi M 2019 Simulation analysis of graphene contacted perovskite solar cells using SCAPS-1D *Opt. Quantum Electron.* **51** 91
[138] Kakavelakis G *et al* 2017 Efficient and highly air stable planar inverted perovskite solar cells with reduced graphene oxide doped PCBM electron transporting layer *Adv. Energy Mater.* **7** 1602120
[139] Peng Y *et al* 2015 Efficient semitransparent perovskite solar cells with graphene electrodes *Adv. Mater.* **27** 3632–38
[140] Bouclé J and Herlin-Boime N 2016 The benefits of graphene for hybrid perovskite solar cells *Synth. Met.* **222** 3–16
[141] Ahmad W *et al* 2024 Revolutionizing photovoltaics: from back-contact silicon to back-contact perovskite solar cells *Mater. Today Electron.* **9** 100106
[142] Clarke W *et al* 2023 IonMonger 2.0: software for free, fast and versatile simulation of current, voltage and impedance response of planar perovskite solar cells *J. Comput. Electron.* **22** 364–82
[143] Sha W *et al* 2024 SolarDesign: an online photovoltaic device simulation and design platform arXiv: 2412.20009
[144] Mann S *et al* 2021 dPV: an end-to-end differentiable solar-cell simulator arXiv: 2105.06305

[145] Calado P *et al* 2022 Driftfusion: an open source code for simulating ordered semiconductor devices with mixed ionic-electronic conducting materials in one dimension *J. Comput. Electron.* **21** 960–91

[146] Kresse G and Furthmüller J 1996 Efficiency of *ab-initio* total energy calculations for metals and semiconductors using a plane-wave basis set *Comput. Mater. Sci.* **6** 15–50

[147] Giannozzi P *et al* 2009 Quantum espresso: a modular and open-source software project for quantumsimulations of materials *J. Phys. Condens. Matter* **21** 395502

[148] Hess B *et al* 2008 Gromacs 4: algorithms for highly efficient, load-balanced, and scalable molecular simulation *J. Chem. Theory Comput.* **4** 435–47

[149] Plimpton S 1995 Fast parallel algorithms for short-range molecular dynamics *J. Comput. Phys.* **117** 1–19

[150] Sabetvand R, Ghazi M and Izadifard M 2020 DFT study of electronic and optical properties of $CH_3NH_3SnI_3$ perovskite *Energy Sources* A **46** 13778–90

[151] Filip M R *et al* 2014 Steric engineering of metal-halide perovskites with tunable optical band gaps *Nat. Commun.* **5** 5757

[152] Yin Y *et al* 2021 Recent progress in defect tolerance and defect passivation in halide perovskite solar cells *Acta Phys.-Chim. Sin* **37** 2008048

[153] Mosconi E, Quarti C and De Angelis F 2016 First principles modeling of perovskite solar cells: interplay of structural, electronic and dynamical effects *Unconventional Thin Film Photovoltaics* (Royal Society of Chemistry)

[154] Rezini B *et al* 2025 Exploring the optoelectronic properties and solar cell performance of $Cs_2SnI_{6-x}Br_x$ lead-free double perovskites: combined DFT and SCAPS simulation *Physics* **7** 3

[155] Salem M S *et al* 2023 Full optoelectronic simulation of lead-free perovskite/organic tandem solar cells *Polymers* **15** 784

IOP Publishing

Perovskites

Fundamentals, properties, preparation and applications

N M Ravindra, Priyanka Singh, Leqi Lin and Pankaj Kumar

Chapter 7

Current, emerging and frontier applications of perovskite materials

This chapter presents a comprehensive overview of the rapidly expanding applications of perovskite materials in photovoltaics, optoelectronics, and emerging technologies. It covers developments in perovskite solar cells (PSCs), including flexible and tandem configurations, as well as advances in perovskite-based light-emitting diodes (PeLEDs) and photodetectors. The chapter further explores emerging applications such as smart wearables, quantum devices, neuromorphic systems, and bioelectronics. It also highlights the role of perovskites in energy storage and emphasizes the importance of encapsulation strategies, from conventional glass and polymeric methods to innovative self-healing approaches, for improving device stability and operational lifespan.

7.1 Introduction

As discussed in chapter 6, the computational and simulation tools are playing an increasingly vital role in designing advanced perovskite materials and device architectures, accelerating their integration into a broad range of technologies. While initially focused on photovoltaics, perovskite semiconductors are now being applied across diverse electronic and optoelectronic platforms, including LEDs, photodetectors, sensors, lasers, and energy storage systems, due to their exceptional optical and electrical properties [1]. Metal halide perovskites, with their tunable bandgaps, strong photoluminescence (PL), and defect-tolerant electronic structures, have emerged as key materials for next-generation optoelectronic technologies [2]. Additionally, their combined ionic and electronic transport properties make them highly attractive for advanced neuromorphic and memory applications, including memristors and artificial synapses [3]. The multifunctionality of perovskite materials is illustrated in figure 7.1, which highlights their structural versatility and performance across a wide range of devices; these devices include solar cells, LEDs,

doi:10.1088/978-0-7503-5427-1ch7

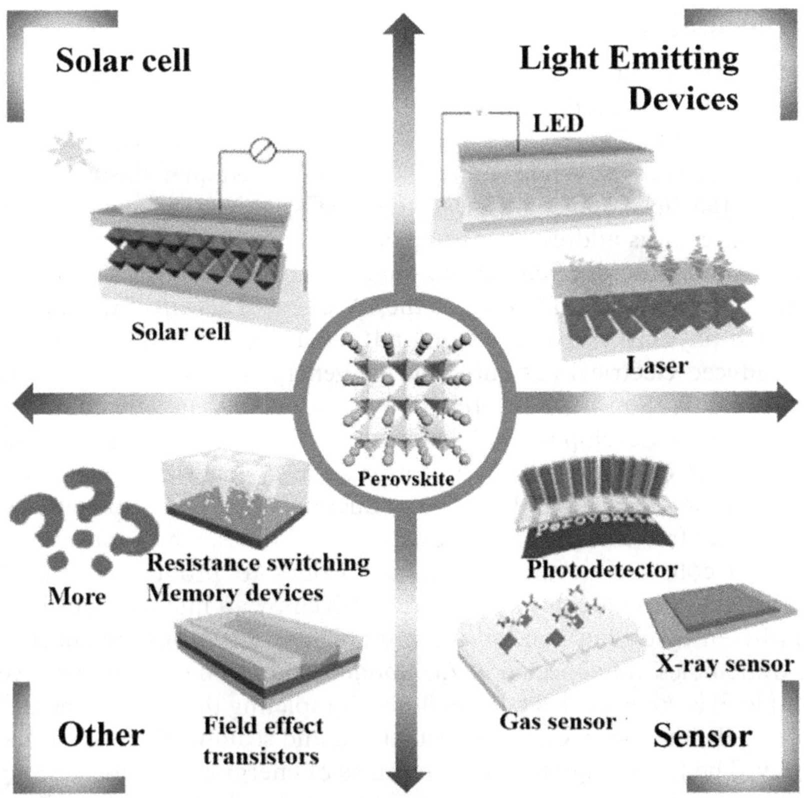

Figure 7.1. Schematic illustration of multifunctional applications of perovskite materials across diverse optoelectronic and electronic devices. (Reproduced from [4]. CC BY 4.0.)

photodetectors, sensors, lasers, memory devices, and field-effect transistors (FETs) [4, 5]. Beyond electronics, perovskite oxides show significant potential in energy storage and electrocatalysis. In particular, perovskites have demonstrated efficacy in electrochemical nitrate reduction to ammonia, presenting new opportunities for sustainable nitrogen fixation technologies [6]. Furthermore, their solution-processable nature supports the development of flexible and wearable electronics, including smart textiles and conformable sensors [7]. Printed PSCs have a wide range of applications spanning from residential, to commercial, and defense sectors. Specific examples include building-integrated photovoltaics (BIPV), portable power sources for consumer electronics, solar-integrated transport vehicles, wearable energy-harvesting textiles, and deployable power systems such as tents and canopies for remote or emergency use. While perovskites are being investigated across a broad range of technological applications, this chapter highlights selected examples that demonstrate both the progress of established technologies and the potential of emerging applications.

7.2 Photovoltaics

PSCs remain the leading application of perovskites, benefiting significantly from device simulation and modeling. The following sections explore the photovoltaic (PV) application of PSCs across various formats, from conventional PSCs to flexible and tandem architectures. While section 7.2.1.2 is focused on mainstream lead-based technologies, the development of lead-free PSCs, driven by environmental and regulatory concerns, is addressed in chapter 9.

PV refers to the process of converting sunlight into electricity using semiconducting materials, a phenomenon known as the PV effect. This effect was first observed by the French physicist Alexandre Edmond Becquerel in 1839, who demonstrated that light-induced electrical current could be generated in an electrochemical cell. Later, in 1873, W Smith discovered photoconductivity in selenium, laying the groundwork for the development of solid-state solar cells. A PV device, commonly known as a solar cell, operates by absorbing the incident photons, which excite electrons from the valence band to the conduction band in a semiconductor. This process generates free charge carriers, electrons and holes, which must be efficiently separated and collected at the opposite electrodes to produce a usable electric current. The separation is achieved via drift (driven by an internal electric field) and diffusion (driven by carrier concentration gradients). Electrons are collected at the cathode, while holes are collected at the anode, establishing a photovoltage. When an external load is connected, a current flows, completing the power generation cycle [8, 9]. Solar cells, therefore, offer a direct, clean, and scalable method for harvesting solar energy. The technology holds the promise of energy independence and a long-term solution to climate change and air pollution. While solar cell principles have been known for decades, the first solid-state device was created by Charles Fritts in 1883, using a thin layer of gold over selenium. This early cell demonstrated a power conversion efficiency (PCE) of approximately 1%. Though incremental improvements followed, the global interest in solar PV remained limited until more advanced semiconductor materials emerged [8, 9].

Silicon (Si) has long been the cornerstone of PV technology due to its excellent semiconducting properties as well as its abundance in nature. It was first utilized in early radio detectors in the 1900s, but the modern Si solar cell emerged in 1940, when Russell Ohl of Bell Laboratories observed the PV effect in a cracked monocrystalline silicon (c-Si) sample, an accidental p–n junction formed by the opposite charges on either side of the crack [10]. Although the efficiency of this primitive device was just $\sim$1%, further advancements led to a landmark achievement in 1954, when Chapin, Pearson, and Fuller, also from Bell Labs, developed a c-Si solar cell with $\sim$6% PCE [11]. Since then, continuous research and refinement have propelled the certified efficiency of Si solar cells to over 26% [12]. The structure and working mechanism of a Si solar cell are illustrated in figure 7.2(a) and figure 7.2(b), respectively. While c-Si solar cells offer high performance and long-term reliability, their high manufacturing costs, complex fabrication processes, and limitations in large-area deployment have encouraged the search for alternative materials and

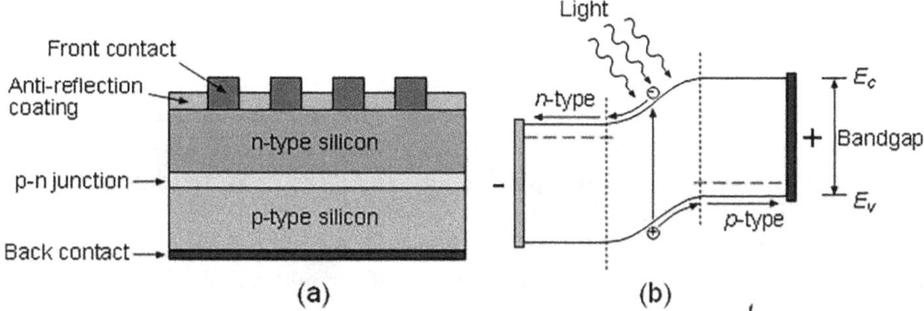

Figure 7.2. (a) Schematic structure and (b) working mechanism of a p–n junction Si solar cell.

architectures. This quest has given rise to the next generations of solar cell technologies.

The first generation includes wafer-based cells such as c-Si and multi-crystalline (mc)-Si, as well as III–V compound semiconductors such as gallium arsenide (GaAs). The second generation shifted towards thin-film technologies using materials such as amorphous silicon (a-Si), cadmium telluride (CdTe), and copper indium gallium diselenide (CIGS). While these offered reduced costs, some posed environmental concerns due to toxic elements such as cadmium and selenium [13]. The third generation introduced novel semiconductors including organic dyes, organic polymers, quantum dots, and most notably, perovskite semiconductors. The fourth generation blends organic and inorganic materials into hybrid architectures, that are aimed at achieving both high efficiency and low production costs. Although many of these emerging technologies remain in the development and demonstration stage, they promise transformative impacts on solar energy accessibility and sustainability [13]. In the context of this historical development, perovskite materials have catalyzed a dramatic resurgence in PV research due to their high absorption coefficients, long carrier diffusion lengths, and tunable bandgaps. The ongoing integration of modeling tools with experimental research continues to accelerate innovation, bringing PSCs closer to widespread commercialization. Despite the material differences and structural variations, all solar cells operate on the same basic principle: the conversion of incident light into electrical energy. The performance of a PV device is generally evaluated using its illuminated current–voltage (I–V) characteristics, and the parameters such as open-circuit voltage (V_{oc}), short-circuit current density (J_{sc}), fill factor (FF), and PCE, as discussed in chapter 6. This chapter focuses specifically on PSC technology and its evolution through different configurations. A comprehensive overview of the historical developments and alternative PV technologies can be found in established literature sources [14, 15].

7.2.1 Perovskite solar cells

PSCs utilize perovskite-structured semiconductors as light-absorbing materials and are rapidly emerging as a next-generation PV technology. Owing to their high PCEs, low production costs, and versatile form factors, PSCs are expected to play a transformative role in the future energy generation across multiple sectors. Unlike conventional Si-based solar cells, PSCs offer several fundamental advantages; they are lightweight, mechanically flexible, semitransparent, and can be manufactured using low-temperature, solution-based printing techniques on flexible substrates [16–20]. These properties enable high-throughput roll-to-roll (R2R) fabrication, making PSCs well-suited for scalable and low-cost production. The global PSC market was valued at USD 64.05 million in 2023, and is projected to grow significantly, reaching USD 105.23 million in 2024 and surging to USD 1760.59 million by 2032, indicating its substantial commercial potential [16–20]. The adoption of PSCs could reduce the dependence on fossil fuels, bolster domestic energy production, and generate significant employment opportunities. Several international companies are leading the development and commercialization of PSCs, including Saule Technologies (Poland), GCL Suzhou Nanotechnology Co., Ltd (China), Hunt Perovskite Technologies (USA), Oxford PV (Germany), and Greatcell Energy (Australia) [21].

7.2.1.1 Structure of perovskite solar cell

Although the structure of PSCs has been discussed in earlier chapters, a brief overview is necessary here to contextualize their applications. PSCs are typically fabricated in two main device architectures: the conventional n–i–p structure and the inverted p–i–n configuration. In the n–i–p structure, the bottom electrode on the substrate acts as the cathode, whereas in the p–i–n architecture, it serves as the anode. These architectural distinctions significantly influence charge transport dynamics and, consequently, the overall device performance. Figure 7.3 illustrates the layer arrangements in both configurations, emphasizing their structural differences. While n–i–p structured PSCs generally exhibit slightly higher PCEs, p–i–n structured devices offer enhanced operational stability. The relatively lower PCEs observed in p–i–n devices are largely attributed to the interfacial energy losses between the perovskite active layer and adjacent charge transport layers.

7.2.1.2 Lead based perovskite solar cell

Lead-based PSCs represent the most widely researched and technologically advanced class of perovskite PV. Metal halide perovskites, generally structured as

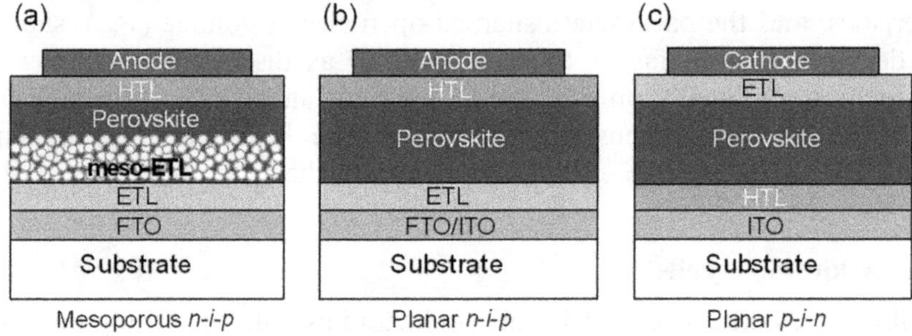

Figure 7.3. Schematic representation of the (a) mesoporous n–i–p, (b) planar n–i–p and (c) planar p–i–n structures of PSCs.

ABX_3 compounds with lead (Pb) occupying the B-site, have demonstrated exceptional PCEs now surpassing those of established Si technologies. While lead-based perovskites continue to dominate the current high-efficiency PV research, the pursuit of safer, more stable, and environmentally responsible alternatives, such as lead-free, two-dimensional, and high-entropy perovskite materials, is rapidly gaining momentum.

Although perovskite materials have been known since 1839, their application in solar cell technologies was not realized until 2009. The development of PSCs traces its roots to dye-sensitized solar cells (DSSCs), which were first invented in 1988 and formally reported in 1991 by Michael Gratzel and Brian O'Regan at the Ecole Polytechnique Federale de Lausanne (EPFL), Switzerland. Commonly referred to as Gratzel cells, DSSCs feature a photo-sensitized anode composed of a dye-coated titanium dioxide (TiO_2) layer deposited on fluorine-doped tin oxide (FTO) glass, paired with a counter electrode generally made of platinum or lead-coated FTO [22]. The device operates via a redox electrolyte, typically iodide/triiodide (I^-/I_3^-), which facilitates charge transport between electrodes. Upon illumination, the dye absorbs photons and generates electron–hole pairs. The photo-generated electrons are transferred to the photo-sensitized anode, whereas the holes are neutralized by electrons from I^- ions in the electrolyte. After the I^- ions donate the electrons, they are oxidized into I_3^- ions, which transport through the electrolyte and reach the counter electrode. The electrons from photo-sensitized anode transport through external circuit and reach the counter electrode where they reduce the I_3^- ions back into I^- ions. The original DSSCs achieved a PCE of ~7%–8% in 1991 [22], which gradually improved to around 12% by 2011 [23]. However, further improvements have been modest, with certified efficiencies peaking near 13% [12], and the technology continues to face challenges related to long-term operational stability.

In 2009, Miyasaka and co-workers pioneered the first perovskite-based solar cells by incorporating synthetic organometal halide perovskites, specifically methylammonium lead iodide ($CH_3NH_3PbI_3$ or $MAPbI_3$) and methylammonium lead bromide ($CH_3NH_3PbBr_3$ or $MAPbBr_3$), as light absorbers in DSSCs, replacing the conventional organic dyes [24]. TiO_2 was used as sensitizer on FTO coated glass substrate and 0.15 M LiI + 0.075 M I_2 in methyoxyacetonitrile and 0.4 M LiBr + 0.4 M Br_2 in acetonitrile were used as electrolyte mediums for the corresponding perovskites. The PSC, incorporating $MAPbI_3$, exhibited PCE of 3.8% whereas the PSC based on $MAPbBr_3$ exhibited PCE of 3.13% [24]. Although these efficiencies were modest, and the cells suffered from severe instability, primarily due to rapid dissolution of the perovskite layer in the liquid electrolyte, $MAPbI_3$ demonstrated important optoelectronic advantages. Notably, it exhibited a broad absorption spectrum across the visible range, high charge carrier mobility, long diffusion lengths, and extended carrier lifetimes. With a bandgap of approximately 1.5 eV, $MAPbI_3$ theoretically allows for a maximum efficiency of ~30% as per the Shockley–Queisser (S–Q) limit [9]. Furthermore, the high solubility of these organometal halide perovskites in organic solvents enabled easy thin-film fabrication via spin coating techniques. In order to improve the stability of DSSCs, Michael Gratzel and his team were working on solid state DSSCs, and were using thin films of high

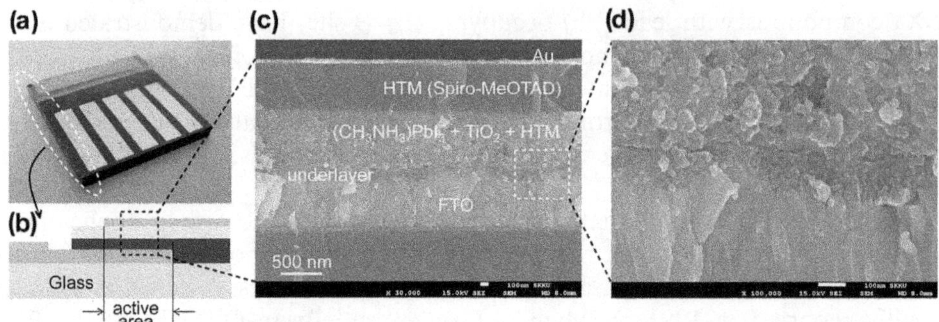

Figure 7.4. (a) Photograph of the first solid-state PSC. (b) Schematic cross-sectional structure and (c) cross-sectional SEM image of the solar cell. (d) Cross-sectional SEM image of the FTO/underlayer/active layer junctions. (Reproduced with permission from [25]. CC BY-NC-SA 3.0.)

mobility hole transport materials in place of liquid electrolyte medium. In 2012, they replaced the organic dye with $MAPbI_3$ and used 2,2′,7,7′-tetrakis(N,N-di-p-methoxyphenyl-amine)−9,9′-spirobifluorene (spiro-OMeTAD) as the hole transport layer (HTL) instead of a liquid electrolyte. This innovation marked the creation of the first solid-state perovskite-sensitized solar cell, which achieved a remarkable PCE of 9.7% along with improved stability [25]. This breakthrough triggered immense global interest in perovskite PV, laying the foundation for a new era in solar energy technologies. Figure 7.4 illustrates the photograph, structure and cross-sectional scanning electron microscope (SEM) images of the first solid-state PSC reported by Gratzel *et al* [25]. In brief, the solar cells were prepared on FTO coated glass substrates. FTO served as the bottom transparent electrode. A thin compact layer of TiO_2 was deposited on FTO by spin coating of TiO_2 precursor, which was followed by deposition of a mesoporous TiO_2 film using TiO_2 nano-crystalline paste. A thin layer of $MAPbI_3$ perovskite was spin coated from its precursor solution and annealed at 100 °C for 15 min on the hotplate to obtain perovskite nanocrystals. A thin layer of spiro-OMeTAD, doped with bis(trifluoromethane)sulfonamide lithium (LiTFSI) and 4-tert-butylpyridine (TBP), was deposited on perovskite film to serve as the HTL. The solar cell was completed by deposition of 60 nm of Au by thermal evaporation as the counter electrode.

Subsequently, Gratzel and his team refined the deposition technique of the perovskite layer to enhance the crystallinity, morphology, and coverage of the perovskite film [26]. They adopted a sequential deposition method (also known as two-step deposition) to grow the perovskite layer over the mesoporous TiO_2 scaffold. In this approach, a solution of lead iodide (PbI_2) in dimethylformamide (DMF) was first infiltrated into the mesoporous TiO_2 layer and allowed to dry. This was followed by immersing the PbI_2-coated substrate into a methylammonium iodide (MAI) solution prepared in 2-propanol. After an immersion time of approximately 20 s, the samples were rinsed with 2-propanol and dried, allowing the PbI_2 to chemically convert into $MAPbI_3$ via *in situ* reaction. This sequential deposition strategy provided superior control over the perovskite film morphology and enabled the formation of perovskite crystals deep within the porous TiO_2

network, something not easily achievable through single-step spin coating. HTL, spiro-OMeTAD, was subsequently deposited onto the perovskite film by spin coating, and the device was finalized by thermal evaporation of a gold (Au) top electrode. This improved fabrication protocol yielded a significant enhancement in performance, achieving a PCE of approximately 15% [26]. The success of this method catalyzed rapid growth in global research interest. Since then, research on PSCs has surged, with an exponential rise in the number of scientific publications and innovations reported annually. Several strategies have been employed to achieve controlled growth of perovskite films with improved morphology and electrical properties for high-efficiency PSCs [27–33]. Molecular engineering has played a crucial role in this progress, with researchers synthesizing multi-cation, mixed-halide, and mixed-metal perovskites to enhance optoelectronic properties and photovoltaic performance [34–37]. Representative examples of such engineered perovskites include $MAPbI_3$, $MAPbBr_3$, formamidinium lead iodide ($FAPbI_3$), cesium lead iodide ($CsPbI_3$), methylammonium lead iodide chloride ($MAPbI_{3-x}Cl_x$), and mixed cation variants such as $FA_xMA_{1-x}PbI_3$. Among these, $FAPbI_3$ has attracted significant attention due to its potential for high efficiency and stability. However, $FAPbI_3$ exists in two phases: the yellow, photoinactive δ-phase and the dark brown, photoactive α-phase. For PSC applications, stabilizing the α-$FAPbI_3$ phase with minimal defects and high crystallinity is essential. Park *et al* [33] demonstrated a successful strategy using alkylammonium chloride (RACl) additives in the $FAPbI_3$ precursor, enabling *in situ* δ-to-α phase conversion during crystallization. This approach resulted in PSCs achieving a remarkable PCE of 26.08% (certified 25.73%) under standard test conditions. The type and amount of RACl played a crucial role in the transition rate from δ-phase to α-phase, surface morphology, crystallinity and crystallographic orientation of the grains in the α-$FAPbI_3$ film. Utilizing these similar innovations in molecular and device engineering, PSCs have reached certified efficiencies as high as 27% [12]. Furthermore, defects, both within the bulk perovskite layer and at its interfaces, critically impact device performance, an issue addressed in chapter 8. To counteract these defect-related losses, researchers have implemented various mitigation strategies, including the incorporation of bulk additives and surface passivation treatments, such as ammonium-based defect-passivating ligands, topics explored further in chapter 9 [38–41].

Lead-based PSCs have demonstrated outstanding PCEs, with certified values high enough to support commercialization. As of 2025, the highest certified PCE for a single-junction lead-based PSC has reached 27%, [12]. Despite this progress, their practical deployment is hindered by rapid degradation under real-world operating conditions, limiting commercial viability [42–46]. One strategy to enhance device reproducibility and performance involves the use of Lewis base adducts with PbI_2 during film formation. Oxygen-containing Lewis bases and iodide ions form adducts with PbI_2, as confirmed by infrared spectroscopy. In a simple one-step spin coating process (shown in figure 7.5(a)), a precursor solution comprising MAI and PbI_2 (1:1 mol%) with DMSO in DMF (50 wt%) was spin coated at 4000 rpm. The resulting $MAPbI_3$ films exhibited rod-shaped crystals; however, incomplete substrate

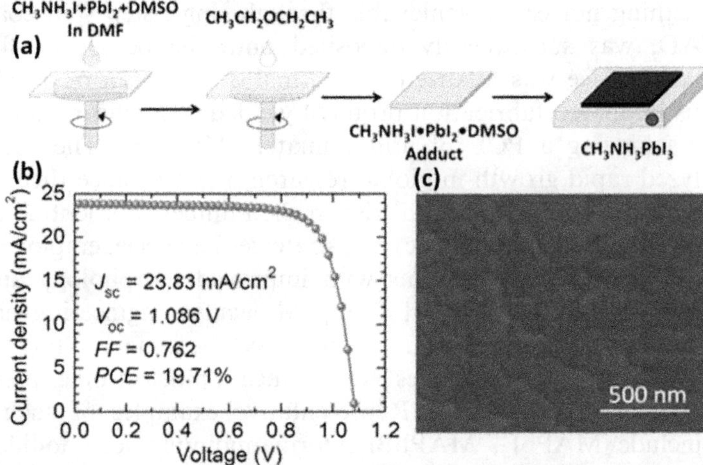

Figure 7.5. (a) Schematic illustration of the fabrication process for $MAPbI_3$ perovskite layers using a one-step spin coating method from a DMF solution containing MAI, PbI_2, with DMSO as an additive. (b) *J*–*V* characteristics of the best-performing PSC among 41 devices, measured under standard AM 1.5 G illumination (100 mW cm^{-2}) using a reverse scan. (c) SEM image of $MAPbI_3$ perovskite layers. (Reproduced with permission from [47]. Copyright 2015 American Chemical Society.)

coverage was observed, likely due to solubility differences between the precursors. Morphological analysis (figure 7.5(c)) revealed that the Lewis base adduct-assisted films had a dense microstructure with uniform grain sizes of 200–300 nm, and the cross-sectional SEM confirmed a well-defined layered device architecture. Devices fabricated using this method achieved an average PCE of 18.3% across 41 samples, with a champion efficiency of 19.71%, characterized by a J_{sc} of 23.83 mA cm^{-2}, V_{oc} of 1.086 V, and a FF of 0.762 (figure 7.5(b)) [47]. These results highlight the effectiveness of adduct engineering in improving film quality and device performance. Despite these advancements, challenges such as material instability and lead toxicity remain significant hurdles to commercial deployment. Ongoing research aims to address these issues through improved material compositions and encapsulation techniques.

7.2.2 Flexible perovskite solar cells

PSCs can be fabricated on a wide range of flexible substrates, including thin glass sheets, polymer films (e.g. PET or PEN), metal foils, fabrics, and even paper, enabling the development of flexible and lightweight PV devices [19, 48]. The use of flexible substrates supports scalable fabrication approaches such as R2R printing, making PSCs attractive for portable, wearable, and BIPV applications. Flexible PSCs are generally manufactured using techniques such as thermal evaporation, spin coating, and printing. Spin coating and thermal evaporation are commonly employed for fabricating small-area devices (~0.1–1 cm^2), while printing techniques (e.g. slot-die, blade, or inkjet printing) are used to process large-area devices (>10 cm^2), especially in R2R manufacturing. Kumar *et al* [16] reported the fabrication of

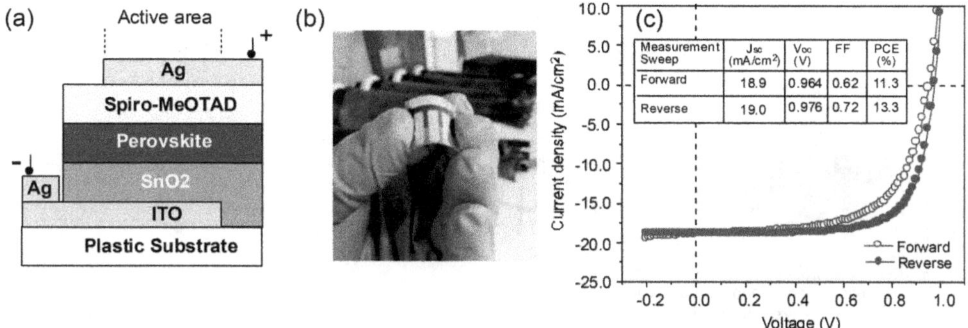

Figure 7.6. (a) Schematic structure and (b) photograph of flexible PSC. (c) *J–V* characteristics of the flexible PSCs in forward and reverse sweeps. (Reproduced with permission from [16]. Copyright 2020 IOP.)

flexible PSCs on ITO-coated PET substrates using an n–i–p device architecture ITO/SnO_2/$CH_3NH_3PbI_{3-x}Cl_x$/spiro-OMeTAD/Ag [16]. In this structure, layers of SnO_2, perovskite, and spiro-OMeTAD were deposited via spin coating, while the Ag electrodes were deposited through thermal evaporation under vacuum. Figure 7.6(a) and (b) show the schematic structure and a photograph of one of the flexible PSCs. Figure 7.6(c) shows the current density–voltage (*J–V*) characteristics of the champion flexible solar cell measured in forward and reverse sweeps. The solar cells exhibited PCE of 11.3% in forward sweep whereas 13.3% in the reverse sweep with a little hysteresis. The hysteresis in the *J–V* characteristics was attributed to ion migration and possible defects in the perovskite film. The mechanical robustness of the device was also tested through repeated bending at a 90° angle. Impressively, no significant degradation in device performance was observed even after 100 bending cycles, highlighting its mechanical durability. However, it is crucial to note that flexible PSCs are multilayered and composed of different materials with varying mechanical properties. During bending, strain can alter interatomic spacing, and if constituent layers are brittle or inelastic, they may crack or deform irreversibly, leading to performance loss or complete device failure.

7.2.3 Tandem solar cells

A tandem solar cell (TSC) is a multi-junction solar cell and generally consists of two subcells, stacked to absorb different portions of the solar spectrum more efficiently. Perovskite TSCs mark a crucial evolution in PV technologies, and offers a pathway to surpass the S–Q efficiency limit (~33%) associated with single p–n junction solar cells [49]. Tandem structures utilize two light-absorbing layers with complementary bandgaps to capture a wider range of the solar spectrum (inset, figure 7.7(a)). Specifically, a wide-bandgap perovskite top cell is stacked over a lower-bandgap bottom cell, enabling more efficient utilization of high-energy and low-energy photons, and thereby significantly reducing losses due to thermalization and non-absorption. TSCs have been explored extensively in the literature [21, 50–53], providing a comprehensive overview of their underlying principles, material systems, and integration strategies. In a single p–n junction solar cell, excess energy from

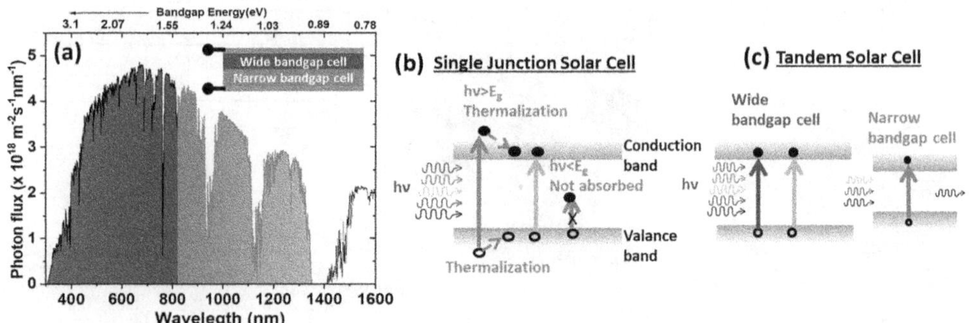

Figure 7.7. (a) Absorption spectra of top (blue) and bottom (red) subcells in a TSC. Solar spectra data taken from ASTM G173–03 (AM 1.5 G). (b) Energy band diagram of a single p–n junction solar cell, showing thermalization losses. (c) TSC with two p–n junctions for improved spectral utilization. (Reproduced with permission from [21]. Copyright 2025 IOP Publishing.)

photons with energies higher than the bandgap (E_g) is generally lost in the form of heat through thermalization processes as depicted through the energy band diagram in figure 7.7(b). TSCs enables the absorption of these high-energy photons and utilize their energy more effectively, reducing thermalization losses and improving efficiency as shown in figure 7.7(c) [21].

7.2.3.1 Processing and device structures of tandem solar cells

Generally, TSCs are constructed using two primary approaches; monolithic integration and mechanical stacking. Monolithic TSCs consist of subcells grown sequentially on a substrate in a continuous process (figure 7.8(a)). The upper subcell, with a wider bandgap, is deposited on the lower narrow-bandgap subcell. A tunnel junction or transparent conductive layer connects the two subcells to enable the charge transfer. Precise lattice matching between layers minimizes defects and enhances performance. While monolithic tandem cells offer high efficiency due to their integrated structure, they require advanced fabrication techniques and equipment for layer-by-layer deposition [52].

In mechanically stacked tandems, subcells are fabricated independently and then physically assembled (figure 7.8(c)) using methods such as epitaxial liftoff, wafer bonding, or transfer printing. Unlike monolithic tandem cells, the layers need not be lattice-matched, and no tunnel junction is required and hence offers greater material flexibility. However, this approach may introduce interface defects, alignment challenges, and increased optical losses due to the need for transparent conducting electrodes (TCEs). While simpler to fabricate, mechanically stacked tandems generally exhibit lower efficiency due to interfacial recombination and optical losses [54].

TSCs can be designed in two-terminal (2T), three-terminal (3T), or four-terminal (4T) structures (figure 7.8), each one having its own benefits and challenges in terms of efficiency, fabrication complexity, and operational stability. As shown in figure 7.8(a), 2T tandem cells feature monolithically integrated top and bottom subcells on a single substrate. The benefit of a 2T tandem cell is its simplicity and ease of integration and such cells can be coupled in parallel or series with minimal space between them [55].

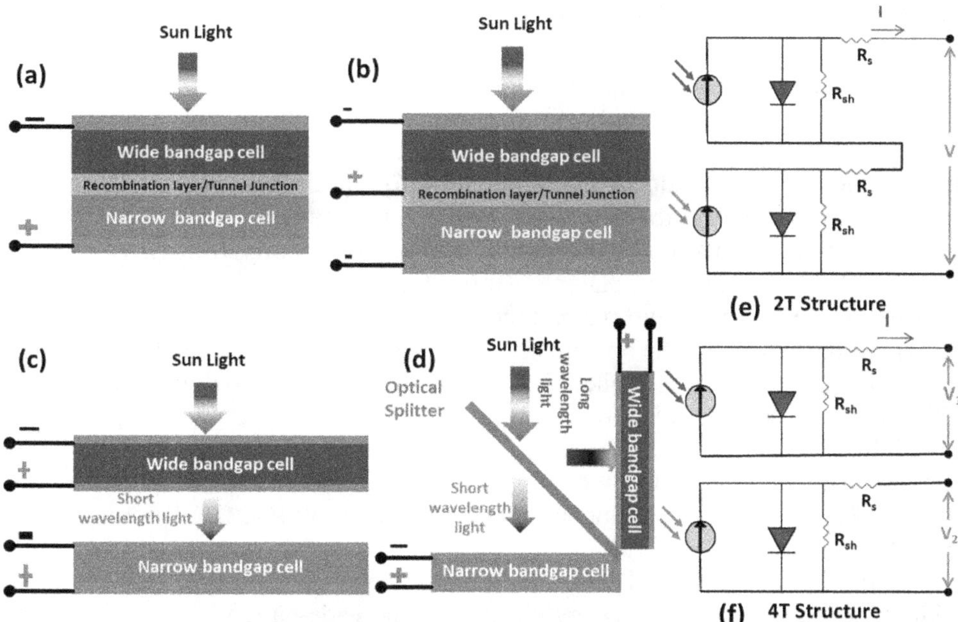

Figure 7.8. Schematic representations of various TSC configurations: (a) 2T monolithic tandem device featuring a tunnel junction or recombination layer between subcells; (b) 3T monolithic tandem device; (c) 4T mechanically stacked tandem device; and (d) 4T tandem device using an optical splitter. (e) Equivalent circuit of a 2T tandem cell, and (f) equivalent circuit of a 4T tandem cell. The top and bottom cells are designed to absorb light in the visible (400–800 nm, blue arrows) and infrared (800–1100 nm, red arrows) regions. ((a), (b), (c) and (d) Reproduced with permission from [21]. Copyright 2025 IOP Publishing. (e) and (f) Adapted from [56]. CC BY 4.0.)

The tunnel junction is an important component, which ensures efficient charge transfer by providing low resistance and high optical transmittance between subcells. The equivalent circuit under illumination is illustrated in figure 7.8(e), where R_s and R_{sh} denote the series and shunt resistances, respectively, and I and V represent the output current and voltage respectively [56]. In 2T tandem devices, the subcells are connected in series, making current matching between the top and bottom cells essential to prevent the overall current from being limited by subcell with lower current. Some of the challenges in fabricating efficient 2T tandems are as follows: achieving precise current matching, developing high-performance recombination junctions, and minimizing efficiency losses under varying spectral conditions [57]. With advances in these areas, 2T tandems are expected to be highly suitable for future energy generation.

In 3T TSCs, subcells are interconnected through wiring, tunnel junctions, or recombination layers [58–61]. As illustrated in figure 7.8(b), a third terminal is introduced at the interface between the top and bottom cells to enable the extraction of additional power by adapting to variations in the incident solar spectrum. The inclusion of a third terminal offers greater flexibility in independently optimizing operation of each subcell, thereby alleviating current-matching constraints seen in 2T devices and lead to improved efficiency and performance.

4T TSCs exist in two configurations: mechanically stacked devices (figure 7.8(c)) and optical splitting designs (figure 7.8(d)). Mechanically stacked 4T tandem devices consist of independently fabricated subcells stacked optically but electrically isolated from one another (figure 7.8(c)). 4T connections are required for each subcell to produce its own electricity. In order to transmit infrared light to the bottom cell, the top cell incorporates bifacial TCEs to minimize optical losses [62, 63]. These structures allow each subcell to function at its maximum efficiency without current-matching constraints. A photon down-conversion strategy can further enhance cell performance by placing a thin film on top of the device that absorbs high-energy blue photons and reemits them at longer wavelengths, thereby enabling better light harvesting by the bottom cell [64]. The equivalent circuit of a 4T device is illustrated in figure 7.8(f). While 4T tandems offer superior design flexibility and independent subcell optimization, they also involve complex interconnections and potential optical/electrical losses due to parasitic absorption and electrode resistance. Despite these challenges, 4T tandem cells have demonstrated high efficiencies, such as >30% for perovskite/silicon and 25.9% for perovskite/CIGS configurations [65, 66], as shown in figure 7.8(a).

In an optical splitting 4T TSCs, a dichroic mirror is used to spectrally divide the incident light, directing high-energy photons to the wide-bandgap top cell and low-energy photons to the narrow-bandgap bottom cell, as shown in figure 7.8(d) [67]. This configuration eliminates the need for additional TCEs and hence has greater flexibility in independently fabricating each subcell. While this design enhances optical management and simplifies electrical isolation, its broader application is limited by the high cost of the optical splitter, impacting overall economic viability [57]. Nonetheless, optical splitting 4T tandems have shown impressive PCE of ~23.26% for all-perovskite TSCs [68].

7.2.3.2 Types of perovskite tandem solar cells

Perovskite TSCs can be further categorized based on the material composition of their subcells, such as perovskite/silicon (pk/Si), perovskite/organic (pk/OPV), all-perovskite (pk/pk) and (pk/CIGS). Each configuration presents distinct advantages and challenges concerning efficiency, stability, and suitability for various applications. Significant efficiency advancements in perovskite TSCs are represented in figure 7.9(a) and (b).

Figure 7.9(a) displays the efficiency achieved for 2T, 3T and 4T TSCs based on pk/Si, pk/pk, pk/CIGS, and pk/OPV [69]. Recently, in response to the increasing focus on perovskite TSCs, NREL introduced a new 'hybrid tandems' category in its official efficiency chart [70], highlighting progress in combinations such as pk/Si, pk/CIGS and pk/OPV tandems, as illustrated in figure 7.9(b). However, fully perovskite-based tandems pk/pk remain classified under the 'emerging PV' category [70].

In the pk/Si tandem configuration, a wide-bandgap perovskite top cell is combined with a silicon bottom cell. The perovskite layer efficiently absorbs high-energy (short-wavelength) photons, while the silicon layer harvests the low-energy (long-wavelength) photons that pass through the perovskite layer. With continued material and interfacial engineering, pk/Si tandems have demonstrated certified

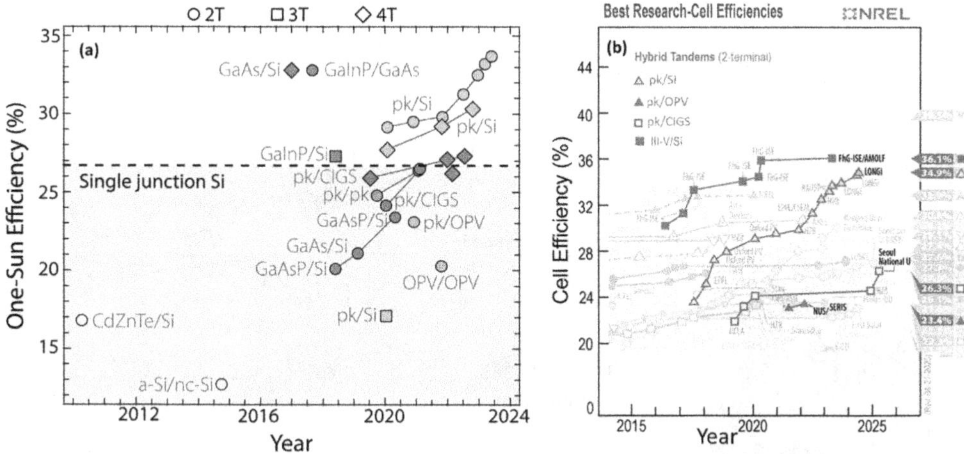

Figure 7.9. (a) Progress in the PCEs of TSCs for 2T, 3T and 4T configurations. The tandems are based on various perovskite combinations; pk/Si, pk/CIGS, pk/OPV and pk/pk. (b) Progress in 2T hybrid tandem technologies, including pk/Si, pk/OPV, pk/CIGS. (a) Reproduced from [69]. CC BY 4.0. (b) Adapted from [12]. Courtesy of the National Renewable Energy Laboratory, Golden, CO.)

PCEs exceeding 33%, making them a strong candidate for next-generation commercial solar modules. A notable early milestone was achieved by Mailoa *et al* in 2015, reporting a 13.7% efficiency for a 2T monolithic pk/Si TSC (1 cm^2) [71]. Since then, research activity has surged, with Oxford PV reporting a certified 2T pk/Si efficiency of 29.5% (figure 7.9(a) and (b)). Most recently, Longi achieved a record 34.9% efficiency, surpassing earlier 33.9% mark [70, 72]. Furthermore, the compatibility of perovskites with existing silicon infrastructure adds to their scalability and cost-effectiveness. An example of a mechanically stacked 4T pk/Si TSC is shown in figure 7.10(a) [73]. The device performance is evaluated by measuring the performance of silicon bottom cell receiving the light filtered through the semitransparent perovskite top cell. The current–voltage curves of the subcells are presented in figure 7.10(b). The original Si heterojunction cell shows a PCE of 23.4% (V_{oc} of 708 mV, J_{sc} of 40.1 mA cm^{-2} and FF of 82.5%). However, in tandem structure, the bottom Si cell exhibited 8.5% efficiency with J_{sc} of 14.5 mA cm^{-2} and other parameters, V_{oc} of 698 mV and FF of 83.5%. A substantial decrease in J_{sc} is observed, as the top perovskite cell absorbs the majority of the high-energy photons, only a small fraction of near-infrared photons is able to enter the Si bottom cell. The corresponding external quantum efficiency (EQE) spectra, shown in figure 7.10(c), indicate integrated photocurrents of 14.1 mA cm^{-2} for the Si bottom cell and 21.3 mA cm^{-2} for the perovskite top cell, closely matching the *J*–*V* results [73].

In pk/pk TSCs, two perovskite subcells with different bandgaps are incorporated, generally a wide-bandgap top cell (∼1.7–1.9 eV) and a narrow-bandgap bottom cell (∼1.1–1.3 eV), to achieve complementary light absorption. These fully perovskite-based tandems are particularly attractive due to their low-temperature solution-processability, lightweight nature, and the potential for flexible and transparent

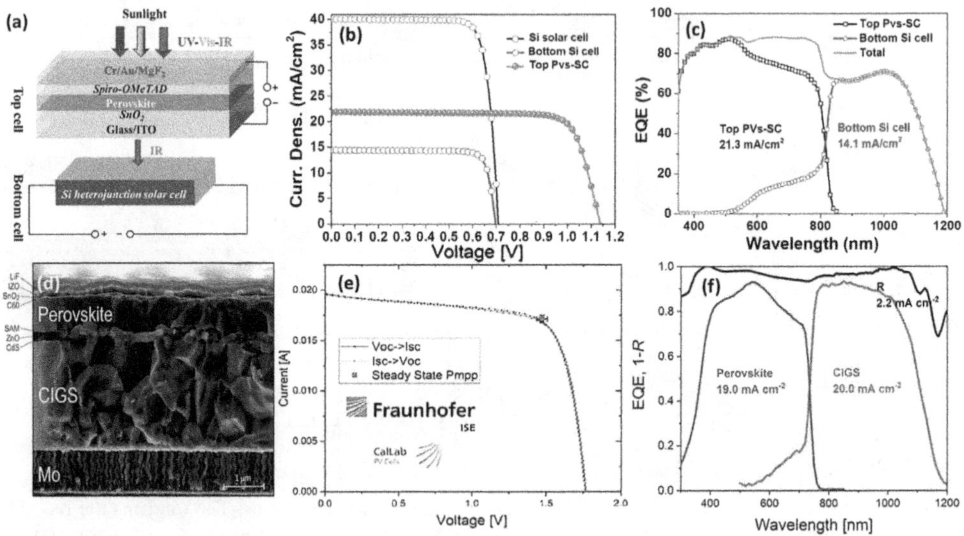

Figure 7.10. (a) Mechanically stacked 4T pk/Si TSC structure, (b) J–V characteristics and (c) corresponding EQEs for top and bottom solar cells. (d) Schematic illustration of a monolithic pk/CIGS TSC with all the layers superimposed on a cross-sectional SEM image. (e) I–V characteristics of the device. (f) EQE and $(1 - R)$ spectra of TSC. Integrated photocurrent densities were derived using the AM 1.5 G spectrum. The perovskite and CIGS subcells have bandgaps of 1.68 eV and 1.1 eV, respectively, determined from the EQE inflection points. ((a), (b) and (c) Reproduced with permission from [73]. Copyright 2021 Elsevier. (d), (e) and (f) Reproduced from [82]. CC BY 4.0.)

devices. A recent study [74] reports a certified PCE of 29.1% for monolithic pk/pk tandem solar cells by improving the crystal orientation of the wide-bandgap (1.78 eV) top subcell. The researchers introduced a thin 2D perovskite layer during film formation, promoting (100) orientation and reducing non-radiative recombination. This resulted in a record V_{OC} of 1.373 V and a high fill factor of 84.7% in the top subcell. This work demonstrates that controlled crystallographic alignment can significantly enhance tandem device performance without compromising charge transport [74]. However, challenges such as long-term operational stability, interfacial losses, and scalable fabrication still need to be addressed before pk/pk tandems can be deployed at scale.

In pk/CIGS TSCs, a perovskite top absorber is integrated with a CIGS bottom cell. Although lower PCEs of 25%–26% are being reported at the moment (as can be seen from figure 7.9), the pk/CIGS tandem configuration is especially appealing for lightweight and flexible photovoltaic modules that are suited for portable electronics, BIPV, and aerospace applications [75]. This device configuration has a substantially smaller carbon footprint per kWh produced. pk/CIGS tandem cells, with a combined radiation hardness of both subcells [76, 77], offer a high-energy yield option for space applications. These benefits suggest that pk/CIGS devices will find use in a wide range of terrestrial and space applications in the future. There are relatively few papers on pk/CIGS tandem solar cells [75–81] in comparison to the many on pk/Si TSCs. This is probably because of the lower PCE, smaller market share of CIGS, and difficult integration of the perovskite subcell on top of a (nano) rough CIGS cell surface. In addition, issues such as parasitic absorption, lattice mismatch, and deposition compatibility between layers present fabrication challenges. In a recent work, a monolithic 2T pk/CIGS tandem device with certified PCE of 24.2% [70, 82], surpassing the previous record [81], is reported. Figure 7.10(d) shows a cross-sectional SEM image of the device. As shown in figure 7.10(e), the I–V characteristics for a 1.04 cm^2 active area demonstrate a J_{sc} of 18.8 mA cm^{-2}, V_{oc} of 1.77 V, and FF of 71.2%. Optical and photocurrent analysis, via EQE and absorbance $(1 - R)$, is presented in figure 7.10(f).

In pk/OPV TSCs, the excellent optoelectronic properties of perovskites are combined with the mechanical flexibility and lightweight nature of organic PV. This tandem architecture features a wide-bandgap perovskite top cell and a narrow-bandgap organic bottom cell. The use of entirely solution-based and low-temperature processing makes pk/OPV tandems particularly suitable for flexible and semitransparent device applications, such as wearable electronics and curved surfaces. Despite these advantages, the overall efficiency of pk/OPV tandems remains modest, with PCEs of ~23.4% (figure 7.9), primarily due to the limited stability and charge transport properties of the organic subcell. Nonetheless, recently, PCE of 26.4% has been achieved in pk/OPV through two isomeric structures of a diammonium molecule that passivate (repair) defects on the surface of the perovskite [83]. Ongoing research is focusing on improving interfacial charge extraction, reducing recombination losses, and enhancing the stability of both subcells under real-world conditions.

In pk/III–V TSCs, perovskite materials are integrated with III–V semiconductors such as GaAs or GaInP, aiming to combine the low-cost processing advantages of perovskites with the high efficiency and radiation hardness of III–V semiconductors [84, 85]. These tandems are particularly attractive for space and concentrator PV applications, where performance outweighs cost constraints. The high V_{oc} and spectral complementarity of III–V semiconductors pair well with the broad absorption range of perovskites. Record efficiencies exceeding 30% have been reported in 2T and 4T configurations for such combinations. However, challenges remain in terms of lattice and thermal expansion mismatch, interface passivation, and the high cost of III–V semiconductor device fabrication, which currently limits the widespread deployment of pk/III–V tandem solar cells for niche applications. Li *et al* demonstrated high-efficiency wide-bandgap pk/GaAs TSCs in both the 2T and 4T configurations. The devices achieved PCEs of 24.27% for 2T and 25.19% for 4T, showing excellent spectral complementarity. Notably, TSC also exhibited mechanical flexibility, surviving over 1000 bending cycles, highlighting its potential for advanced flexible PV [85].

Tandem PSCs offer high efficiency-to-weight ratios, making them well-suited for space applications, BIPV, and portable power solutions. Scalable fabrication methods such as blade coating and R2R printing enhance their commercial viability. Moreover, the tandem architecture not only drives research innovation but also facilitates integration into the next-generation multi-junction solar modules.

7.3 Optoelectronic devices

Beyond their prominent role in PV, perovskite materials have shown immense potential in a wide range of optoelectronic devices applications such as LEDs, photodetectors, lasers, and sensors [86]. Among these, perovskite-based LEDs and photodetectors have attracted considerable attention in recent years. Sections 7.4.1 and 7.4.2 explore these technologies, highlighting their device architectures and recent performance achievements.

7.3.1 Light-emitting diodes

LEDs are semiconductor devices that emit light through radiative recombination of electrons and holes under an applied electric field. Perovskite light-emitting diodes (PeLEDs) have emerged as promising candidates for the next-generation display and lighting technologies due to the exceptional optoelectronic properties of metal halide perovskites, including high photoluminescence quantum yield, narrow emission line widths, and tunable emission wavelengths across the visible and near-infrared (NIR) spectrum [87]. Similar to PSC structure, PeLED consists of multiple functional layers stacked on a transparent conductive substrate such as indium tin oxide (ITO). The standard architecture includes an HTL, a perovskite emissive layer, and an ETL, sandwiched between the anode and cathode (figure 7.11(A)). The perovskite emissive layer acts as the active region in which injected electrons and holes recombine to emit light. Recent developments, particularly in quasi-2D perovskite engineering, film crystallization, and interfacial design, have driven external quantum efficiencies (EQEs) beyond 20% in green and NIR devices. The successful demonstration of bright PeLEDs operating at room temperature [88] initiated a surge in research activity. Contemporary state-of-the-art PeLEDs now achieve EQEs ranging from 18% to 30% depending on the emission color [89–95], with notable progress in NIR PeLED exhibiting exceptional efficiency along with improved operational stability. A notable example is the use of a dipolar molecular stabilizer, sulfobetaine 10 (SFB10), which enabled NIR PeLEDs to operate continuously for over 3600 h (5 months) at J_{sc} of 5 mA cm^{-2} without significant radiance degradation (figures 7.11(A) and (B)) [95]. Accelerated lifetime testing of 62 stabilized PeLEDs revealed an estimated T_{50} lifetime of 32 675 h (~3.7 years) at an initial radiance of 2.1 W sr^{-1} m^{-2} (3.2 mA cm^{-2}), which is comparable to the performance of green organic LEDs operating at 1000 cd m^{-2} (figure 7.11(B)) [87].

These findings highlight the potential of PeLEDs for high-performance and long-lasting LED applications. However, several challenges remain particularly with respect to the long-term stability of blue and lead-free PeLEDs. Degradation mechanisms such as ion migration, phase segregation, and interface degradation, similar to those observed in PSCs, must be addressed. Continued progress in material stabilization, compositional engineering, and device encapsulation is crucial for the commercial potential of PeLEDs.

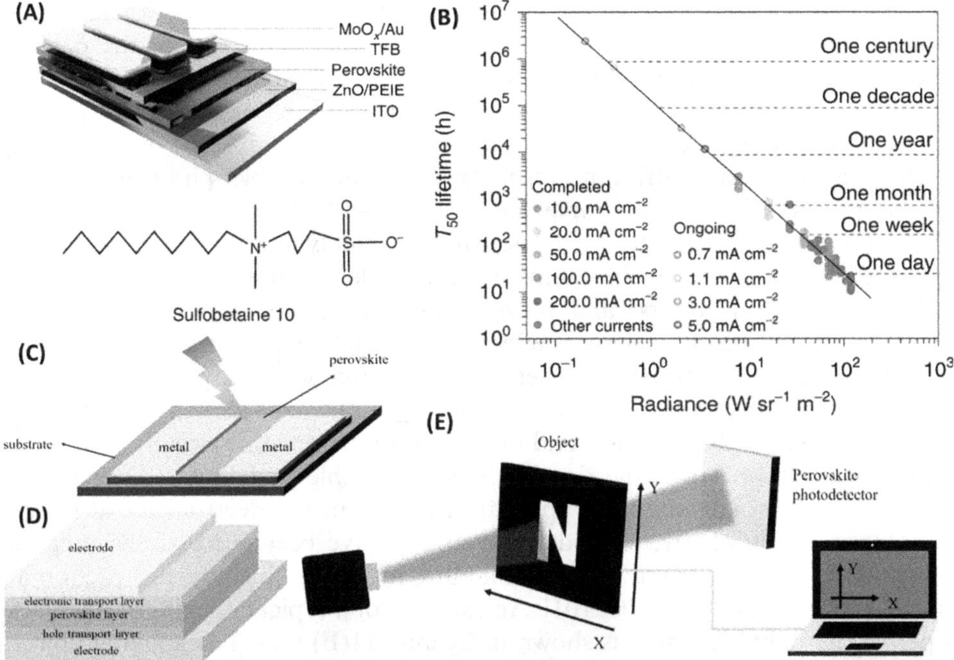

Figure 7.11. (A) Schematic of the ultra-stable NIR PeLED device architecture along with the chemical structure of the SFB10 stabilizer. (B) T_{50} lifetime results of SFB10-stabilized PeLEDs. The solid line represents a fit of 62 lifetime data points using the equation $R_0^n \times T_{50}$ = constant (with $n = 1.86$). Solid dots indicate completed measurements, while open circles represent ongoing tests. (C) Structure of a perovskite-based photoconductor, (D) p–i–n type perovskite photodiode and (E) schematic illustration of the perovskite-based imaging system. ((A)and (B) Reproduced with permission from [95]. Copyright 2022 Springer Nature. (C), (D) and (E) Reproduced from [97]. CC BY 4.0.)

7.3.2 Photodetectors

Photodetectors convert incident light into electrical signals, playing a crucial role in diverse applications such as imaging, environmental monitoring, biomedical sensing, and optical communication systems. Conventional photodetectors are primarily based on Si or other inorganic semiconductors such as gallium phosphide (GaP) and lead sulfide (PbS), which, despite their performance advantages, often require high-temperature and vacuum-based fabrication processes, leading to high production costs and limited scalability. In contrast, emerging photodetectors based on hybrid perovskite materials offer low-cost, solution-processable alternatives with superior optoelectronic properties [96, 97], exhibiting broad absorption spectra that extend from the ultraviolet (UV) to NIR regions. As a result, perovskite-based photodetectors have achieved remarkable performance metrics, including specific detectivities exceeding 10^{13} jones, fast photoresponse times, and broadband spectral tunability [98, 99].

Structurally, perovskite photodetectors can be realized in several architectures. One common configuration is the metal–semiconductor–metal (MSM) structure, as shown in figure 7.11(C) [97]. In this structure, metal electrodes (Au, Ag, Al, or Cu) are deposited in a coplanar layout on the perovskite layer. When the device is illuminated with photons of energy greater than the bandgap ($h_{\nu} > E_g$), electrons are excited from the valence band to the conduction band and electron–hole pairs are generated. An external bias facilitates the separation and transport of these charge carriers via drift and diffusion, resulting in a measurable photocurrent. This architecture benefits from simple fabrication and fast response, making it suitable for low-cost large-area sensors. Another widely adopted design is the p–i–n photodiode structure, which consists of a wide perovskite intrinsic (i) layer sandwiched between p-type and n-type transport layers (figure 7.11(D)). The extended depletion region enhances effective photoactive area and reduces junction capacitance, thereby improving the sensitivity and frequency response of the device by allowing better charge separation and collection. This design is particularly effective for weak-light detection and integrated optoelectronic circuits [97]. Recent advancements have led to the development of flexible and large-area perovskite photodetectors, broadening their applications in wearable electronics and imaging technologies. Additionally, photodetector arrays have been successfully integrated into scanning-based imaging systems, enabling real-time object reconstruction and pattern recognition [96, 97, 100, 101]. An example of a typical imaging system based on perovskite photodetectors is shown in figure 7.11(E) [97]. The imaging object is fixed on the *X–Y* scanning platform between the laser and the perovskite photodetector. The *X–Y* moving platform is controlled by computer software programming, and the object can move in the horizontal and vertical directions. The photodetector is connected to the phase-locked amplifier. By moving the object, the current signal matching the object position and the corresponding position coordinates can be collected. The measured dark current value represents the background noise level of each pixel, and the photocurrent value represents the light intensity level. Finally, the data are transferred to the computer and converted into grayscale values to realize the image presentation [97].

In order to further boost the performance of perovskite-based photodetectors, particularly in terms of sensitivity and photoconductive gain, several innovative strategies have been explored. One prominent approach involves the use of TiO_2/perovskite bilayer photoconductors, in which engineered porous or compact TiO_2 blocking layers facilitate hole injection into the perovskite valence band, thereby significantly enhancing photosensitivity [102, 103]. Another effective method is the hole-injection strategy via interface charge modulation, where the accumulation of negative charges at the interface lowers the electrode work function and improves the overall charge carrier conductivity [104]. Trap-state engineering has also shown to be a promising approach. By careful controlling of trap states, devices can transit from diode-like behavior to strong photoconductive response under illumination, enabled by multiplicative charge injection mechanisms [105, 106]. Additionally, in MSM photodetectors, tuning the Schottky barrier at the metal–perovskite interface has been demonstrated to enhance electron injection and overall photocurrent response [107]. These techniques highlight the versatility and tunability of perovskite materials and device interfaces, paving the way for high-performance, application-specific photodetector designs [101, 108]. These features position perovskite photodetectors as promising candidates for next-generation photonic and imaging technologies.

7.4 Emerging applications

As the scope of application of perovskite materials continues to expand, the subsequent section presents representative cases, such as wearables, smart windows, quantum photonics, neuromorphic computing, bioelectronics, and sustainable systems [109–117] to illustrate their growing significance in next-generation technological advancements.

7.4.1 Wearables and smart windows

Perovskite materials are gaining considerable attention in the field of smart and wearable electronics [110–113]. A prominent demonstration of this potential is illustrated by the development of a flexible, self-powered smart bracelet that integrates a PSC with a quasi-solid-state zinc-ion battery (ZIB), as shown in figure 7.12(A) [111]. In this design, high-performance ZIBs were fabricated using MnO_{2x} nanosheets grown on carbon cloth (MnO_{2-x} @CC), where lithium treatment introduced oxygen vacancies to enhance ion transport by expanding the interlayer spacing. With a high mass loading of 25.5 mg cm^{-2}, the resulting cathode

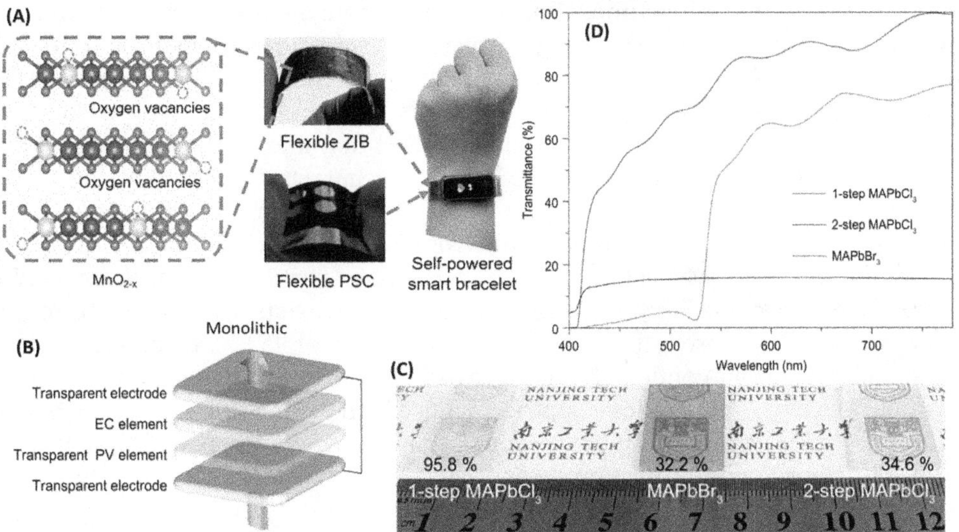

Figure 7.12. (A) Schematic of a flexible, self-powered wearable smart bracelet with integrated flexible PSC and a ZIB. (B) 2T monolithic, integrated PVCD architecture which requires only a single external circuit to connect the top and bottom terminals, representing an ideal design for achieving a full-frame active area along with high transmittance simultaneously. (C) Photographs of one-step $MAPbCl_3$, $MAPbBr_3$, and two-step $MAPbCl_3$ films, respectively. (D) Transmittance spectra of one-step $MAPbCl_3$, $MAPbBr_3$, and two-step $MAPbCl_3$ films. ((A) Reproduced with permission from [111]. Copyright 2021 American Chemical Society. (B), (C) and (D) Adapted from [112]. CC BY 4.0.)

exhibited a remarkable specific capacity of 3.63 mAh cm^{-2} at 3.93 mA cm^{-2} and sustained cycling stability over 5000 cycles. This ZIB delivered an impressive energy density of 5.11 mWh cm^{-2} (59.42 mWh cm^{-3}) while maintaining flexibility and safety across various operating conditions. When integrated with flexible PSCs, the device successfully powered a commercial smart bracelet, representing a significant step toward the next-generation self-sustaining wearable systems.

Beyond wearables, perovskite-based technologies are also showing great promise in smart window applications, where they contribute to both energy generation and dynamic light regulation. Transparent PV technology offers a promising route for developing power-generating smart windows, BIPV, and greenhouse applications [110–113]. Photovoltachromic devices (PVCD), which integrate fully transparent perovskite photovoltaic layers with ion-gel-based electrochromic elements in a vertical tandem architecture (figure 7.12(B)), offer a novel solution to meet the increasing demand for building energy efficiency [112]. These devices eliminate the need for intermediate electrodes and, through fine-tuned halide ion exchange, can achieve a high initial transmittance of up to 76%, an excellent color-rendering index of 96, and over 30% contrast ratio in visible-light transmission (400–780 nm). Furthermore, their self-adaptive transmittance properties enable dynamic indoor light and thermal regulation based on ambient solar irradiance. Chloride-based perovskites have seen limited use in PV applications compared to their iodide and bromide counterparts. This is primarily due to challenges in solution-processability and excessive crystallinity, which result in poor film morphology and significant grain boundary recombination. For example, methylammonium lead chloride ($MAPbCl_3$) films, fabricated using a one-step spin coating method (one-step $MAPbCl_3$), display a frosted, hazy appearance attributed to poor crystallinity (figure 7.12(C)). A halide-exchange strategy is developed to convert a solution-processed methylammonium lead bromide ($MAPbBr_3$) film into a fully transparent perovskite, referred to as two-step $MAPbCl_3$, characterized by low haze and enhanced optical transmittance (figure 7.12(C)). As shown in figure 7.12(D), the transparent cut-off wavelength blue-shifts from 525 nm in pristine $MAPbBr_3$ to 407 nm in the two-step $MAPbCl_3$, while the average visible transmittance (AVT) increases from 44% to 76% [112]. In contrast, the one-step $MAPbCl_3$ film retains a low AVT of 15%, correlating with its high haze level of 95.8%. Comparative haze measurements further confirm this, showing 32.2% for $MAPbBr_3$ and 34.6% for the two-step $MAPbCl_3$ film. These improvements highlight the importance of halide-exchange strategies and morphology control in achieving transparent and efficient perovskite films for multifunctional window and display technologies.

7.4.2 Quantum photonic devices

Perovskite quantum dots (PQDs), particularly $CsPbBr_3$, are rapidly emerging as leading candidates for quantum photonic applications due to their excellent optoelectronic properties and solution-processability [118, 119]. This progress builds upon the broader success of colloidal quantum dots (QDs), such as InP and CdSe, in which precise control over size, shape, and composition has enabled electronically well-defined nanomaterials. Larger PQDs, with sizes equal to or exceeding the Bohr diameter (~7 nm for $CsPbBr_3$), have demonstrated remarkable properties including long coherence times [120], bright single-photon emission [121], superlattice self-assembly [122, 123], superfluorescence [122, 124] and high efficiency in display and LED technologies [65, 125]. However, these properties remain relatively unexplored in strongly confined lead-halide PQDs, largely due to previous challenges in achieving high chemical stability and size monodispersity below ~7 nm [119]. Addressing this limitation, a recent study demonstrates the postsynthetic treatment of ~5 nm $CsPbBr_3$ PQDs using didodecyldimethylammonium bromide ligands, resulting in chemically stable and highly uniform QDs (monodispersity ~7.5% ± 2.0%) [119]. These QDs self-assemble into rhombic superlattices with long-range order and emit narrow-band cyan light. Most notably, they function as room-temperature single-photon sources, showing a single-photon purity of 73% and minimal blinking (78% 'on' fraction), highlighting their suitability for scalable quantum photonic devices [119]. Figure 7.13(A) illustrates the structural and photonic characteristics of 5 nm $CsPbBr_3$ quantum dots (QDs).

Experimental investigations into such strongly confined PQDs are crucial for understanding exciton, phonon interactions and Coulomb many-body effects [126,

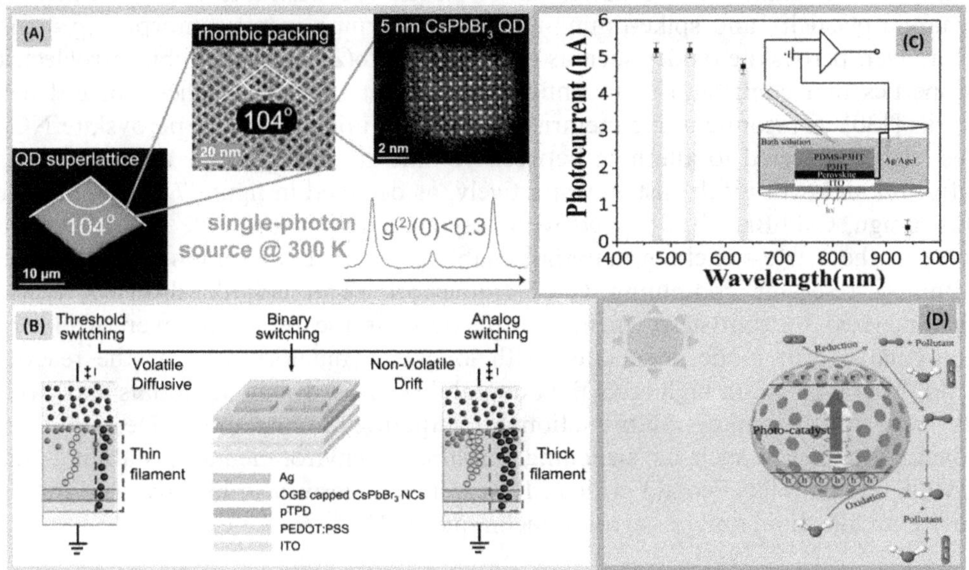

Figure 7.13. (A) Illustration of structural characteristics of 5 nm $CsPbBr_3$ QD. (B) Depiction of a reconfigurable halide perovskite memristor based on $CsPbBr_3$ NCs with OGB ligands, enabling both diffusive (volatile, short-term memory) and drift (non-volatile, long-term memory) modes for integrated reservoir computing. (C) Photocurrent response of the perovskite-based (ITO/$CH_3NH_3PbI_3$/PDMS:P3HT and ITO/$CH_3NH_3PbI_3$/P3HT/PDMS:P3HT) biointerface under varying illumination and aCSF conditions. Inset: Schematic of the photocurrent measurement set-up. (D) Schematic illustration of the photocatalytic degradation process of organic pollutants. ((A) Adapted from [119]. CC BY 4.0. (B) Adapted from [115]. CC BY 4.0. (C) Reproduced with permission from [116]. Copyright 2019 The Authors. Published by John Wiley and Sons. CC-BY-NC-ND. (D) Adapted from [117]. CC BY 4.0.)

127]. These insights guide the design of quantum photonic systems across regimes, from high-efficiency cyan LEDs (weak coupling) to single-photon nonlinear devices (strong coupling) [128].

Furthermore, spectral tunability across perovskite compositions, including CsPb (Cl–Br)$_3$, $CsPbBr_3$, $FAPbBr_3$, $CsPbI_3$, and $PaPbI_3$, combined with their sharp emission peaks, adjustable photoluminescence, and short to moderate lifetimes (0.16–25 ns), makes them versatile sources for quantum light emission [114]. Their integration into photonic platforms such as waveguides, microlasers, and interferometers further strengthens their role in the next-generation quantum photonic circuits.

7.4.3 Neuromorphic systems

The tunable ionic conductivity and resistive switching capabilities of halide perovskites enable their use in memristors—key components of neuromorphic computing architectures [109, 129]. Halide perovskites such as $MAPbI_3$ and $CsPbBr_3$ demonstrate non-volatile resistive switching, low operating voltages, and analogue signal processing—essential features for synaptic emulation and edge AI systems. These materials support multi-level conductance states, low operation voltages, and analogue switching, all of which are desirable for implementing synaptic plasticity and spike-timing-dependent learning in neuromorphic systems. Moreover, perovskite oxides such as $BaTiO_3$ and Pb(Zr, Ti)O_3 exhibit ferroelectric properties that contribute to reconfigurable weight states in artificial neural networks [130]. A reconfigurable memristor platform, utilizing halide perovskite NCs, has been engineered to alternate between volatile and non-volatile modes through diffusion and drift mechanisms, respectively, as depicted in figure 7.13(B) [115]. In this design, $CsPbBr_3$ NCs capped with oleylguanidinium bromide (OGB) ligands serve as the active switching material, enabling two distinct computational modes within a reservoir computing (RC) framework. The volatile diffusive mode, characterized by short-term memory, functions as the reservoir layer, while the non-volatile drift mode, associated with long-term memory, acts as the readout layer. This innovation highlights the adaptability of perovskite materials and paves the way for dynamic, multifunctional computing architectures. Despite these advantages, issues such as switching variability, environmental instability, and CMOS compatibility remain critical. Research is underway to improve endurance, device architecture, and passivation techniques [131].

7.4.4 Bioelectronics, hybrid systems, and environmental interfaces

The mechanical compliance and hybrid ionic–electronic conduction of perovskites have enabled their applications in bioelectronic interfaces, including artificial skin, neural stimulators, and implantable biosensors. Their biocompatibility and optoelectronic responsiveness make them suitable for direct interaction with biological systems. For example, the optoelectronic performance of a perovskite-based biointerface, structured as ITO/$CH_3NH_3PbI_3$/P3HT, is enabled by a device architecture consisting of an ITO-coated glass substrate as the conductive layer, a

$CH_3NH_3PbI_3$ perovskite film acting as the absorptive photoactive layer, a biocompatible P3HT layer functioning as HTL, and a PDMS:P3HT composite acting as the encapsulation layer. Photocurrent measurements were carried out in aqueous artificial cerebrospinal fluid (aCSF) using an electrophysiology amplifier, under varying light intensities and wavelengths [116]. The results (figure 7.13(C)) demonstrate strong photocurrent responses across the visible spectrum, attributed to interband transitions in the perovskite layer. Control devices, comprising ITO/PDMS:P3HT and ITO/P3HT/PDMS:P3HT, exhibited negligible signals, confirming that the enhanced response originates from the perovskite material rather than the polymeric matrix. In hybrid optoelectronic systems, perovskites facilitate multifunctional operations-such as sensing, energy harvesting, and light emission-in platforms such as smart textiles, wearables, and self-powered IoT nodes.

Perovskites, beyond their electronic applications, are increasingly used in environmental photocatalysis and electrocatalysis for CO_2 reduction, nitrogen fixation, water splitting, and wastewater treatment. Their strong visible-light absorption, tunable band edges, and structural versatility enable efficient redox reactions. A recent report examines the emerging role of perovskite-based photocatalysts in the removal of persistent organic pollutants from wastewater [117]. Under illumination, perovskites generate electron–hole pairs that produce reactive oxygen species, which degrade pollutants into harmless byproducts such as CO_2 and H_2O (figure 7.13(D)).

7.5 Energy storage

Metal halide perovskites are now emerging as promising materials for the next-generation energy storage and self-powered systems. Recent research highlights their potential in lithium-ion batteries (LIBs), supercapacitors, and integrated photo-rechargeable devices due to their excellent ionic conductivity, tunable structure, and compatibility with solution processing. Han *et al* [132] discussed the use of halide perovskites as electrode materials in LIBs and formed the basis of hybrid systems via integration with PSCs to achieve efficient energy conversion and storage. Similarly, a review by Yang *et al* [133] elaborates on PSC-based self-charging power packs, emphasizing their application in portable electronics and the importance of interface engineering and architectural design. Zhang *et al* [134] summarized recent advances in halide perovskite materials for energy storage, underlining the role of factors such as morphology, defect density, and doping in significantly impacting the performance of batteries and supercapacitors. Nonetheless, in spite of the current challenges in stability and scalability, these studies collectively suggest that halide perovskites hold great promise for integrated, high-efficiency, and flexible energy storage devices.

7.6 Encapsulation technologies

Encapsulation plays a critical role in ensuring the environmental stability and operational durability of perovskite devices by protecting them from moisture, oxygen, UV light, and thermal degradation [135]. Adapting the approaches from matured silicon and organic PV technologies, encapsulation methods in PSCs have

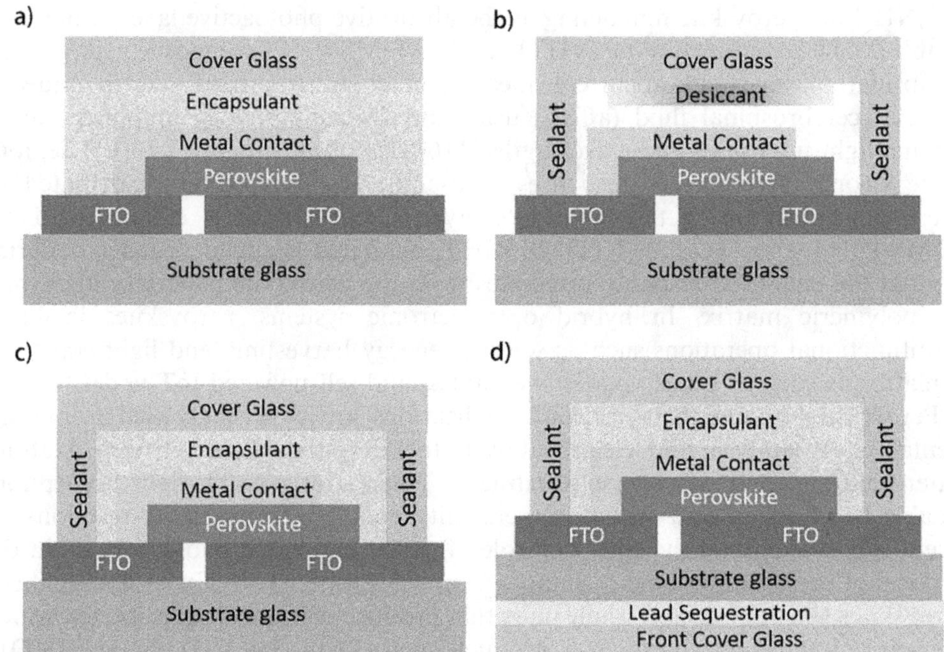

Figure 7.14. Schematic representations of various cover glass encapsulation strategies for PSCs: (a) blanket encapsulation, (b) edge encapsulation, (c) combined blanket encapsulation with edge sealing, and (d) sealed-edge blanket encapsulation incorporating lead containment. In all configurations, the metal contacts remain fully enclosed within the encapsulated region. (Reproduced with permission from [135]. Copyright 2022 The Author(s). Published by the American Chemical Society. CC-BY-NC-ND 4.0.)

evolved to include both standalone thin-film coatings and cover glass (or flexible cover) architectures. As depicted in figure 7.14(a)–(d), these configurations include blanket encapsulation, edge sealing, and hybrid approaches, some of which incorporate functionalities such as lead leakage suppression. Various encapsulation strategies are developed to enhance the environmental stability and operational durability of perovskite devices, including glass and polymeric barriers, lamination and thin-film coatings, laser-assisted and UV-curable sealants, as well as emerging self-healing encapsulation approaches.

Rigid glass-to-glass encapsulation remains a benchmark for long-term stability, sandwiching the PSC between two glass plates with edge-sealing adhesives such as polyisobutylene (PIB) [136]. This configuration offers excellent resistance to moisture and oxygen, with studies showing retention of over 90% of initial PCE under ambient storage conditions. However, its bulk and rigidity limit flexibility, making it less suited for lightweight or wearable devices.

In contrast, polymer-based encapsulation using materials such as ethylene-vinyl acetate (EVA), UV-curable resins, and thermoplastic polyolefins (TPO) offers flexible, lightweight alternatives [137]. These polymers support R2R manufacturing and can adapt to a wide range of device geometries. UV-curable resins, in particular, enable rapid curing and strong adhesion, which helps minimize thermal stress and improve device durability. More recently, there has been increasing interest in polyolefin elastomer (POE) based encapsulants, as well as other EVA alternatives such as ionomers, polyvinyl butyral (PVB), and thermoplastic polyurethane (TPU) [138].

For example, the TPO encapsulated devices (figure 7.15(A)) exhibited superior environmental resistance, maintaining >95% of initial PCE after one month of ambient storage and showing no degradation after water immersion for one week, in stark contrast to the EVA and POE encapsulated counterparts, which dropped below 20% within 8 days. The encapsulation process used in figure 7.15(A) (I)–(IV) involved p–i–n PSCs (FTO/NiO_x/perovskite/PCBM/BCP/Au) with heat-activated adhesive copper ribbons. These were sealed in a multi-layer structure and laminated at 150 °C [137]. Additionally, a low-temperature glass-to-glass encapsulation using UV-curable epoxy (Vitralit, Panacol) further demonstrated improved stability while avoiding the thermal stress associated with Surlyn-based methods [139] (figure 7.15 (B)). Devices sealed at low temperatures (26 °C–48 °C during UV curing) retained 83% PCE after 30 days, with a T_{80} lifetime of 70 days, outperforming Surlyn-based seals that required heating at 125 °C for sealing and led to early-stage thermal degradation (figure 7.15(B)) [140–142].

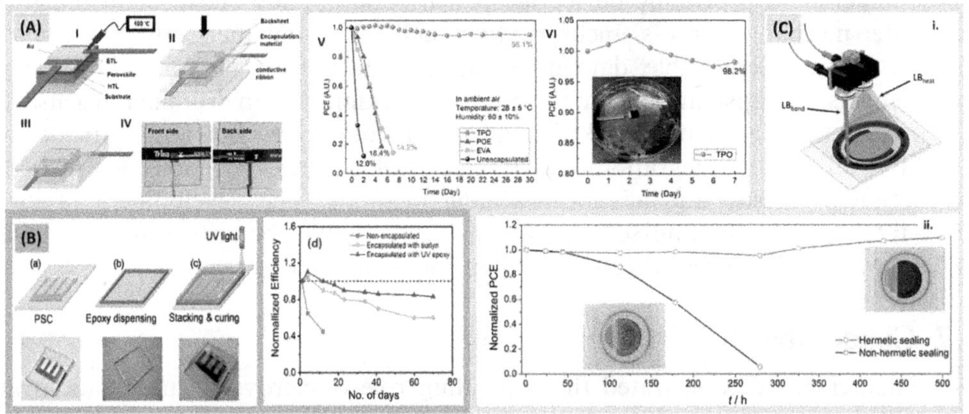

Figure 7.15. (A) Schematic illustration of (I)–(III) the PSCs encapsulation process, (IV) photographic images of the PSC after encapsulation, (V) stability comparison of PSCs encapsulated with various materials under ambient conditions and (VI) normalized PCE variation of PSC encapsulated with TPO during water immersion test. (B) Top row: Schematic illustration of the glass-to-glass encapsulation process for a PSC using a UV-curable epoxy edge sealant: (a) PSC on a FTO glass substrate, (b) cover glass with patterned epoxy edge sealant, and (c) stacking of layers followed by selective UV curing from the cover glass side. Bottom row: Corresponding photographs of each encapsulation step. (d) Performance variations of PSCs stored in the dark at 30 °C and 50% relative humidity (RH). The dashed line indicates the initial performance value as a reference. (C) (i) Schematic of the laser-assisted glass frit sealing process utilizing a dual-laser beam configuration. (ii) Normalized PCE of devices with hermetic and non-hermetic encapsulation after 500 h of exposure to humid air. ((A) Adapted from [137]. CC BY 4.0. (B) Reproduced with permission from [139]. Copyright 2019 Elsevier. (C) Reproduced with permission from [144]. Copyright 2020 Royal Society of Chemistry.)

Lamination involves thermally or pressure-activated bonding of protective films over PSCs and is widely used in scalable module manufacturing. Recent innovations employ semi-solid adhesives and viscoelastic polymers to improve device adhesion, mechanical strength, and resistance to environmental factors. Lamination offers compatibility with flexible architectures and enables fast processing. Complementing lamination are thin-film coating techniques which include atomic layer deposition (ALD) and chemical vapor deposition (CVD), which apply ultrathin, conformal barrier layers (e.g. Al_2O_3, SiO_2) directly onto device surfaces. These layers act as effective diffusion barriers against water and oxygen ingress. Despite challenges that are related to vacuum processing and thermal sensitivity, low-temperature ALD processes show great promise for preserving perovskite integrity [143].

Furthermore, dual-laser beam sealing techniques (figure 7.15(C)) enable localized bonding of encapsulant layers at moderate temperatures [144]. A combination of pre-heating at 100 °C–120 °C and targeted bonding at approximately 380 °C resulted in over 80% reproducibility of hermetic sealing, with devices sustaining high performance for more than 200 h. In contrast, non-hermetically sealed devices degraded to 50% efficiency within the same period, mainly due to seal defects or organic contamination (figure 7.15(C)).

Recent research explores self-healing encapsulation materials that autonomously repair cracks and defects under external stimuli such as light or heat [145, 146]. These materials can prolong device lifespan by maintaining encapsulant integrity even after mechanical stress. Incorporating such smart polymers into PSCs holds promise for field-deployable, damage-resistant modules with minimal maintenance requirements. Encapsulation is important in addressing the environmental sensitivity of PSCs. Research has markedly enhanced both short and long-term device stability through the integration of rigid and flexible barrier technologies, scalable lamination processes, and advanced sealing methods. The development of low-temperature, lead-safe, and self-healing encapsulants will be critical to advancing PSCs toward reliable commercialization and widespread deployment.

7.7 Conclusion

This chapter has demonstrated the expanding role of perovskite materials, from high-performance photovoltaics to innovative optoelectronic and cross-disciplinary applications. With growing progress in flexible devices, tandem architectures, and reliable encapsulation, perovskites are steadily moving from lab-scale research to real-world deployment across a range of advanced technologies.

References

[1] Xu X *et al* 2018 1D organic–inorganic hybrid perovskite micro/nanocrystals: fabrication, assembly, and optoelectronic applications *Small Methods* **2** 1700340

[2] Zhang L *et al* 2023 Advances in the application of perovskite materials *Nano-Micro Lett.* **15** 177

[3] Vats G *et al* 2023 Optical memory, switching, and neuromorphic functionality in metal halide perovskite materials and devices *Adv. Mater.* **35** 2205459

[4] Cheng M *et al* 2024 Progress and application of halide perovskite materials for solar cells and light emitting devices *Nanomaterials* **14** 391

[5] Korde V B *et al* 2024 Review: Perovskite nanostructures materials versatile platform for advance biosensor applications *Sens. Actuators Rep.* **7** 100201

[6] Ullah S *et al* 2024 Emerging applications of perovskite oxides in electrochemical reduction of carcinogenic nitrate to ammonia: a recent review *Inorg. Chem. Front.* **11** 7204–37

[7] Pandiyan A *et al* 2023 A comprehensive review on perovskite and its functional composites in smart textiles: progress, challenges, opportunities, and future directions *Prog. Mater. Sci.* **140** 101206

[8] Luque A and Hegedus S 2011 *Handbook of Photovoltaic Science and Engineering* (Wiley)

[9] Sze S M, Li Y and Ng K K 2021 *Physics of Semiconductor Devices* (Wiley)

[10] Ohl R S 1946 Light-sensitive electric device *Patent* USA US2402662A

[11] Chapin D M, Fuller C S and Pearson G L 1954 A new silicon p–n junction photocell for converting solar radiation into electrical power *J. Appl. Phys.* **25** 676–7

[12] NREL 2025 Best Research-Cell Efficiency Chart *US Department of Energy, Office of Energy Efficiency and Renewable Energy* https://www2.nrel.gov/pv/cell-efficiency

[13] Sharma P and Mishra R K 2025 Comprehensive study on photovoltaic cell's generation and factors affecting its performance: a review *Mater. Renew. Sustain. Energy* **14** 21

[14] Kumar P 2016 *Organic Solar Cells: Device Physics, Processing, Degradation, and Prevention* (CRC Press)

[15] Dittrich T 2018 *Materials Concepts for Solar Cells* (World Scientific)

[16] Kumar P and Chauhan A K 2020 Highly efficient flexible perovskite solar cells and their photo-stability *J. Phys.* D: Appl. Phys. **53** 035101

[17] Yang T-Y, Kim Y Y and Seo J 2021 Roll-to-roll manufacturing toward lab-to-fab-translation of perovskite solar cells *APL Mater.* **9** 110901

[18] Weerasinghe H C *et al* 2024 The first demonstration of entirely roll-to-roll fabricated perovskite solar cell modules under ambient room conditions *Nat. Commun.* **15** 1656

[19] Kutsarov D I *et al* 2025 Progress in flexible perovskite solar cells: paving the way for scalable manufacturing *Adv. Mater. Technol.* **10** 2401834

[20] Aftab S *et al* 2023 Advances in flexible perovskite solar cells: a comprehensive review *Nano Energy* **120** 109112

[21] Ravindra N M, Lin L and Singh P 2025 *Recent Advances in Solar Cells* (IOP Publishing)

[22] O'Regan B and Grätzel M 1991 A low-cost, high-efficiency solar cell based on dye-sensitized colloidal TiO_2 films *Nature* **353** 737–40

[23] Yella A *et al* 2011 Porphyrin-sensitized solar cells with cobalt (II/III)–based redox electrolyte exceed 12 percent fficiency *Science* **334** 629–34

[24] Kojima A *et al* 2009 Organometal halide perovskites as visible-light sensitizers for photovoltaic cells *J. Am. Chem. Soc.* **131** 6050–1

[25] Kim H-S *et al* 2012 Lead iodide perovskite sensitized all-solid-state submicron thin film mesoscopic solar cell with efficiency exceeding 9% *Sci. Rep.* **2** 591

[26] Burschka J *et al* 2013 Sequential deposition as a route to high-performance perovskite-sensitized solar cells *Nature* **499** 316–9

[27] Liu M, Johnston M B and Snaith H J 2013 Efficient planar heterojunction perovskite solar cells by vapour deposition *Nature* **501** 395–8

[28] Lee M M *et al* 2012 Efficient hybrid solar cells based on meso-superstructured organometal halide perovskites *Science* **338** 643–7

[29] Lee K-M *et al* 2019 Effect of anti-solvent mixture on the performance of perovskite solar cells and suppression hysteresis behavior *Org. Electron.* **65** 266–74

[30] Jiang J *et al* 2020 Improvement in solar cell efficiency based on the $MAPbI_3$ films extracted by a mixed anti-solvent *Appl. Phys. Lett.* **117** 203901

[31] Chao L *et al* 2021 Solvent engineering of the precursor solution toward large-area production of perovskite solar cells *Adv. Mater.* **33** 2005410

[32] Park S M *et al* 2023 Engineering ligand reactivity enables high-temperature operation of stable perovskite solar cells *Science* **381** 209–15

[33] Park J *et al* 2023 Controlled growth of perovskite layers with volatile alkylammonium chlorides *Nature* **616** 724–30

[34] Saliba M *et al* 2016 Incorporation of rubidium cations into perovskite solar cells improves photovoltaic performance *Science* **354** 206–9

[35] Mateen M *et al* 2020 High-performance mixed-cation mixed-halide perovskite solar cells enabled by a facile intermediate engineering technique *J. Power Sources* **448** 227386

[36] Yang M *et al* 2023 Mixed-halide inorganic perovskite solar cells: opportunities and challenges *Adv. Opt. Mater.* **11** 2301052

[37] McMeekin D P *et al* 2016 A mixed-cation lead mixed-halide perovskite absorber for tandem solar cells *Science* **351** 151–5

[38] Seshaiah K V and Kim J H 2024 Nature of defects and their passivation engineering for advancements in perovskite solar cells *Chem. Eng. J.* **492** 152370

[39] Gao W *et al* 2020 A review of flexible perovskite oxide ferroelectric films and their application *J. Materiomics* **6** 1–16

[40] Zhao W *et al* 2021 A special additive enables all cations and anions passivation for stable perovskite solar cells with efficiency over 23% *Nano-Micro Lett.* **13** 169

[41] Tingare Y S *et al* 2024 Charged hole-transporting materials based on imidazolium for defect passivation in inverted perovskite solar cells *Sol. RRL* **8** 2300817

[42] Chauhan A and Kumar P 2017 Degradation in perovskite solar cells stored in different environmental conditions *J. Phys.* D: Appl. Phys. **50** 325105

[43] Chauhan A K and Kumar P 2019 Photo-stability of perovskite solar cells with Cu electrode *J. Mater. Sci., Mater. Electron.* **30** 9582–92

[44] Chauhan A K *et al* 2017 Air-processed organo-metal halide perovskite solar cells and their air stability *J. Mater. Sci.* **52** 10886–97

[45] Zhang D *et al* 2022 Degradation pathways in perovskite solar cells and how to meet international standards *Commun. Mater.* **3** 58

[46] Miah M H *et al* 2025 Key degradation mechanisms of perovskite solar cells and strategies for enhanced stability: issues and prospects *RSC Adv.* **15** 628–54

[47] Ahn N *et al* 2015 Highly reproducible perovskite solar cells with average efficiency of 18.3% and best efficiency of 19.7% fabricated via Lewis base adduct of lead(II) iodide *J. Am. Chem. Soc.* **137** 8696–9

[48] Aftab S *et al* 2024 Advances in flexible perovskite solar cells: a comprehensive review *Nano Energy* **120** 109112

[49] Shockley W and Queisser H J 1961 Detailed balance limit of efficiency of p–n junction solar cells *J. Appl. Phys.* **32** 510–9

[50] Li H and Zhang W 2020 Perovskite tandem solar cells: from fundamentals to commercial deployment *Chem. Rev.* **120** 9835–950

[51] Lal N N *et al* 2017 Perovskite tandem solar cells *Adv. Energy Mater.* **7** 1602761

[52] Jošt M *et al* 2020 Monolithic perovskite tandem solar cells: a review of the present status and advanced characterization methods toward 30% efficiency *Adv. Energy Mater.* **10** 1904102

[53] Duan L *et al* 2023 Stability challenges for the commercialization of perovskite–silicon tandem solar cells *Nat. Rev. Mater.* **8** 261–81

[54] Todorov T, Gunawan O and Guha S 2016 A road towards 25% efficiency and beyond: perovskite tandem solar cells *Mol. Syst. Des. Eng.* **1** 370–6

[55] Leijtens T *et al* 2018 Opportunities and challenges for tandem solar cells using metal halide perovskite semiconductors *Nat. Energy* **3** 828–38

[56] Cheng Y and Ding L 2021 Perovskite/Si tandem solar cells: fundamentals, advances, challenges, and novel applications *SusMat* **1** 324–44

[57] Werner J, Niesen B and Ballif C 2018 Perovskite/silicon tandem solar cells: marriage of convenience or true love story?—An overview *Adv. Mater. Interfaces* **5** 1700731

[58] Park I J *et al* 2019 A three-terminal monolithic perovskite/Si tandem solar cell characterization platform *Joule* **3** 807–18

[59] Gota F *et al* 2020 Energy yield advantages of three-terminal perovskite–silicon tandem photovoltaics *Joule* **4** 2387–403

[60] Tayagaki T *et al* 2019 Three-terminal tandem solar cells with a back-contact-type bottom cell bonded using conductive metal nanoparticle arrays *IEEE J. Photovolt.* **10** 358–62

[61] Tockhorn P *et al* 2020 Three-terminal perovskite/silicon tandem solar cells with top and interdigitated rear contacts *ACS Appl. Energy Mater.* **3** 1381–92

[62] Guo F *et al* 2015 Fully printed organic tandem solar cells using solution-processed silver nanowires and opaque silver as charge collecting electrodes *Energy Environ. Sci.* **8** 1690–7

[63] Tyagi B *et al* 2022 High-performance, large-area semitransparent and tandem perovskite solar cells featuring highly scalable a-ITO/Ag mesh 3D top electrodes *Nano Energy* **95** 106978

[64] Lamanna E *et al* 2020 Mechanically stacked, two-terminal graphene-based perovskite/silicon tandem solar cell with efficiency over 26% *Joule* **4** 865–81

[65] Kim S *et al* 2021 Over 30% efficiency bifacial 4-terminal perovskite-heterojunction silicon tandem solar cells with spectral albedo *Sci. Rep.* **11** 15524

[66] Kim D H *et al* 2019 Bimolecular additives improve wide-band-gap perovskites for efficient tandem solar cells with CIGS *Joule* **3** 1734–45

[67] Uzu H *et al* 2015 High efficiency solar cells combining a perovskite and a silicon heterojunction solar cells via an optical splitting system *Appl. Phys. Lett.* **106** 013506

[68] Yao Y *et al* 2020 Highly efficient Sn–Pb perovskite dolar vell and high-performance all-perovskite four-terminal tandem solar cell *Sol. RRL* **4** 1900396

[69] Alberi K *et al* 2024 A roadmap for tandem photovoltaics *Joule* **8** 658–92

[70] NREL 2024 Best Research-Cell Efficiency Chart by NREL *US Department of Energy, Office of Energy Efficiency and Renewable Energy* https://www2.nrel.gov/pv/cell-efficiency

[71] Mailoa J P *et al* 2015 A 2-terminal perovskite/silicon multijunction solar cell enabled by a silicon tunnel junction *Appl. Phys. Lett.* **106** 121105

[72] Shaw V 2023 Longi claims 33.9% efficiency for perovskite-silicon tandem solar cell *PV Magazine—Photovoltaics Markets and Technology* 3 November https://www.pv-magazine.com/2023/11/03/longi-claims-33-9-efficiency-for-perovskite-silicon-tandem-solar-cell/

[73] Yang D *et al* 2021 28.3%-efficiency perovskite/silicon tandem solar cell by optimal transparent electrode for high efficient semitransparent top cell *Nano Energy* **84** 105934

[74] Liu Z *et al* 2025 All-perovskite tandem solar cells achieving >29% efficiency with improved (100) orientation in wide-bandgap perovskites *Nat. Mater.* **24** 252–9

[75] Fu F *et al* 2019 Flexible perovskite/Cu(In,Ga)Se_2 monolithic tandem solar cells arXiv:1907.10330

[76] Lang F *et al* 2019 Efficient minority carrier detrapping mediating the radiation hardness of triple-cation perovskite solar cells under proton irradiation *Energy Environ. Sci.* **12** 1634–47

[77] Lang F *et al* 2020 Proton radiation hardness of perovskite tandem photovoltaics *Joule* **4** 1054–69

[78] Todorov T *et al* 2015 Monolithic perovskite-CIGS tandem solar cells via *in situ* band gap engineering *Adv. Energy Mater.* **5** 1500799

[79] Han Q *et al* 2018 High-performance perovskite/Cu(In,Ga)Se_2 monolithic tandem solar cells *Science* **361** 904–8

[80] Jošt M *et al* 2019 21.6%-efficient monolithic perovskite/Cu(In,Ga)Se_2 tandem solar cells with thin conformal hole transport layers for integration on rough bottom cell surfaces *ACS Energy Lett.* **4** 583–90

[81] Al-Ashouri A *et al* 2019 Conformal monolayer contacts with lossless interfaces for perovskite single junction and monolithic tandem solar cells *Energy Environ. Sci.* **12** 3356–69

[82] Jošt M *et al* 2022 Perovskite/CIGS tandem solar cells: from certified 24.2% toward 30% and beyond *ACS Energy Lett.* **7** 1298–307

[83] Jiang X *et al* 2024 Isomeric diammonium passivation for perovskite–organic tandem solar cells *Nature* **635** 860–6

[84] Kamat P V 2018 Hybrid perovskites for multijunction tandem solar cells and solar fuels. A virtual issue *ACS Energy Lett.* **3** 28–9

[85] Li Z *et al* 2020 Wide-bandgap perovskite/gallium arsenide tandem solar cells *Adv. Energy Mater.* **10** 1903085

[86] Rogalski A *et al* 2025 The perovskite optoelectronic devices—a look at the future *Small Methods* **9** 2400709

[87] Zhao B *et al* 2024 Highly stable perovskite light-emitting diodes *Matter* **7** 772–93

[88] Tan Z-K *et al* 2014 Bright light-emitting diodes based on organometal halide perovskite *Nat. Nanotechnol.* **9** 687–92

[89] Lin K *et al* 2018 Perovskite light-emitting diodes with external quantum efficiency exceeding 20 per cent *Nature* **562** 245–8

[90] Cao Y *et al* 2018 Perovskite light-emitting diodes based on spontaneously formed submicrometre-scale structures *Nature* **562** 249–53

[91] Chiba T *et al* 2018 Anion-exchange red perovskite quantum dots with ammonium iodine salts for highly efficient light-emitting devices *Nat. Photonics* **12** 681–7

[92] Bai W *et al* 2023 Perovskite light-emitting diodes with an external quantum efficiency exceeding 30% *Adv. Mater.* **35** 2302283

[93] Kim J S *et al* 2022 Ultra-bright, efficient and stable perovskite light-emitting diodes *Nature* **611** 688–94

[94] Xu W *et al* 2019 Rational molecular passivation for high-performance perovskite light-emitting diodes *Nat. Photonics* **13** 418–24

[95] Guo B *et al* 2022 Ultrastable near-infrared perovskite light-emitting diodes *Nat. Photonics* **16** 637–43

[96] Li G *et al* 2022 Research progress of high-sensitivity perovskite photodetectors: a review of photodetectors: noise, structure, and materials *ACS Appl. Electron. Mater.* **4** 1485–505

[97] Wang H *et al* 2022 A review of perovskite-based photodetectors and their applications *Nanomaterials* **12** 4390

[98] Chen G *et al* 2019 Stable α-$CsPbI_3$ perovskite nanowire arrays with preferential crystallographic orientation for highly sensitive photodetectors *Adv. Funct. Mater.* **29** 1808741

[99] Li C *et al* 2021 High-performance photodetectors based on nanostructured perovskites *Nanomaterials* **11** 1038

[100] Ma T *et al* 2024 Recent progress in photodetectors: from materials to structures and applications *Micromachines* **15** 1249

[101] Li C *et al* 2020 Advances in perovskite photodetectors *InfoMat* **2** 1247–56

[102] Moehl T *et al* 2014 Strong photocurrent amplification in perovskite solar cells with a porous TiO_2 blocking layer under reverse bias *J. Phys. Chem. Lett.* **5** 3931–6

[103] Chen H-W *et al* 2015 A switchable high-sensitivity photodetecting and photovoltaic device with perovskite absorber *J. Phys. Chem. Lett.* **6** 1773–9

[104] Domanski K *et al* 2015 Working principles of perovskite photodetectors: analyzing the interplay between photoconductivity and voltage-driven energy-level alignment *Adv. Funct. Mater.* **25** 6936–47

[105] Dong R *et al* 2015 High-gain and low-driving-voltage photodetectors based on organolead triiodide perovskites *Adv. Mater.* **27** 1912–8

[106] Liu C *et al* 2016 PbS quantum dots-induced trap-assisted charge injection in perovskite photodetectors *Nano Energy* **30** 27–35

[107] Hu X *et al* 2014 High-performance flexible broadband photodetector based on organolead halide perovskite *Adv. Funct. Mater.* **24** 7373–80

[108] Li Y *et al* 2022 Multifunctional histidine cross-linked interface toward efficient planar perovskite solar cells *ACS Appl. Mater. Interfaces* **14** 47872–81

[109] Li H *et al* 2024 Recent advances in artificial neuromorphic applications based on perovskite composites *Mater. Horiz.* **11** 5499–532

[110] Andrew B *et al* 2022 Perovskite fiber-shaped optoelectronic devices for wearable applications *J. Mater. Chem.* C **10** 6957–91

[111] Zhao J *et al* 2021 A safe flexible self-powered wristband system by integrating defective MnO_{2-x} nanosheet-based zinc-ion batteries with perovskite solar cells *ACS Nano* **15** 10597–608

[112] Liu Y *et al* 2021 Full-frame and high-contrast smart windows from halide-exchanged perovskites *Nat. Commun.* **12** 3360

[113] Traverse C J *et al* 2017 Emergence of highly transparent photovoltaics for distributed applications *Nat. Energy* **2** 849–60

[114] Castelletto S, Angelis F D and Boretti A 2022 Prospects and challenges of quantum emitters in perovskites nanocrystals *Appl. Mater. Today* **26** 101401

[115] John R A *et al* 2022 Reconfigurable halide perovskite nanocrystal memristors for neuromorphic computing *Nat. Commun.* **13** 2074

[116] Aria M M *et al* 2019 Perovskite-based optoelectronic biointerfaces for non-bias-assisted photostimulation of cells *Adv. Mater. Interfaces* **6** 1900758

[117] Ajit D, Dipankar M and Mrinal Kanti A 2021 Perovskite based photocatalyst for wastewater treatment: green approach of environmental sustainability *Am. J. Biol. Environ. Stat.* **7** 1–8

[118] Aftab S *et al* 2024 From energy to light: advancements in perovskite quantum dots for optoelectronics *J. Mater. Chem.* C **12** 17789–801

[119] Boehme S C *et al* 2023 Strongly confined $CsPbBr_3$ quantum dots as quantum emitters and building blocks for rhombic superlattices *ACS Nano* **17** 2089–100

[120] Utzat H *et al* 2019 Coherent single-photon emission from colloidal lead halide perovskite quantum dots *Science* **363** 1068–72

[121] Park Y-S *et al* 2015 Room temperature single-photon emission from individual perovskite quantum dots *ACS Nano* **9** 10386–93

[122] Cherniukh I *et al* 2021 Perovskite-type superlattices from lead halide perovskite nanocubes *Nature* **593** 535–42

[123] van der Burgt J S *et al* 2018 Cuboidal supraparticles self-assembled from cubic $CsPbBr_3$ perovskite nanocrystals *J. Phys. Chem.* C **122** 15706–12

[124] Krieg F *et al* 2021 Monodisperse long-chain sulfobetaine-capped $CsPbBr_3$ nanocrystals and their superfluorescent assemblies *ACS Cent. Sci.* **7** 135–44

[125] Fakharuddin A *et al* 2022 Perovskite light-emitting diodes *Nat. Electron.* **5** 203–16

[126] Cho K *et al* 2021 Luminescence fine structures in single lead halide perovskite nanocrystals: size dependence of the exciton–phonon coupling *Nano Lett.* **21** 7206–12

[127] Mondal A *et al* 2018 Ultrafast exciton many-body interactions and hot-phonon bottleneck in colloidal cesium lead halide perovskite nanocrystals *Phys. Rev.* B **98** 115418

[128] Zasedatelev A V *et al* 2021 Single-photon nonlinearity at room temperature *Nature* **597** 493–7

[129] Kim S J *et al* 2025 Linearly programmable two-dimensional halide perovskite memristor arrays for neuromorphic computing *Nat. Nanotechnol.* **20** 83–92

[130] Mikolajick T *et al* 2023 From ferroelectric material optimization to neuromorphic devices *Adv. Mater.* **35** 2206042

[131] Kim S-Y, Zhang H and Rubio-Magnieto J 2024 Operating mechanism principles and advancements for halide perovskite-based memristors and neuromorphic devices *J. Phys. Chem. Lett.* **15** 10087–103

[132] Han B *et al* 2023 Energy storage research of metal halide perovskites for rechargeable batteries *Nano Energy* **115** 108646

[133] Yang Y *et al* 2022 Perovskite solar cells based self-charging power packs: fundamentals, applications and challenges *Nano Energy* **94** 106910

[134] Zhang L *et al* 2020 Halide perovskite materials for energy storage applications *Adv. Funct. Mater.* **30** 2003653

[135] Wang Y *et al* 2022 Encapsulation and stability testing of perovskite solar cells for real life applications *ACS Mater. Au* **2** 215–36

[136] Wang F *et al* 2022 Recent progress of scalable perovskite solar cells and modules *Energy Rev.* **1** 100010

[137] Xu Y *et al* 2023 A facile approach for the encapsulation of perovskite solar cells *Energies* **16** 598

[138] Institution B S 2019 *Measurement Protocols for Photovoltaic Devices Based on Organic, Dye-sensitized or Perovskite Materials* (British Standards Institution)

[139] Ramasamy E *et al* 2019 Glass-to-glass encapsulation with ultraviolet light curable epoxy edge sealing for stable perovskite solar cells *Mater. Lett.* **250** 51–4

[140] Domanski K *et al* 2018 Systematic investigation of the impact of operation conditions on the degradation behaviour of perovskite solar cells *Nat. Energy* **3** 61–7

[141] Islavath N *et al* 2017 Effect of hole-transporting materials on the photovoltaic performance and stability of all-ambient-processed perovskite solar cells *J. Energy Chem.* **26** 584–91

[142] Hwang I *et al* 2015 Enhancing stability of perovskite solar cells to moisture by the facile hydrophobic passivation *ACS Appl. Mater. Interfaces* **7** 17330–6

[143] Park H H 2021 Inorganic materials by atomic layer deposition for perovskite solar cells *Nanomaterials* **11** 88

[144] Martins J *et al* 2020 Novel laser-assisted glass frit encapsulation for long-lifetime perovskite solar cells *J. Mater. Chem.* A **8** 20037–46

[145] Yu Y, Zhang F and Yu H 2020 Self-healing perovskite solar cells *Sol. Energy* **209** 408–14

[146] Pallotta R *et al* 2024 Smart materials to empowering perovskite solar cells with self-healing capability *Small Struct.* **5** 2300448

Chapter 8

Challenges, limitations, and sustainability considerations

Perovskite solar cells (PSCs) offer high-efficiency and low-cost processing but face critical challenges that inhibit their commercialization. This chapter explores the limitations affecting their stability, scalability, and sustainability. Important issues include sensitivity to moisture, heat, UV light, ion migration, and interfacial degradation, as well as lead toxicity and poor recyclability. The lack of standardized protocols and low technology readiness levels (TRLs) further restrict their industrial deployment. Sustainability approaches, including circular design and eco-friendly materials, are explored. Addressing these multifaceted barriers is essential for enabling stable, safe, and scalable perovskite technologies for the next-generation photovoltaic (PV) applications.

8.1 Introduction

Metal halide perovskites (MHPs) have attracted global attention for their outstanding optoelectronic properties, with high power conversion efficiencies (PCEs) in solar cells, low manufacturing costs, and solution-based processing. Perovskites have emerged as strong candidates for the next-generation PV, advanced optoelectronics, sensors, and neuromorphic computing devices [1–3]. Chapter 7 highlighted these wide-ranging possibilities and the rapid progress that has been made in improving the device performance. However, despite their potential applications, several critical challenges, particularly related to long-term stability, continue to limit their commercial deployment. Exposure to moisture, heat, light, or oxygen can rapidly degrade perovskite materials, reducing device efficiency and operational lifetimes [4, 5]. Phenomena such as ion migration and phase segregation, especially in mixed halide perovskite systems, further compromise device performance,

doi:10.1088/978-0-7503-5427-1ch8

introducing current–voltage (*I–V*) hysteresis and device instability [6, 7]. Another major challenge is environmental safety, especially the use of toxic lead (Pb) in most high-efficiency perovskite devices. Lead leakage poses a risk to human health and the environment, prompting growing concern over the lifecycle and safe disposal of these materials and devices [8, 9]. Moreover, lifecycle assessments (LCA) for PSCs are still in the early stages, with limited data on resource use, emissions, and end-of-life (EoL) impacts compared to well-established technologies based on inorganic semiconductors such as silicon (Si) [10]. Scaling up perovskite production also presents obstacles, as many lab-scale methods are not easily transferable to industrial manufacturing. Challenges in reproducibility, process standardization, certification, and long-term performance monitoring still need to be addressed before perovskites can enter the mainstream solar market [11–13]. Furthermore, as perovskite research moves forward, sustainability and circular economy strategies will be essential. This includes developing recycling and resource recovery methods for perovskite modules and solvents, as well as designing materials and device structures with EoL reuse and eco-friendly supply chains [14, 15]. This chapter explores these challenges, beginning with intrinsic material instabilities, followed by environmental and toxicological concerns. This chapter also addresses the challenges associated with the manufacturing, scalability, and sustainability of perovskite devices. Understanding and overcoming these barriers is critical to unlock the full potential of perovskite technologies for safe, stable, and scalable applications.

Although perovskite materials are being explored for a wide array of applications, this chapter focuses primarily on PSCs, for which most degradation and stability studies have been investigated systematically. The mechanisms, such as ion migration, moisture sensitivity, thermal instability, and defect dynamics, are not only critical for PV operations but are also important for other devices based on halide perovskites. As such, the insights presented here extend beyond solar applications and remain broadly relevant to the long-term performance and sustainability of halide perovskite materials for general applications.

8.2 Material instability

One of the foremost challenges impeding the long-term viability of perovskite-based devices is their instability under real-world conditions. Despite achieving remarkable efficiencies, PSCs encounter major challenges in sustaining their performance over time, primarily due to intrinsic and extrinsic degradation mechanisms, as illustrated in figure 8.1 [16, 17]. Extrinsic degradation results from exposure to environmental factors such as moisture, oxygen, heat, and UV light (figure 8.1). While these can be partially mitigated through high-barrier encapsulation layers and UV filtering strategies [16], the bigger issue is the intrinsic degradation caused by internal issues with the perovskite materials and their interfaces. This includes crystal defects, ion migration, interfacial reactions, lattice strain, mechanical degradation and phase segregation (figure 8.1) [17]. Addressing such intrinsic issues requires smart material design and careful device engineering.

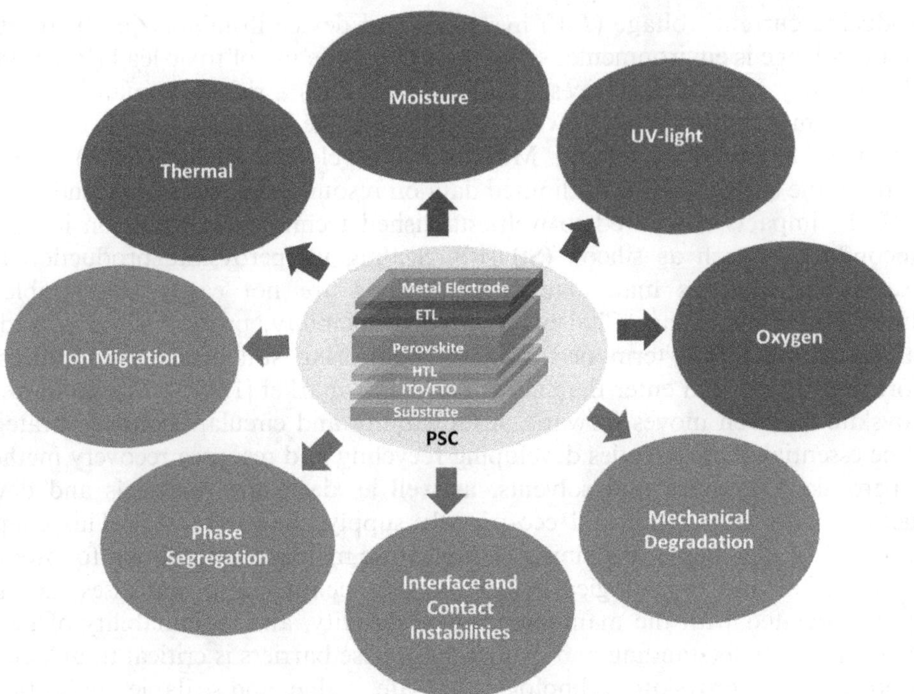

Figure 8.1. Conceptual diagram illustrating the extrinsic (in blue circles) and intrinsic (in red circles) degradation factors affecting the stability of PSCs.

8.2.1 Moisture, oxygen, thermal, and UV sensitivity

MHPs, particularly hybrid organic–inorganic halide types, are highly sensitive to environmental conditions such as moisture, oxygen, heat, and UV light. Moisture can penetrate into the perovskites and trigger hydrolysis reactions and break down the material into lead iodide (PbI_2), making the perovskite lose its optical and electronic properties [18]. At temperatures above 85 °C, thermal instability can cause phase transitions or volatilization of organic components, leading to degradation in device performance [19]. Moreover, UV light, especially in the presence of electron transport layer such as TiO_2, can generate reactive oxygen species that catalyze photodegradation of the perovskite layer [4, 20]. Various strategies have been attempted to enhance their stability, such as compositional tuning by replacing volatile methylammonium (MA^+) with more thermally stable ions such as formamidinium (FA^+) or cesium (Cs^+) [21]. $MAPbI_3$ ($CH_3NH_3PbI_3$), a widely studied material due to its ideal bandgap and superior optoelectronic characteristics, is especially susceptible to humidity. Exposure to humidity causes it to degrade and change from black to yellow, as it converts into PbI_2 via the release of volatile CH_3NH_2 and HI (equation (8.1)) [22–26], as explained in equations (8.1)–(8.4) and shown in figure 8.2(A) [27]:

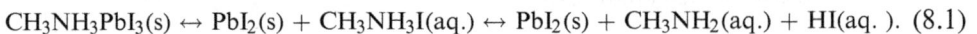

$$CH_3NH_3PbI_3(s) \leftrightarrow PbI_2(s) + CH_3NH_3I(aq.) \leftrightarrow PbI_2(s) + CH_3NH_2(aq.) + HI(aq.). \quad (8.1)$$

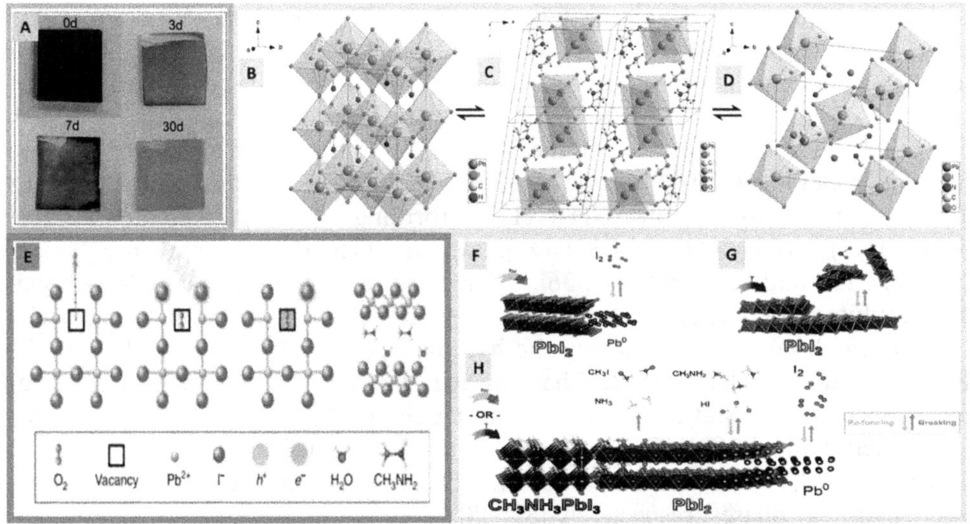

Figure 8.2. (A) A pictorial view of perovskite films for different numbers of days of exposure to air. Schematic illustration of crystal structures of (B) $CH_3NH_3PbI_3$ in its cubic phase, (C) monohydrate phase, $CH_3NH_3PbI_3$. H_2O and (D) the dihydrate phase, $(CH_3NH_3)_4PbI_6{\cdot}2H_2O$. In (B) and (D), the positions of the hydrogen atoms in $(CH_3NH_3)^+$ ions and the water molecules are not specified. (E) Schematic patterns of the reaction steps of O_2 with $MAPbI_3$. (F) PbI_2 decomposition process by visible light (< 530 nm) above its bandgap at 40 °C–60 °C. (G) PbI_2 evaporation at ~70 °C in the dark and (H) $MAPbI_3$ photodecomposition and thermal degradation, producing irreversible $CH_3I + NH_3$ release, reversible $CH_3NH_2 + HI$ formation, and reversible I_2 and Pb generation under illumination or mild heat. The irreversibility of $CH_3I + NH_3$ release is shown by the one-way arrow. ((A) Reproduced from [27]. CC BY 4.0. (B), (C) and (D) Reproduced with permission from [22]. Copyright 2015 American Chemical Society. (E) Reproduced with permission from [41]. Copyright 2022 Elsevier. (F), (G) and (H) Reproduced from [44]. CC BY 3.0.)

Moreover, exposure of $CH_3NH_3PbI_3$ to moisture can lead to the formation of hydrated intermediate phases:

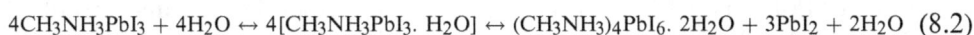

$$4CH_3NH_3PbI_3 + 4H_2O \leftrightarrow 4[CH_3NH_3PbI_3.\ H_2O] \leftrightarrow (CH_3NH_3)_4PbI_6.\ 2H_2O + 3PbI_2 + 2H_2O \quad (8.2)$$

$$(CH_3NH_3)_4PbI_6.\ 2H_2O \leftrightarrow 4CH_3NH_3I + PbI_2 + 2H_2O \quad (8.3)$$

$$CH_3NH_3I \leftrightarrow CH_3NH_2 + HI. \quad (8.4)$$

Perovskites are soluble in polar solvents such as water, leading to rapid degradation upon exposure to moisture [28]. Structural stability is governed by hydrogen bonding between organic and inorganic components, which is disrupted by water [29]. While an appropriate percentage of moisture can aid the crystallization in perovskites and enhance film quality and grain growth [22, 30–33], excessive moisture disrupts the perovskite lattice, leading to irreversible decomposition or formation of unstable hydrates.

Water molecules bond with both organic and inorganic sublattices, forming hydrated intermediates [34, 35]. For instance, when $MAPbI_3$ reacts with H_2O, it first breaks down into CH_3NH_3I (aqueous) and solid PbI_2 via hydrolysis. CH_3NH_3I then further decomposes into volatile CH_3NH_2 and HI (equations (8.2)–(8.4)). As a result, moisture exposure can cause MAI, PbI_2, CH_3NH_2, and HI to coexist within the perovskite film. Additionally, defects in the crystal structure (discussed in section 8.3.2) serve as pathways for water penetration, accelerating degradation by disrupting the bonding of alkylamine cations [7]. Similarly, Walsh *et al* proposed that water degrades $MAPbI_3$ via a simple acid–base reaction, in which H_2O abstracts a proton from $CH_3NH_3^+$, forming an intermediate complex $[(CH_3NH_3^+)_{n-1}(CH_3NH_2)_nPbI_3][H_3O^+]$, which then decomposes into volatile CH_3NH_2, HI, and solid PbI_2 [34, 36]. It has been reported that the hydrated phase of perovskite materials can be reversibly converted back to the original perovskite structure (without H_2O) by storing the hydrated perovskite in dry air for 48 h, as shown in figures 8.2(B)–(D) [22]. The presence of water molecules causes significant distortion of the crystal lattice by separating the $[PbI_6]^{4-}$ octahedra as shown in figures 8.2(C) and 8.2(D). This makes the perovskite structure transform from a 3D network of $[PbI_6]^{4-}$ octahedra to a 1D and 0D chain of octahedra for monohydrated and dehydrated phases, respectively [37, 38].

The degradation of perovskite materials induced by moisture, as given in equations (8.2)–(8.4), can be significantly accelerated in the presence of UV light and oxygen [23, 26, 39, 40] and can be expressed as

$$4HI(aq.) + O_2 \leftrightarrow 2I_2 + 2H_2O \quad (8.5)$$

$$2HI(aq.)h\nu \leftrightarrow H_2(g) + I_2. \quad (8.6)$$

Equations (8.5) and (8.6) illustrate that moisture-induced degradation in perovskite is intensified in the presence of oxygen and light. Oxygen oxidizes HI to form I_2 and H_2O, while UV light decomposes HI into H_2 and I_2 vapors.

Figure 8.2(E) depicts schematic patterns of the reaction steps of O_2 with $MAPbI_3$ under white light illumination [41]. It has been shown that the generation of superoxide (O_2^-) can deprotonate MA^+ of photo-excited MA^*, accelerating the formation of PbI_2, H_2O, MA, and I_2 [42]. PbI_2 can further degrade into metallic Pb and I_2 gas [19, 43]. Figure 8.2(F)–(H) [44] highlight three photodecomposition pathways of $MAPbI_3$ under simulated sunlight and operating temperatures from 40 °C to 80 °C: (1) reversible release of CH_3NH_2 and HI, (2) irreversible formation of NH_3 and CH_3I, and (3) reversible Pb + I_2(g) reaction. If the irreversible pathway (pathway (2)) is avoided, and only the reversible processes (pathways (1) and (3)) occur, encapsulated $MAPbI_3$ can be regenerated during dark periods, supporting the potential for self-healing under controlled conditions [44].

The important chemical and physical mechanisms derived from degradation and recovery experiments on PbI_2 powders and thin films are outlined by equation (8.9) and visualized in figure 8.2(F). When exposed to light with photon energies above 2.34 eV (<530 nm), PbI_2 undergoes photodecomposition. However, this process is reversible if Pb is exposed again to I_2(g) [44]:

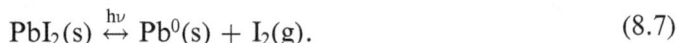

$$PbI_2(s) \overset{h\nu}{\leftrightarrow} Pb^0(s) + I_2(g). \quad (8.7)$$

Under dark conditions, at moderate temperatures (~70 °C) in a high vacuum (~10^{-6} torr), PbI_2 sublimates as molecules or clusters, as shown in figure 8.2(G) and described by equation (8.8):

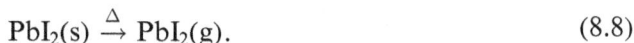

$$PbI_2(s) \xrightarrow{\Delta} PbI_2(g). \tag{8.8}$$

Importantly, no free I_2 gas release is detected under these conditions.

The chemical processes observed during degradation and recovery experiments on $MAPbI_3$ powders and thin films are summarized by equations (8.9) and (8.10) and illustrated in figure 8.2(H). Under illumination or dark and mild heating conditions relevant to PV operation, the photodecomposition and thermal decomposition reactions can be given as [44]

$$2CH_3NH_3PbI_3(s) \xleftrightarrow{h\nu \text{ or } \Delta} PbI_2(s) + Pb^0(s) + I_2(g) + CH_3NH_2(g) + CH_3I(g) + HI(g) + NH_3(g) \tag{8.9}$$

$$CH_3NH_3PbBr_3(s) \xleftrightarrow{h\nu \text{ or } \Delta} PbBr_2(s) + CH_3NH_2(g) + HBr(g). \tag{8.10}$$

CH_3X/NH_3 (X = I or Br) molecules are reported to be thermodynamically driven products of degradation of CH_3NH_3X (MAX) [45] and this degradation is irreversible. In contrast, the release of CH_3NH_2 and HX is considered to be reversible due to high reactivity of these gases, which readily recombine to form MAX and $MAPbX_3$ in the presence of PbX_2 [46–48]. In fact, CH_3NH_2 and HI have been successfully used to directly synthesize $MAPbI_3$ [47, 48]. Under encapsulated conditions, this reversibility supports a dynamic equilibrium between degradation and reformation, making CH_3NH_2 + HI a benign degradation route (figure 8.2(H)). However, in open systems (non-encapsulated), the permanent escape of these volatile gases causes the irreversible reaction. It is notable that $MAPbBr_3$ exhibits greater thermal and photochemical stability than $MAPbI_3$, particularly at moderate temperatures (40 °C–80 °C), as the detrimental degradation pathway involving CH_3Br and NH_3 release is not observed under these conditions (equation (8.10)). As a result, Br-based methylammonium perovskites are expected to demonstrate improved stability compared to their I-based counterparts when encapsulated and operated near ambient conditions [44].

8.2.2 Ion migration and phase segregation

Ion migration is the major contributor to degradation in PSCs and is caused by both external and internal factors. Some of the external factors that are responsible for ion migration are electric field, light exposure, and thermal stress. In mixed halide compositions, this leads to phase segregation and bandgap instability. Mobile halide ions, such as I^- and Br^-, can accumulate at interfaces, forming interface barriers and reducing the device efficiency. Ion migration can also deteriorate optoelectronic properties of perovskites, and lead to device failure during electrical bias or under prolonged exposure to light, moisture and oxygen [7, 49–51]. As shown in figure 8.3(A) [50], uniform halide distributions in mixed Br^-/I^- perovskites become unstable under illumination, resulting in phase segregation into I-rich (low bandgap) and Br-rich (high

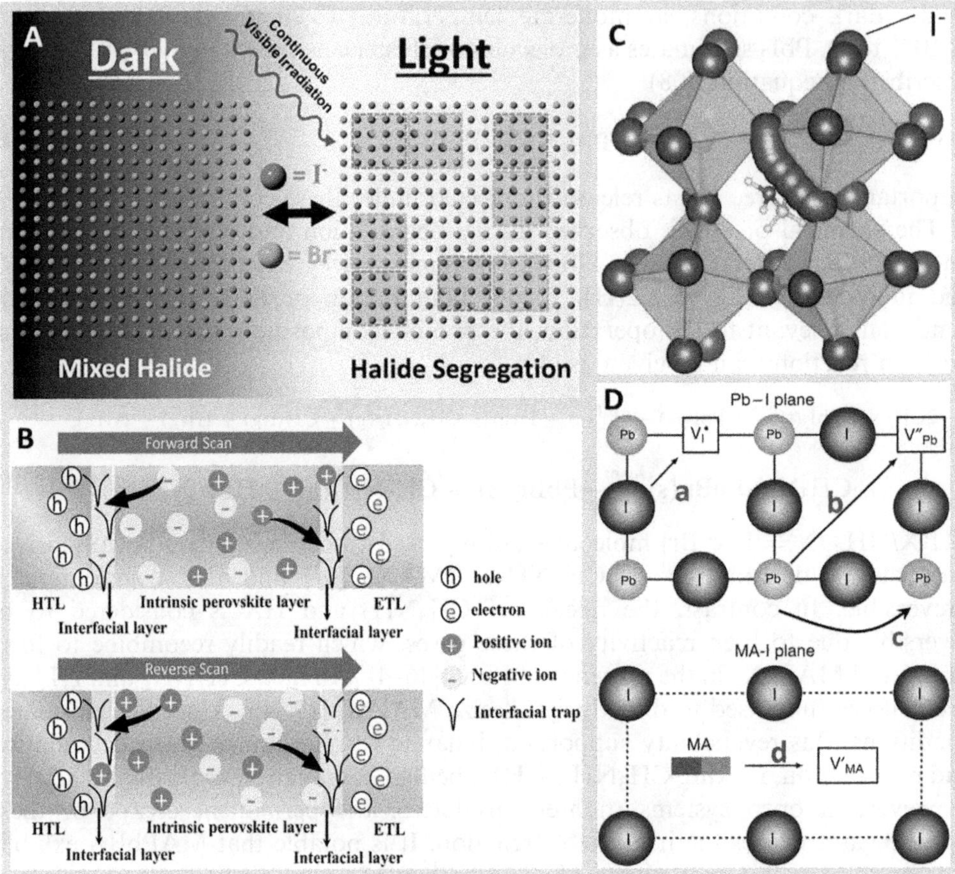

Figure 8.3. (A) Schematic illustration of light-induced halide segregation in mixed halide perovskites. (B) Representation of ion migration and re-orientation in the perovskite layer under forward and reverse scans. (C) Calculated migration path of I^- ion along the I^-–I^- edge indicating a slightly curved path and local relaxation/tilting of the octahedra and (D) schematic illustration of migration paths of the three ions: I^- (path a), Pb^{2+} (path b, c) in the Pb–I plane, and MA^+ (path d) in the MA–I plane. ((A) Reproduced with permission from [50]. Copyright 2018 American Chemical Society. CC BY-NC 4.0. (B) Reproduced from [54]. CC BY 3.0. (C) and (D) Reproduced from [52]. CC BY 4.0.)

bandgap) domains. This segregation induces spectral instability and hysteresis in the I–V characteristics [7, 51–53]. Consequently, the PV parameters, namely open-circuit voltage (V_{oc}), short-circuit current density (J_{sc}), fill factor (FF), and PCE are adversely affected.

Beyond halide segregation, broader ion migration within the perovskite lattice, affecting both cations (e.g. MA^+, FA^+) and anions (e.g. I^-, Br^-), poses severe challenges to their structural and electrical stability. Figure 8.3(B) illustrates multiple ion and defect migration pathways that are present within a standard PSC device under forward and reverse bias conditions. Migration of iodide ions and vacancies at interfaces can initiate interfacial degradation, leading to progressive performance loss and failure [55, 56]. Hoke *et al* have confirmed phase separation in $MAPbBr_xI_{3-x}$ under illumination, reaffirming ion migration as a destabilizing factor [7, 57]. These mobile ions distort the crystal structure, increase trap state density, and trigger defect accumulation, ultimately undermining device integrity [58]. Ion migration is not confined to the active layer only, but the ions can also diffuse into adjacent layers such as hole transport layer (HTL), electron transport layer (ETL), and even up to the electrodes. Besleaga *et al* demonstrated iodide migration into the HTL and Ag electrode, forming AgI and compromising interfacial contact [59]. Reactions of mobile ions (I^-, MA^+) with organic charge transport materials degrade their conductivity and hinder charge transport. For example, Zhao's group found that MA^+ infiltration into Spiro-OMeTAD reduces its hole mobility and accelerates device degradation [60], while Wang *et al* reported that iodide exposure caused p-type conductivity loss in both Spiro-OMeTAD and PTAA, resulting in significant loss in device efficiency [61]. Further insight into ion migration was provided by R O Nughays *et al* through combined experimental and DFT studies of FA/MA-based perovskites [62]. Experimental analysis revealed that FA-rich compositions ($FA_{0.6}MA_{0.4}PbI_3$) exhibit longer carrier lifetimes, whereas MA-rich films ($FA_{0.4}MA_{0.6}PbI_3$) suffer from faster recombination due to increased surface defects. DFT simulations on $FAPbI_3$, $MAPbI_3$, and $FA_{0.5}MA_{0.5}PbI_3$ revealed that FA^+ ions exhibit the lowest energy barrier for migration from bulk to the surface, while MA^+ ions have the highest energy barrier. Iodide migration barriers also varied, $FAPbI_3$ showed the lowest energy for bulk-to-surface migration (0.26 eV) but higher energy (0.41 eV) from surface-to-vacuum, while $MAPbI_3$ showed a higher bulk energy (0.383 eV) but a low surface energy (0.185 eV), encouraging vacancy formation [17, 62].

An important issue in understanding ion transport in $MAPbI_3$ is the identification of primary migrating species. While I^- is widely regarded as the most mobile ion, Pb^{2+} and MA^+, and even other mobile species such as H^+, may also migrate under certain conditions [63]. Migration rates are governed by activation energy (E_A). The migration rate (r_m) is influenced by E_A as, $r_m \propto \exp(-E_A/k_BT)$, where k_B and T are Boltzmann constant and temperature, respectively, and k_BT represents the thermal activation energy. Furthermore, E_A is influenced by factors such as ionic radius, ion charge, crystal structure, and jump distance. In $MAPbI_3$, the I^- ions at the edges of PbI_6^{4-} octahedra travel a shorter path (~4.46 Å) to adjacent I^- vacancies compared to MA^+ or Pb^{2+} (~6.28 Å), as shown in figures 8.3(C) and (D) [52]. Despite its relatively large radius (0.206 nm), I^- ion exhibits the lowest E_A for migration (0.58 eV) along the I–I edge (path a in figure 8.3(D)), as calculated by Eames *et al* [52]. MA^+ migrates along <100> (path d in figure 8.3(D)) with $E_A \approx 0.84$ eV, while Pb^{2+} requires $E_A \approx 2.31$ eV along the <110> (path b in figure 8.3(D)). Azpiroz *et al* reported similar trends with activation energies of 0.08 eV (I^-), 0.80 eV (Pb^{2+}(path c in figure 8.3(D)), and 0.46 eV (MA^+) [55] suggesting that while I^- migrates rapidly, the slower ionic responses observed experimentally may stem from MA^+ or Pb^{2+} [64]. Among all species, H^+ is the only one speculated to migrate via Frenkel defects due to its minimal size. Egger *et al* calculated an E_A of 0.29 eV for H^+ ion migration along a transient hydrogen bond connecting two equatorial iodide ions [63].

Ion transport is facilitated not only through point defects but also through structural features such as grain boundaries (GBs) and surface defects, as illustrated in figure 8.4(a)–(g). Schottky (figure 8.4(a)) and Frenkel (figure 8.4(b)) defects serve

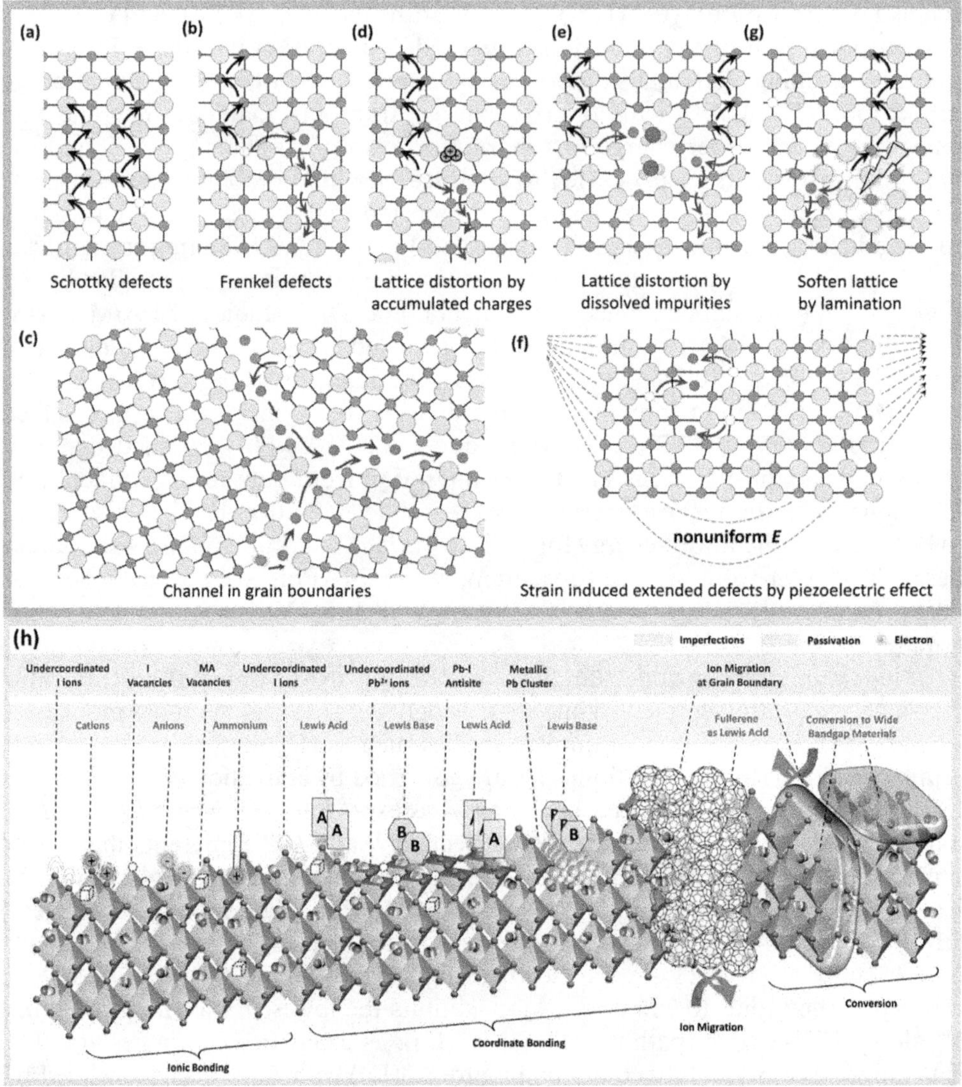

Figure 8.4. Schematic illustration of ion migration pathways in perovskites enabled by: (a) Schottky defects, (b) Frenkel defects, (c) open space and wrong bonds at the grain boundary, (d) lattice distortion from charge accumulation, (e) lattice distortion by dissolved impurities, (f) non-uniform strain caused by piezoelectric effect, and (g) light-induced lattice softening due to bond weakening. (h) Common imperfections in MHPs films addressed by passivation through ionic and coordinate bonding, bandgap widening through compositional modification, and inhibition of ion migration at extended defect sites. ((a)–(g) Reproduced with permission from [65]. Copyright 2016 American Chemical Society. (h) Reproduced with permission from [66]. Copyright 2019 The Royal Society of Chemistry.)

as common pathways. Kim *et al* estimated Pb–I Schottky defect formation energies as low as 27–73 meV, supporting a highly 'soft' lattice [67]. This softness correlates with the relatively low E_A for ion migration in $MAPbI_3$ (0.36 eV), much lower than in oxide perovskites such as $LaFeO_3$ (0.77 eV) or $LiNbO_3$ (0.75 eV) [67]. GBs (figure 8.4(c)), due to their open structure and lower coordination, present low-barrier pathways for ion movement. Xiao *et al* showed that ion-drift in the perovskite layer can lead to the formation of reversible p–i–n structures, resulting in field-switchable photovoltaic effect [64]. Other factors that contribute to ion migration include local lattice distortions induced by scaffold confinement [68], interfacial charge accumulation (figure 8.4(d)) [69], absorbed molecular clusters (figure 8.4(e)) [70], and mechanical strain from piezoelectric effects (figure 8.4(f)). Additionally, photoinduced weakening of bonding of MA^+ ion with inorganic ions can soften the lattice and promote migration, as predicted by Gottesman *et al* (figure 8.4(g)) [71].

Halide perovskite films also exhibit a wide range of intrinsic and extrinsic defects, including vacancies (e.g. I^-, MA^+), antisite defects (e.g. $PbI_3{}^-$), undercoordinated ions, and metallic Pb clusters (figure 8.4(h)) [66]. Most of the defects in perovskite films arise during film formation or post-formation, due to the factors such as non-stoichiometric precursors [72], lattice mismatch [73], growth conditions [74], and nucleation mechanisms [75]. While shallow traps may have minimal effect, the deep-level traps, often located at surfaces or GBs, act as recombination centers that reduce device performance [76, 77]. These defects and ion migration result in hysteresis and instability in perovskite devices [55, 78–80].

Passivation strategies are essential for mitigating such defects. Deep-level traps can be neutralized through Lewis acid–base coordination [81, 82] or ionic interactions with suitable cations [80, 83, 84]. Similar to the passivation layers used in silicon PV devices (e.g. SiO_2, Si_3N_4, Al_2O_3) [82, 85, 86], wide-bandgap coatings have been employed to stabilize perovskite surfaces [66]. However, defect formation remains sensitive to fabrication conditions, including precursor ratios, residual PbI_2, and annealing protocols [77, 83, 87, 88]. Incorporation of small isovalent cations has been shown to suppress vacancies and improve structural stability [89] but many such methods rely on external additives. The development of *in situ* passivation strategies that maintain intrinsic composition is thus critical for advancing the long-term stability of PSCs [90].

8.2.3 Interface and contact instabilities

In addition to bulk ion migration, interfacial instability plays a pivotal role in the performance decay of PSCs. Interfaces between the perovskite layer and adjoining charge transport layers (e.g. Spiro-OMeTAD, PCBM) or electrodes are particularly susceptible to chemical and structural degradation. Unwanted reactions at these junctions, such as metal diffusion (e.g. Ag, Au) from the electrode into the perovskite can trigger interfacial breakdown and the formation of shunting paths [91]. These issues are especially pronounced in inverted (p–i–n) device architectures, in which degradation at the interface and increased recombination are driven by the accumulation of structural and chemical defects [92].

As shown in figures 8.5(A) and (B) [93], interfacial degradation mechanisms are complex and multifaceted. Defects originating in the perovskite, charge transport, and metal oxide layers can serve as recombination centers and accelerate ion migration. Oxygen vacancies in metal oxides react with migrating Pb^{2+} ions to form PbO, destabilizing the interface and promoting the decomposition of organic cations

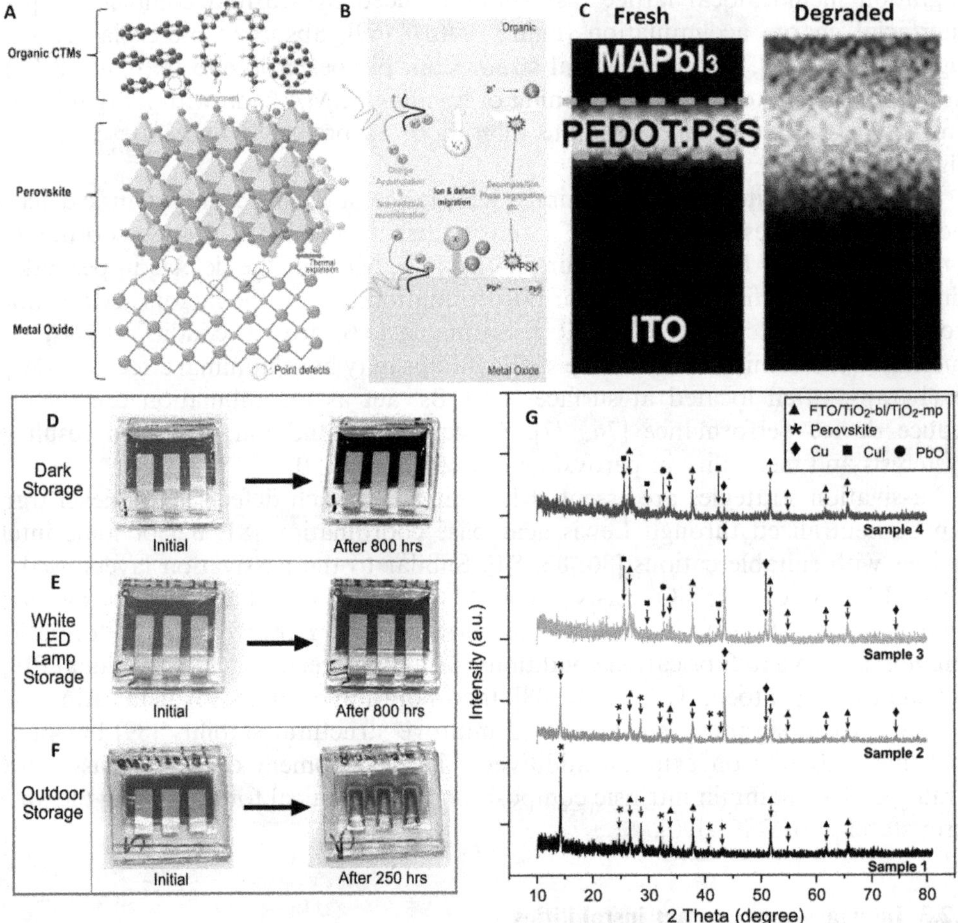

Figure 8.5. (A) Metal oxide–perovskite–organic material interfaces in PSCs with various point defects and (B) illustration of typical interfacial degradation processes occurring in PSCs. (C) ToF-SIMS analysis of ITO/HTL/$MAPbI_3$ samples, illustrating the spatial distribution of characteristic markers for the HTL material PEDOT:PSS. Photographs of the solar cells stored in (D) dark (E) under white LED lamp and (F) under direct sunlight and (G) XRD graphs of the samples, stored in dark (sample 1 and sample 2) and in direct sunlight (sample 3 and sample 4), that helped in understanding the degradation mechanisms in PSCs. ((A) and (B) Reproduced with permission from [93]. Copyright 2023 John Wiley and Sons. (C) Reproduced with permission from [100]. Copyright 2020 American Chemical Society. (D)–(G) Reproduced with permission from [101]. Copyright 2019 Springer Nature.)

into volatile species, thereby degrading both the absorber and charge transport layers [94]. Furthermore, moisture not only degrades the perovskite layer but also affects charge transport layers, especially HTL. Spiro-OMeTAD, widely used in n–i–p PSCs, requires doping with additives such as lithium bis(trifluoromethane sulfonyl) imide (LiTFSI) to enhance its conductivity. However, LiTFSI is hygroscopic and promotes moisture uptake, leading to HTL and device degradation. Additionally, Li^+ migration facilitates moisture infiltration, accelerating deterioration in both HTL and perovskite layers [95–97]. ETLs are generally more moisture-resistant but the commonly used ETLs such as TiO_2 are hydrophilic and prone to adsorbing ambient moisture. Even PCBM, a more hydrophobic ETL used in inverted PSCs, can undergo water-induced degradation and fails to adequately passivate undercoordinated Pb^{2+}ions, thus compromising device durability [98, 99].

Figure 8.5(C) presents a time-of-flight secondary ion mass spectrometry (ToF-SIMS) depth profile analysis comparing fresh and 1000 h aged ITO/PEDOT:PSS/$MAPbI_3$ samples under O_2^+ ion beam bombardment. While fresh devices showed a well-defined interface (evidenced by strong C_2^- ions), aged samples revealed significant interfacial diffusion of C_2^- and S^- ions into the perovskite, indicating intermixing and morphological changes due to prolonged light exposure [100].

In order to further investigate interface-driven degradation under different environmental conditions, Kumar *et al* studied devices with the architecture FTO/TiO_2-bl/TiO_2-mp/$MAPbI_3$/P3HT/Cu using ISOS protocols [101]. The solar cells were stored in a variety of conditions such as dark, under continuous illumination of white LED lamp, and outdoor under natural sunlight as per ISOS-D-1, ISOS-L-1 and ISOS-O-1 protocols [101]. Figure 8.5(D)–(F) show the photographs of the PSCs stored in different environmental conditions. The solar cells were tested for ~800 h. Devices stored in dark exhibited minimal performance loss. Under continuous LED illumination, significant degradation was noted, primarily due to decreased J_{sc} and FF, while the V_{oc} remained stable. However, cells exposed to direct sunlight showed a rapid degradation in V_{oc} and FF which led to rapid degradation in PCE. After ~120 h, all the PV parameters decreased very rapidly and became zero. The degradation mechanisms in PSCs, under direct sunlight, were different from those under LED lamp or dark storage and the reason for this was more harsh environmental conditions under direct sunlight. As shown in figure 8.5(F), the degradation was visibly apparent, with the Cu electrode reacting with the underlying layers and eventually disappearing. The degradation mechanisms in the solar cells were understood with the help of x-ray diffraction (XRD) studies (figure 8.5(G)). XRD analyses confirmed that perovskite layers stored in the dark remained chemically stable, whereas those under sunlight decomposed and reacted with Cu, forming copper iodide (CuI) (equations (8.4) and (8.11)–(8.13)). Additionally, photo-bleaching was faster under sunlight compared to LED light; this can be attributed to photo-oxidation accelerated by UV content and elevated temperatures:

$$2PbI_2 + O2 \rightarrow 2PbO + 2I_2\uparrow \quad (8.11)$$

$$2Cu + 2HI \rightarrow 2CuI + H_2\uparrow \quad (8.12)$$

$$2Cu + I_2 \rightarrow 2CuI. \tag{8.13}$$

As discussed above, the decomposition pathway of $CH_3NH_3PbI_3$ included its conversion into CH_3NH_2, PbI_2, and HI, which subsequently transformed into CO_2 and PbO. These photochemical processes compromised the electrode interface and severely degraded device performance. Similar degradation patterns were observed for PSCs using Ag electrodes, which reacted with the perovskite to form AgI [102]. Due to the high reactivity of common electrode materials such as Cu, Ag, and Al, there is growing interest in chemically inert alternatives. Carbon-based electrodes have emerged as promising candidates, offering excellent environmental stability due to their inertness and lack of reactivity with perovskite components [103, 104]. Devices employing carbon electrodes have achieved PCEs exceeding 20%, while demonstrating superior resistance to moisture, oxygen, and light-induced degradation [104–107].

8.2.4 Mechanical degradation

Beyond chemical and interfacial degradation, mechanical failure presents a critical limitation to the long-term performance of flexible PSCs. Due to the intrinsic brittleness of MHPs, flexible devices are particularly prone to mechanical deterioration. Repeated bending or mechanical stress can induce microcracks, delamination of layers, or overall structural weakening, leading to rapid performance loss [57]. Even in rigid architectures, cyclic thermal expansion and contraction can create stresses that result in film cracking or layer separation, highlighting the broader need for mechanical stability strategies across all PSC types.

Flexible PSCs are inherently subjected to mechanical strain, which significantly affects both device performance and operational stability. Strain engineering in MHPs has revealed that mechanical strain alters the optoelectronic properties of perovskites. Specifically, tensile strain increases the bandgap, while compressive strain decreases it, due to differential effects on the Pb–X bond and associated band structure [108]. Compressive strain also enhances hole mobility by reducing the effective mass [109].

However, this same strain can also affect material stability. Tensile strain, in particular, lowers the activation energy required for ion migration, thus accelerating degradation and facilitating the decomposition of MHPs into PbI_2 [110]. Figure 8.6 illustrates the effect of mechanical strain on perovskite lattice distortion and energy band modulation. For example, in $FAPbI_3$ films, residual tensile strain has shown to drive the transformation from the photoactive α-phase to the photo-inactive δ-phase even at room temperature, undermining device efficiency and stability [109]. Conversely, in $CsPbI_3$, tensile strain can help stabilize the photoactive black γ-phase, suppressing its transition to the less desirable yellow δ-phase and thereby enhancing thermal stability [111]. Nonetheless, strain-induced failures such as crack formation remain a serious concern. As depicted in figure 8.6(b), cracks can develop at GBs, within the grains themselves, or at the interfaces between the perovskite and charge transport layers [112]. These cracks disrupt charge transport pathways,

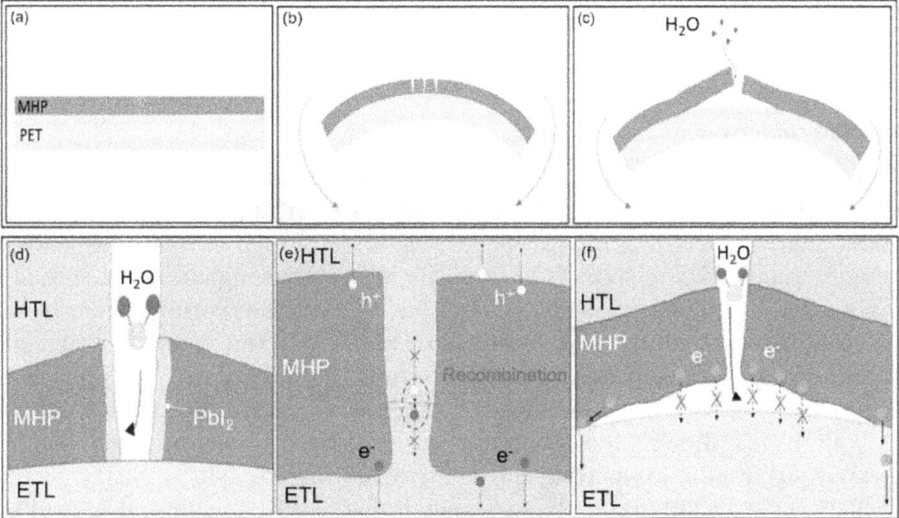

Figure 8.6. Schematic illustration of mechanical failure modes in MHP thin films: (a) MHP on polyethylene terephthalate (PET) without external stress (bending), (b) bending-induced cracks, (c) cracks allow moisture-induced degradation, (d) moisture degrades MHP to PbI_2; (e) HTLs infiltrate cracks, causing undesirable charge recombination as indicated by the red dashed circle (the blue dots and white dots represent electrons and holes respectively), (f) delamination disrupts electrical contact, leading to increased series resistance and device failure, also exposes the film to further moisture-induced degradation. (Reproduced with permission from [112]. Copyright 2022 Optica Publishing Group.)

increase non-radiative recombination, and promote further environmental degradation. Dai *et al* [113] demonstrated that grain size plays a pivotal role in crack propagation: devices with smaller grains are prone to horizontal cracks along grain boundaries, while larger grains tend to localize cracking at the MHP/ETL interfaces. Cracks within grains have also been reported under stress or irradiation [114, 115].

These observations highlight the need for mechanical engineering approaches, such as flexible encapsulants, strain-tolerant layers, and GBs reinforcement, to enhance the mechanical resilience of flexible PSCs.

8.3 Toxicity and environmental impact

The aforementioned degradation mechanisms highlight the complex and interconnected nature of perovskites and the stability challenges that are faced in device applications. While incremental progress has been made in the stabilization of perovskites through compositional tuning, passivation layers, and robust encapsulation, there still exist several other serious issues such as toxicity and environmental impacts that need to be understood and addressed. Processing of perovskite devices may release some toxic materials such as lead and organic solvents into the environment. Lead might also be released into the environment if the device breaks down. Even safe disposal of the modules after their end of life also raises serious concerns about safety and environment impacts. The following section explores the toxicity, environmental sustainability, and regulatory considerations that must be addressed along with technical improvements in device stability and performance of perovskites.

8.3.1 Lead toxicity

Even though lead-based PSCs are very efficient, lead can be very harmful to the environment and human health. Lead in PSCs is often treated similar to the lead found in regular electronics. However, new research shows that the lead from broken or damaged PSCs spreads more easily in the environment and is more likely to be absorbed by living organisms. Studies have shown that perovskite-derived lead is absorbed by plants up to ten times more efficiently than the lead from sources such as mining, industrial or e-waste. In one such study, mint plants grown in perovskite-contaminated soil exhibited drastically higher lead accumulation in their roots, stems, and leaves, though the amount of lead in the soil was not very high [116] (figure 8.7). Over a 20 day period, this led to symptoms of acute toxicity, including tissue blackening and plant death. As shown in figure 8.7, the amount of lead taken up by the plants was up to 366 times higher than expected, showing that lead from perovskites is especially dangerous for the environment.

These findings indicate that existing safety regulations, such as the EU Restriction of Hazardous Substances (RoHS), may require updates to adequately address the specific risks that are associated with PSCs. New safety limits specifically for PSCs should be created, considering how easily lead can leak into the environment and be absorbed. In order to reduce the risk of lead leaking into the soil or water and harming nature, better solutions are required. To lower this risk, stronger encapsulation materials, real-time leaching tests, and safer lead alternatives such as tin (Sn^{2+}) for perovskites must be implemented. Innovations in lead-free perovskites, including Sn- and Bi-based alternatives, are progressing to reduce toxicity without compromising device performance.

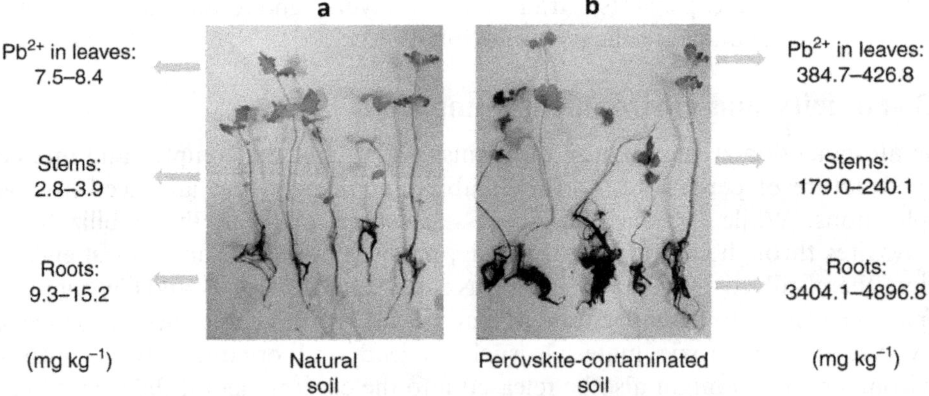

Figure 8.7. Photograph of mint plants cultivated in (a) natural soil and (b) soil contaminated with 250 mg kg^{-1} Pb^{2+} from perovskite material. Lead concentrations detected in the leaves, stems, and roots are indicated along with each image. (Adapted from [116]. CC BY 4.0.)

8.3.2 Challenges in recycling and disposal

Managing PSCs at the end of their life is difficult because no standard recycling processes exist. Traditional recycling often uses toxic chemicals, which are harmful to the environment. Some safer recycling methods have been introduced [117, 118], which recover valuable materials without hazardous solvents. However, recycling solutions for PSCs are still under development and need further validation for industrial use. Unlike Si solar panels that benefit from well-established recycling infrastructure, PSCs pose unique challenges due to their multi-layered architectures and the presence of toxic elements such as lead. If improperly disposed of, particularly in humid or acidic environments, lead leakage from degraded perovskite layers can occur uncontrollably, posing risks to soil and water systems [116]. These concerns are particularly urgent due to the limited regulatory and recycling frameworks currently available for PSCs. The complex material structure and chemical instability of perovskites require careful EoL handling strategies to prevent environmental contamination. Many materials used in PSCs, such as organic charge transport layers and electrodes, are not biodegradable and may be harmful. Currently, there are no clear rules for collecting, separating, and treating PSC waste, making circular design essential for sustainability.

As the solar industry continues its rapid expansion, the demand for raw materials is increasing significantly. While recycling can alleviate some of this demand [117, 119], it is essential to plan dedicated recycling infrastructure from the early stages of deployment. A comprehensive recycling system should encompass three main stages: module collection, material separation, and purification for reuse [120]. Since the PV industry is now the largest segment of the semiconductor sector, it benefits from uniform material types, which makes recycling easier. However, old solar panels are generally bulky and have a low value-to-weight ratio; thus, transporting them is expensive. Despite several challenges, PSCs offer various advantages over traditional silicon solar panels when it comes to enabling circular economy strategies. PSC manufacturing requires relatively low capital investment, allowing for the development of distributed and local recycling centers that reduce transportation burden and material waste. Unlike silicon modules, where ethylene-vinyl acetate (EVA) encapsulation makes recovery of pure materials difficult and costly [121], PSC layers can often be disassembled using organic solvents at room temperature [122]. This facilitates component separation and reuse, supporting 'design-for-recycling' approaches for future scalable perovskite technologies [123]. An aqueous-based, green solvent recycling method has been developed recently, which is capable of recovering all device components, including perovskite absorbers, substrates, charge transport layers, and electrodes, while maintaining performance and achieving a 96.6% reduction in resource depletion and a 68.8% drop in human toxicity impacts [118]. Figure 8.8(A) [118] presents a schematic of a proposed sustainable perovskite PV system designed with holistic recycling strategies. In this model, the solar farm not only generates and supplies energy to meet societal needs but also integrates EoL management by enabling layer-by-layer repair and recycling of used modules. The recovered materials are then reused for the fabrication of new

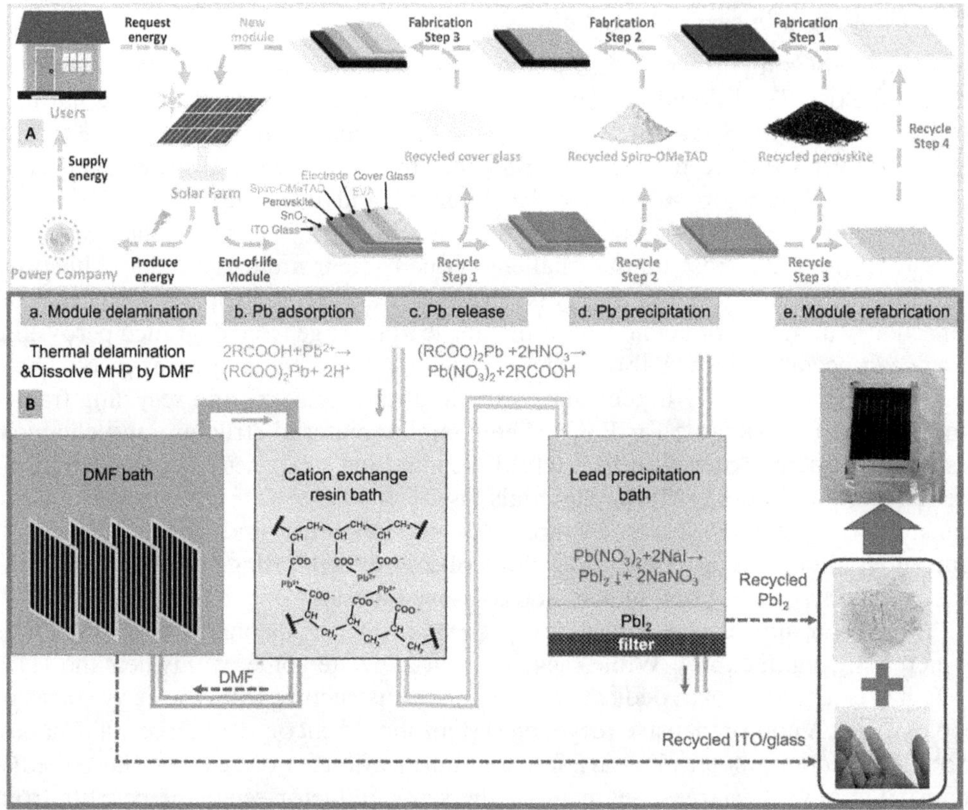

Figure 8.8. (A) Illustration of sustainable perovskite PVs incorporating circular recycling pathways. (B) Roadmap for recycling of perovskite solar modules. (a) Encapsulated perovskite solar modules are delaminated, and the MHP is dissolved using DMF, (b) lead ions in the DMF solution are extracted using a carboxylic acid-based cation-exchange resin, (c) the adsorbed lead ions on resin are released to aqueous solution by resin-regeneration process via HNO_3, (d) PbI_2 is recovered through precipitation by adding NaI to the $Pb(NO_3)_2$ solution and (e) module refabrication using recycled materials. ((A) Reproduced from [118]. CC BY 4.0. (B) Reproduced from [119]. CC BY 4.0.)

solar modules, which are used to establish new solar farms, creating a circular and regenerative system [118]. Theoretically, even the organic materials in PSCs could be recovered because they dissolve in different solvents. But in practical situations, problems such as damage from oxygen make this difficult. Additionally, EVA encapsulation can release by-products that harm the perovskite layer during recycling [124], highlighting the need for new encapsulants [38, 124]. One idea is to design encapsulants as 'release layers', which separate easily when heated or treated with chemicals [123]. Complementarily, green recycling techniques have achieved recovery efficiencies of over 99% using non-toxic solvents [15]. During manufacturing, solvent recovery systems such as condensers and distillation units can recover up to 71% of the used solvents [125]. However, not all materials in PSCs can be recovered with solvents. For example, transparent conductive oxides (TCOs) such as fluorine-doped tin oxide (FTO) cannot be recovered via leaching. So, when FTO is used as the TCO, it becomes necessary to recycle the whole piece of FTO glass. To improve PSC sustainability, a recycling plan on safely recovering toxic lead and valuable glass from used PSCs has been proposed; this is shown in figure 8.8(B). In this process, first the top lamination layers of the solar module are separated via thermal delamination and then the perovskite layer is dissolved using dimethylformamide (DMF). Lead ions are adsorbed by a carboxylic acid-based cation-exchange resin, then released using HNO_3 solution to form water-soluble $Pb(NO_3)_2$. To regenerate usable material, $Pb(NO_3)_2$ is converted into PbI_2, preferred for PSC refabrication, via precipitation with NaI. This method avoids creating hard-to-dissolve waste products such as lead chloride ($PbCl_2$) or lead sulfate ($PbSO_4$) and makes it easier to recycle and reuse lead efficiently [119]. Recyclability, modularity, and use of low-toxicity materials are emerging as critical design criteria for scalable PSCs manufacturing. Akulenko *et al* [126] have highlighted that carbon-based contacts improve recyclability and enable profitable recovery of noble metals.

8.3.3 Life cycle assessment limitations

Life cycle assessments (LCAs) are important tools that are used to understand the total environmental impact of PSCs, from extracting raw materials to throwing them away at the end of their use. However, currently LCAs face problems such as limited data, inconsistent methods, and lack of standard comparison metrics. These issues make it difficult to compare PSCs with other PV technologies. Although early LCA studies suggest that PSCs have lower energy payback times (EPBT) and lower carbon footprints compared to older solar cell technologies, they often ignore factors such as long-term lead leakage, recyclability of toxic parts, and real-world performance of PSCs in different weather conditions over time. LCAs have shown that recycling and reuse can greatly reduce energy consumption and harmful emissions compared to traditional disposal methods [127]. Also, many studies only focus on improving efficiency without considering material toxicity or long-term durability, making it difficult to assess the true sustainability of PSCs.

A recent study evaluated the environmental impact of low-dimensional (LD) material-based PSCs by performing a detailed LCA [128]. Integrating LD materials such as graphene, reduced graphene oxide (rGO), graphene quantum dots (GQDs), molybdenum disulfide (MoS_2), and black phosphorus (BP), can enhance both stability and PCE of PSCs. Among these, rGO, graphene, and MoS_2 show the lowest environmental impact per unit mass. Two alternative LD-PSC designs (FTO/ETL/perovskite/HTL/back contact) were assessed Alt-1, which integrates LD materials with minimal environmental impact in each layer, and Alt-2, which utilizes LD materials with comparatively higher environmental impact across the device structure. Alt-1 utilizes graphene in the ETL, BP in the absorber layer, and rGO in the HTL, while Alt-2 incorporates GQDs in the ETL, rGO in the absorber layer, and BP in the HTL. Alt-1 showed ~25% lower, and Alt-2 ~15% higher environmental impact than reference PSCs.

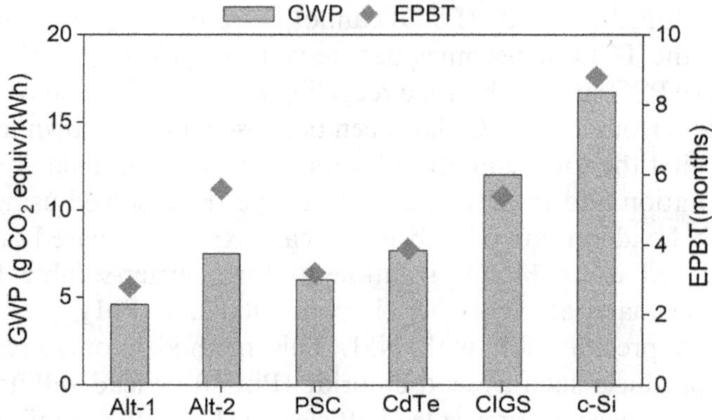

Figure 8.9. Comparison of GWP and EPBT for Alt-1 and Alt-2 lD-PSCs, a reference lead-based PSC, and various commercial PV technologies such as CdTe, CIGS and c-Si. Calculations assume an average global solar irradiation of 1700 kWh m^{-2} $year^{-1}$ and a performance ratio of 0.75. (Reproduced with permission from [128]. CC BY-NC 3.0.)

The global warming potential (GWP), also referred to as the carbon footprint, and energy pay back time (EPBT) of electricity generated from LD-PSCs have been analysed and compared to reference PSCs and established commercial PV technologies, such as crystalline silicon (c-Si), copper indium gallium selenide (CIGS), and cadmium telluride (CdTe); this is shown in figure 8.9. The figure shows that Alt-1 lD-PSCs have the lowest GWP (4.5 g $CO_{2eq.}$ per kWh) and shortest EPBT. They are 20%–40% better than reference PSCs and up to 75% better than commercial PVs. Leccisi *et al* [129] reported GWPs of 5.3, 9.0, and 18.5 g $CO_{2eq.}$ per kWh for PSC, CdTe, and CIGS, respectively, using spray coating with CuSCN/Mo-Al contacts. In contrast, the study, which used spiro-OMeTAD/Ag and slot-die coating, resulted in a ~15% higher GWP [128]. Fthenakis *et al* [130] reported a GWP of 22 g $CO_{2eq.}$ per kWh for c-Si, ~35% higher than the study by Leccisi *et al* mostly due to differences in electricity sources. Lead-based LD-PSCs show EPBTs between 3 and 6 months which outperform many current commercial PV technologies that typically require 4 to 9 months. Alt-1, a lead-free design, had the shortest EPBT of just 2.8 months, while c-Si takes around 9 months. Minor differences between studies come from changes in fabrication methods and electricity mix assumptions [128].

These findings highlight the dual environmental and performance advantages of optimized LD-PSCs, but also highlight concerns around toxicity and long-term sustainability.

8.4 Bottlenecks in commercialization

As PSCs move closer to commercialization, long-term performance and reliability have become major concerns. Despite rapid improvements in PCE and initial stability, several practical barriers still limit their large-scale deployment. This section highlights the issues such as manufacturing consistency, testing standardization, device scalability, and commercial readiness of PSCs.

8.4.1 Manufacturing, scalability, and certification challenges

Scaling up PSCs from laboratory prototypes to industrial-scale production remains a major challenge. Lab-scale fabrication techniques such as spin-coating and thermal evaporation enable high-efficiency small-area devices, but they are inherently limited by low throughput, poor reproducibility, and incompatibility with large-area manufacturing. To overcome these limitations, several scalable deposition methods have been developed, including slot-die coating, spray coating, inkjet printing, and roll-to-roll (R2R) processing. A wide range of additional scalable deposition strategies has also emerged, including doctor blading [131], screen printing [131], electro-deposition [132], vapor-assisted deposition [133, 134], dip coating [135], soft cover deposition [136], R2R and sheet-to-sheet (S2S) methods [137, 138]. These techniques are more suitable for industrial production due to their compatibility with high-throughput and continuous manufacturing. However, they still face persistent technical challenges, such as non-uniform film formation, sensitivity to ambient conditions (e.g. moisture and oxygen), and reduced PCEs (~16%–18%) compared to optimized lab-scale devices.

Printing-based techniques, which are well-established in the field of organic photovoltaics, are particularly promising for PSCs due to their cost-effectiveness and industrial compatibility [139]. These scalable techniques have enabled the fabrication of larger-area modules while maintaining reasonable efficiency [140]. Notably spray coating [141, 142], slot-die coating [143], and inkjet printing have shown promising levels of reproducibility in fabricating large-area devices, further supporting their potential for commercial adoption.

Beyond deposition challenges, lack of standard testing protocols is another major hurdle for commercialization. Although many research groups report high efficiencies, such values are often obtained under idealized lab conditions that do not reflect real environmental conditions. Environmental factors such as temperature variations, UV exposure, humidity, and mechanical stress are rarely included, and the test methods vary lab to lab. This makes it hard to compare results and predict real performance. To bridge this gap and gain industry and consumer trust, it is essential to implement certified testing protocols, such as those established under the IEC 61215 standard. Moreover, long-term outdoor stability testing is necessary to reveal degradation pathways and develop realistic benchmarks for device lifetime and reliability.

Furthermore, reproducibility remains an important concern throughout the fabrication process. Small variations in precursor composition, solvent system, or thermal annealing can lead to significant deviations in efficiency and stability. Achieving consistent performance across batches and devices is crucial for transitioning PSCs from the lab to market. The scalable techniques have enabled the fabrication of larger-area modules while maintaining reasonable efficiency. Gujar *et al* [144] reported minimal performance variation across 35 out of 36 fabricated devices. These findings are supported by work from Leyden *et al* [145] and Borchert *et al* [146]. Odabaşı *et al* [11] have analysed a large number of PSCs using a statistical method called pooled variance (Sp^2) to compare different materials and techniques.

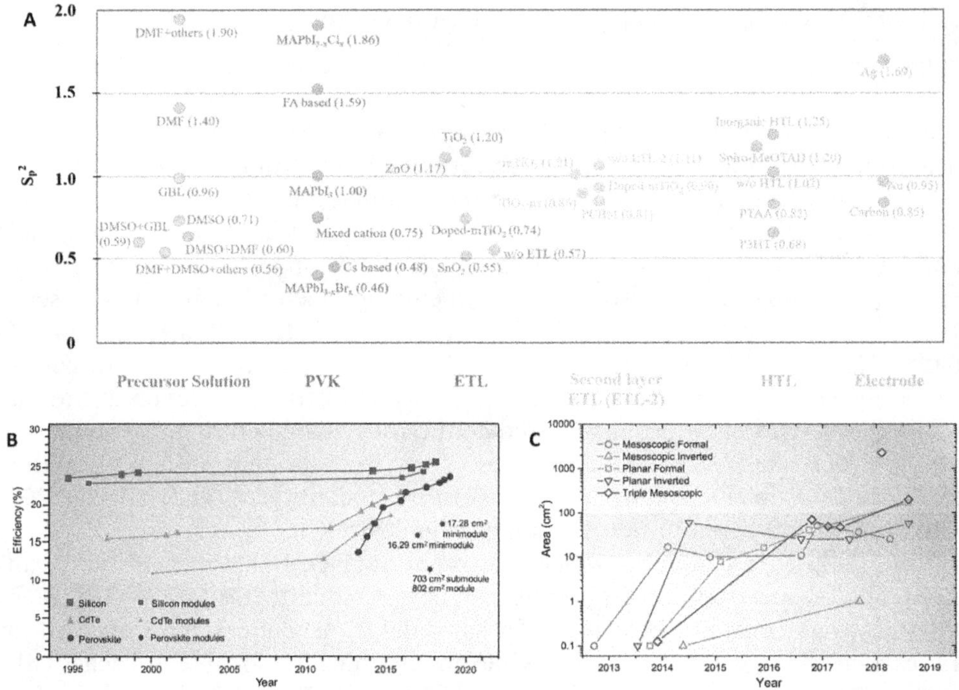

Figure 8.10. (A) The Sp^2 value of materials for each layer of PSCs. The data come from [11]. (B) Historical progression of PCE across various PV technologies over recent decades and (C) technological advancements in perovskite solar modules, highlighting the transition between different device architectures. ((A) Reproduced with permission from [147]. CC BY-NC 3.0. (B) and (C) Reproduced with permission from [148]. Copyright 2019 Elsevier.)

The results, presented in figure 8.10(A), show that lower Sp^2 values correspond to better reproducibility. According to their findings, compact SnO_2 is the most reliable option for ETL. For HTLs, P3HT and PTAA performed better than spiro-MeOTAD. Interestingly, even HTL-free devices performed more consistently than those using some inorganic HTLs. Among electrode materials, carbon stood out for its stable and repeatable performance [11]. Overall, ensuring consistency in materials, deposition methods, and device design is essential for the commercial success of PSCs. Figure 8.10(A) provides a helpful reference for selecting materials and layers that offer high reproducibility, which is a crucial step towards making PSCs a dependable and scalable technology [147].

Despite several challenges, PSCs have demonstrated rapid advancements, within a relatively short timeframe, compared to other technologies like c-Si and CdTe, as shown in figure 8.10(B) [148]. The efficiency of PV technologies such as c-Si and CdTe generally declines as device area increases, largely due to ohmic losses arising from the high sheet resistance ($\sim 10\ \Omega/\square$) of transparent electrodes [149]. This results in a near-linear increase in series resistance and a PCE reduction that follows an inverse scaling law, in which a ten-fold increase in area leads to a $\sim 0.8\%$ decrease in PCE [150]. In contrast, PSCs currently deviate from this trend, primarily because of challenges in achieving uniform large-area deposition of all solution-processed functional layers, such as the perovskite absorber and charge transport layers.

There are five main PSC device architectures: mesoscopic n–i–p, mesoscopic p–i–n, planar n–i–p, planar p–i–n, and triple mesoscopic [151]. Mesoporous structures offer better long-term stability and lower hysteresis [105, 133], while planar designs allow easier, low-temperature processing [152, 153]. The triple mesoscopic design uses screen-printed TiO_2/ZrO_2/carbon layers and allows scalable production by post-deposition perovskite infiltration [154–156]. Figure 8.10(C) [137, 142, 157–159] illustrates the progression in device and module areas for PSCs, highlighting different scalable architectures.

8.4.2 Roadblocks in commercial deployment

Beyond technical limitations, PSCs face challenges in achieving high technology readiness levels (TRLs) [160]. TRL is a standardized scale used to assess the maturity of a technology, ranging from basic research (TRL 1) to full commercial deployment (TRL 9). Most PSC technologies remain below TRL 6, indicating that they have not yet demonstrated robust functionality in relevant operational environments and have only been tested in laboratories or limited industrial environments. Achieving widespread commercialization (TRL 8–9) will require rigorous large-area testing and long-term stability under outdoor conditions. The PEPPERONI project represents a key initiative aiming to advance the TRL of perovskite/Si tandem cells from 4–5 to 7. By focusing on material optimization, pilot-scale production, and supply chain integration, the project targets system-level demonstrations by 2026, paving the way for commercial rollout shortly thereafter [161].

Material supply also poses a significant challenge for large-scale manufacturing of PSCs. For instance, many high-performing PSCs use HTMs such as Spiro-OMeTAD, PTAA, and P3HT. These materials are not only costly and complex to produce but also made at very small scales. To support HTM demands of 100, 1320, and 1430 t/a/TWp (tonnes per annum per terawatt-peak), using PTAA, P3HT, or Spiro-OMeTAD respectively, production would need to scale up by factors of approximately 12 800, 168 955, and 182 778 (figure 8.11(A)) [117]. These HTMs are currently synthesized only in gram-scale batches, with low yields (17.8% for Spiro-OMeTAD; 43.2% for PTAA), lengthy processing times (>45 h and >12 h, respectively), and complex multi-step routes (seven steps for Spiro-OMeTAD; 4 steps for PTAA), involving various solvents, educts, and acids [162, 163]. Advancing industrial-scale synthesis methods is thus critical to enable the use of these materials in multi-TW-scale perovskite PV manufacturing. Another challenge is solvent use. Figure 8.11(B) [117] compares solvent demand for 1 TWp/year with current production levels, showing that while the supply for contact layers is sufficient, processing of perovskite absorbers may require a moderate increase in solvent production, potentially by as much as 20%. Some solvents that are used in PSC fabrication are also toxic and energy-intensive to manufacture. However, figure 8.11(B) shows that the solvent demand for multi-TWp production is manageable and can be reduced further through on-site recycling.

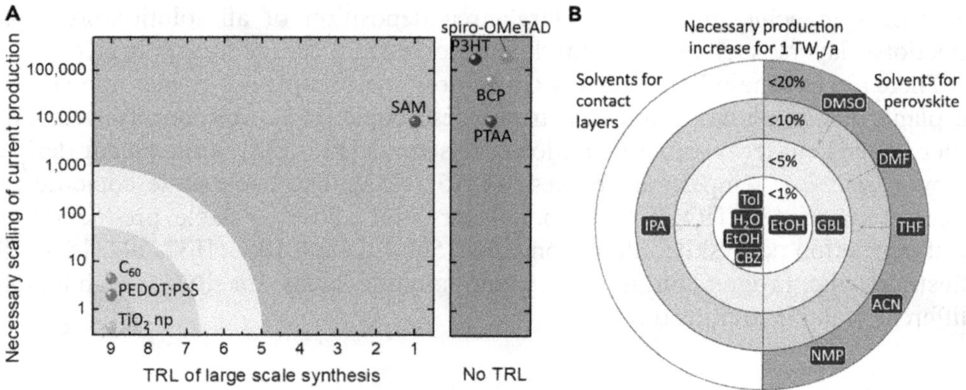

Figure 8.11. Scaling requirements for synthetic contact materials and solvents. (A) Projected production scale-up and TRL advancement needed for the industrial synthesis of synthetic materials to support 1 TWp/a perovskite PV manufacturing. No TRL indicates the absence of a defined industrial production pathway. (B) Estimated increase in solvent demand for achieving 1 TWp/a perovskite PV output. Dashed arrows represent the potential to reduce consumption through effective recycling strategies. (Reproduced with permission from [117]. Copyright 2024 The Author(s). Published by Elsevier. CC BY-NC-ND 4.0.)

All these technical hurdles are worsened by limited collaboration between research institutions and industry. In order to be able to move forward, interdisciplinary teamwork among scientists, manufacturers, and government agencies is needed. This includes better production methods, stable encapsulation techniques, and clear recycling systems. Only then can PSCs become a truly reliable, scalable, and environmentally safe energy technology [164].

8.4.3 Circular economy approaches

As PSCs move closer to commercial use, applying circular economy strategies in both device design and large-scale production is becoming essential. Unlike the conventional linear 'take-make-dispose' model, circular approaches aim to extend the functional lifetime of materials through design for recyclability, reuse, and responsible EoL management [15]. Circularity involves more than just recycling; it begins with eco-design, which includes using modular structures such as thermally releasable encapsulants and carbon-based back contacts. These features make it easier to take apart the device at the end of its life and reduce the energy needed during processing [15, 165, 166]. Such design strategies support the recycling and disposal approaches described in section 8.4.2 and are further strengthened by green solvent-based recovery methods. These techniques have achieved high recovery rates, 100.0% for ITO/SnO_2, 99.1 ± 0.3% for PbI_2, and 89 ± 4% for Spiro-OMeTAD, respectively [167]. In parallel, dynamic material flow modeling and LCA, discussed in section 8.4.3, quantify material demands (lead, silver, indium). These studies are useful for supporting smarter, more sustainable design choices.

In addition to technical solutions, policies and regulations play an important role—rules that require proper EoL and recycling, encourage eco-friendly designs, and provide incentives for sustainable innovation. These are important for building a

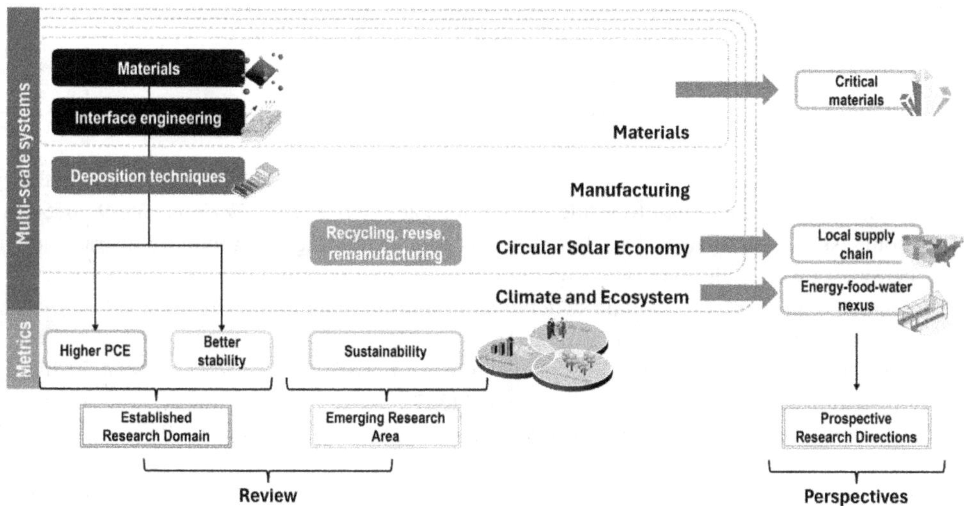

Figure 8.12. Schematic representation of a circular solar economy framework applied to perovskite photovoltaics. (Reproduced from [168]. CC BY3.0.)

circular PSC industry, reflecting the views reported by Charles *et al* [15]. Collaboration between universities, companies, national labs and recycling facilities can help test these ideas, making it easier to scale them up and prepare for market readiness.

Figure 8.12 illustrates a multi-layered circular economy framework for perovskite PVs, integrating materials research, manufacturing processes, recycling ecosystems, and broader climate and ecosystem considerations [168]. The study also highlights the importance of LCA and supportive policies in guiding sustainable technological development. It emphasizes the need for coordinated efforts among researchers, industry stakeholders, and policymakers to make PSCs more reliable, eco-friendly, and ready for commercial deployment.

8.5 Conclusion

Despite being very promising, the commercial viability of PSCs is limited by poor long-term stability, environmental concerns, and manufacturing inconsistencies. Degradation from environmental stressors, interfacial instability, and toxic components such as lead must be addressed to ensure their reliable performance and safe use. Current recycling and certification frameworks are inadequate, and most PSCs remain below TRL 6. Future success depends on interdisciplinary strategies that combine material innovation, lifecycle planning, and scalable fabrication. Implementation of design-for-recycling, lead mitigation, and reproducible processing will be key to transforming perovskites from lab breakthroughs into commercially viable and sustainable solar cell technology.

References

[1] NREL 2025 Best Research-Cell Efficiency Chart *US Department of Energy, Office of Energy Efficiency and Renewable Energy* https://www2.nrel.gov/pv/cell-efficiency

[2] Snaith H J 2013 Perovskites: the emergence of a new era for low-cost, high-efficiency solar cells *J. Phys. Chem. Lett.* **4** 3623–30

[3] Dou L *et al* 2014 Solution-processed hybrid perovskite photodetectors with high detectivity *Nat. Commun.* **5** 5404

[4] Bryant D *et al* 2016 Light and oxygen induced degradation limits the operational stability of methylammonium lead triiodide perovskite solar cells *Energy Environ. Sci.* **9** 1655–60

[5] Domanski K *et al* 2016 Not all that glitters is gold: metal-migration-induced degradation in perovskite solar cells *ACS Nano* **10** 6306–14

[6] Yuan Y *et al* 2015 Photovoltaic switching mechanism in lateral structure hybrid perovskite solar cells *Adv. Energy Mater.* **5** 1500615

[7] Hoke E T *et al* 2015 Reversible photo-induced trap formation in mixed-halide hybrid perovskites for photovoltaics *Chem. Sci.* **6** 613–7

[8] Babayigit A *et al* 2016 Toxicity of organometal halide perovskite solar cells *Nat. Mater.* **15** 247–51

[9] Wang M *et al* 2021 Lead-free perovskite materials for solar cells *Nano-Micro Lett.* **13** 62

[10] Gong J, Darling S B and You F 2015 Perovskite photovoltaics: life-cycle assessment of energy and environmental impacts *Energy Environ. Sci.* **8** 1953–68

[11] Odabaşı Ç and Yıldırım R 2020 Assessment of reproducibility, hysteresis, and stability relations in perovskite solar cells using machine learning *Energy Technol.* **8** 1901449

[12] Khenkin M V *et al* 2020 Consensus statement for stability assessment and reporting for perovskite photovoltaics based on ISOS procedures *Nat. Energy* **5** 35–49

[13] Zhu P *et al* 2024 Toward the commercialization of perovskite solar modules *Adv. Mater.* **36** 2307357

[14] Wu X *et al* 2023 Eco-friendly perovskite solar cells: from materials design to device processing and recycling *EcoMat* **5** e12352

[15] Charles R G *et al* 2023 Circular economy for perovskite solar cells—drivers, progress and challenges *Energy Environ. Sci.* **16** 3711–33

[16] Zhang D *et al* 2022 Degradation pathways in perovskite solar cells and how to meet international standards *Commun. Mater.* **3** 58

[17] Miah M H *et al* 2025 Key degradation mechanisms of perovskite solar cells and strategies for enhanced stability: issues and prospects *RSC Adv.* **15** 628–54

[18] Niu G, Guo X and Wang L 2015 Review of recent progress in chemical stability of perovskite solar cells *J. Mater. Chem.* A **3** 8970–80

[19] Aristidou N *et al* 2017 Fast oxygen diffusion and iodide defects mediate oxygen-induced degradation of perovskite solar cells *Nat. Commun.* **8** 15218

[20] Domanski K *et al* 2017 Migration of cations induces reversible performance losses over day/night cycling in perovskite solar cells *Energy Environ. Sci.* **10** 604–13

[21] Saliba M *et al* 2016 Incorporation of rubidium cations into perovskite solar cells improves photovoltaic performance *Science* **354** 206–9

[22] Leguy A M A *et al* 2015 Reversible hydration of $CH_3NH_3PbI_3$ in films, single crystals, and solar cells *Chem. Mater.* **27** 3397–407

[23] Niu G *et al* 2014 Study on the stability of $CH_3NH_3PbI_3$ films and the effect of post-modification by aluminum oxide in all-solid-state hybrid solar cells *J. Mater. Chem.* A **2** 705–10

[24] Yang J *et al* 2015 Investigation of $CH_3NH_3PbI_3$ degradation rates and mechanisms in controlled humidity environments using *in situ* techniques *ACS Nano* **9** 1955–63

[25] Raval P *et al* 2023 Examining a year-long chemical degradation process and reaction kinetics in pristine and defect-passivated lead halide perovskites *Chem. Mater.* **35** 2904–17
[26] Kore B P, Jamshidi M and Gardner J M 2024 The impact of moisture on the stability and degradation of perovskites in solar cells *Mater. Adv.* **5** 2200–17
[27] Islam M A *et al* 2021 Degradation of perovskite thin films and solar cells with candle soot C/Ag electrode exposed in a control ambient *Nanomaterials* **11** 3463
[28] Habisreutinger S N *et al* 2016 Research update: strategies for improving the stability of perovskite solar cells *APL Mater.* **4** 091503
[29] Liu J *et al* 2014 A dopant-free hole-transporting material for efficient and stable perovskite solar cells *Energy Environ. Sci.* **7** 2963–7
[30] Docampo P and Bein T 2016 A long-term view on perovskite optoelectronics *Acc. Chem. Res.* **49** 339–46
[31] Eperon G E *et al* 2015 The importance of moisture in hybrid lead halide perovskite thin film fabrication *ACS Nano* **9** 9380–93
[32] Bass K K *et al* 2014 Influence of moisture on the preparation, crystal structure, and photophysical properties of organohalide perovskites *Chem. Commun.* **50** 15819–22
[33] You J *et al* 2014 Moisture assisted perovskite film growth for high performance solar cells *Appl. Phys. Lett.* **105** 183902
[34] Frost J M *et al* 2014 Atomistic origins of high-performance in hybrid halide perovskite solar cells *Nano Lett.* **14** 2584–90
[35] Zhu Z *et al* 2016 Interaction of organic cation with water molecule in perovskite $MAPbI_3$: from dynamic orientational disorder to hydrogen bonding *Chem. Mater.* **28** 7385–93
[36] Tiep N H, Ku Z and Fan H J 2016 Recent advances in improving the stability of perovskite solar cells *Adv. Energy Mater.* **6** 1501420
[37] Chi W and Banerjee S K 2021 Achieving resistance against moisture and oxygen for perovskite solar cells with high efficiency and stability *Chem. Mater.* **33** 4269–303
[38] Ma S *et al* 2022 Development of encapsulation strategies towards the commercialization of perovskite solar cells *Energy Environ. Sci.* **15** 13–55
[39] Yu X, Qin Y and Peng Q 2017 Probe decomposition of methylammonium lead iodide perovskite in N_2 and O_2 by *in situ* infrared spectroscopy *J. Phys. Chem.* A **121** 1169–74
[40] Philippe B *et al* 2015 Chemical and electronic structure characterization of lead halide perovskites and stability behavior under different exposures—a photoelectron spectroscopy investigation *Chem. Mater.* **27** 1720–31
[41] Wang S, Wang A and Hao F 2022 Toward stable lead halide perovskite solar cells: a knob on the A/X sites components *iScience* **25** 103599
[42] Zu F S *et al* 2017 Impact of white light illumination on the electronic and chemical structures of mixed halide and single crystal perovskites *Adv. Opt. Mater.* **5** 1700139
[43] Akbulatov A F *et al* 2017 Probing the intrinsic thermal and photochemical stability of hybrid and inorganic lead halide perovskites *J. Phys. Chem. Lett.* **8** 1211–8
[44] Juarez-Perez E J *et al* 2018 Photodecomposition and thermal decomposition in methylammonium halide lead perovskites and inferred design principles to increase photovoltaic device stability *J. Mater. Chem.* A **6** 9604–12
[45] Latini A, Gigli G and Ciccioli A 2017 A study on the nature of the thermal decomposition of methylammonium lead iodide perovskite, $CH_3NH_3PbI_3$: an attempt to rationalise contradictory experimental results *Sustain. Energy Fuels* **1** 1351–7

[46] Juarez-Perez E J *et al* 2016 Thermal degradation of $CH_3NH_3PbI_3$ perovskite into NH_3 and CH_3I gases observed by coupled thermogravimetry–mass spectrometry analysis *Energy Environ. Sci.* **9** 3406–10

[47] Raga S R, Ono L K and Qi Y 2016 Rapid perovskite formation by CH_3NH_2 gas-induced intercalation and reaction of PbI_2 *J. Mater. Chem.* A **4** 2494–500

[48] Pang S *et al* 2016 Transformative evolution of organolead triiodide perovskite thin films from strong room-temperature solid–gas interaction between $HPbI_3$-CH_3NH_2 precursor pair *J. Am. Chem. Soc.* **138** 750–3

[49] Slotcavage D J, Karunadasa H I and McGehee M D 2016 Light-induced phase segregation in halide-perovskite absorbers *ACS Energy Lett.* **1** 1199–205

[50] Brennan M C *et al* 2018 Light-induced anion phase segregation in mixed halide perovskites *ACS Energy Lett.* **3** 204–13

[51] Tress W *et al* 2015 Understanding the rate-dependent *J–V* hysteresis, slow time component, and aging in $CH_3NH_3PbI_3$ perovskite solar cells: the role of a compensated electric field *Energy Environ. Sci.* **8** 995–1004

[52] Eames C *et al* 2015 Ionic transport in hybrid lead iodide perovskite solar cells *Nat. Commun.* **6** 7497

[53] Kang D H and Park N G 2019 On the current–voltage hysteresis in perovskite solar cells: dependence on perovskite composition and methods to remove hysteresis *Adv. Mater.* **31** 1805214

[54] Rahman M B *et al* 2023 Selection of a compatible electron transport layer and hole transport layer for the mixed perovskite $FA_{0.85}Cs_{0.15}Pb\ (I_{0.85}Br_{0.15})_3$, towards achieving novel structure and high-efficiency perovskite solar cells: a detailed numerical study by SCAPS-1D *RSC Adv.* **13** 17130–42

[55] Azpiroz J M *et al* 2015 Defect migration in methylammonium lead iodide and its role in perovskite solar cell operation *Energy Environ. Sci.* **8** 2118–27

[56] Elumalai N K and Uddin A 2016 Hysteresis in organic–inorganic hybrid perovskite solar cells *Sol. Energy Mater. Sol. Cells* **157** 476–509

[57] Zhang M *et al* 2024 Mechanism and regulation of tensile-induced degradation of flexible perovskite solar cells *Energy Adv.* **3** 1431–8

[58] Zai H *et al* 2021 Ion migration in halide perovskite solar cells: mechanism, characterization, impact and suppression *J. Energy Chem.* **63** 528–49

[59] Besleaga C *et al* 2016 Iodine migration and degradation of perovskite solar cells enhanced by metallic electrodes *J. Phys. Chem. Lett.* **7** 5168–75

[60] Zhao Y *et al* 2017 Mobile-ion-induced degradation of organic hole-selective layers in perovskite solar cells *J. Phys. Chem.* C **121** 14517–23

[61] Wang Y *et al* 2019 Stabilizing heterostructures of soft perovskite semiconductors *Science* **365** 687–91

[62] Nughays R O *et al* 2024 Mapping surface-defect and ions migration in mixed-cation perovskite crystals *Adv. Sci.* **11** 2404468

[63] Egger D A, Kronik L and Rappe A M 2015 Theory of hydrogen migration in organic–inorganic halide perovskites *Angew. Chem. Int. Ed.* **54** 12437–41

[64] Xiao Z *et al* 2015 Giant switchable photovoltaic effect in organometal trihalide perovskite devices *Nat. Mater.* **14** 193–8

[65] Yuan Y and Huang J 2016 Ion migration in organometal trihalide perovskite and its impact on photovoltaic efficiency and stability *Acc. Chem. Res.* **49** 286–93

[66] Chen B *et al* 2019 Imperfections and their passivation in halide perovskite solar cells *Chem. Soc. Rev.* **48** 3842–67
[67] Kim J *et al* 2014 The role of intrinsic defects in methylammonium lead iodide perovskite *J. Phys. Chem. Lett.* **5** 1312–7
[68] Choi J J *et al* 2014 Structure of methylammonium lead iodide within mesoporous titanium dioxide: active material in high-performance perovskite solar cells *Nano Lett.* **14** 127–33
[69] Wu B *et al* 2015 Charge accumulation and hysteresis in perovskite-based solar cells: an electro-optical analysis *Adv. Energy Mater.* **5** 1500829
[70] Jeon N J *et al* 2014 Solvent engineering for high-performance inorganic–organic hybrid perovskite solar cells *Nat. Mater.* **13** 897–903
[71] Gottesman R *et al* 2014 Extremely slow photoconductivity response of $CH_3NH_3PbI_3$ perovskites suggesting structural changes under working conditions *J. Phys. Chem. Lett.* **5** 2662–9
[72] Kumar V *et al* 2018 Stoichiometry-dependent local instability in $MAPbI_3$ perovskite materials and devices *J. Mater. Chem.* A **6** 23578–86
[73] Jamali A *et al* 2021 A new and simple method for simulation of lattice mismatch on the optical properties of solar cells: a combination of DFT and FDTD simulations *Sol. Energy* **230** 166–76
[74] Sakai N *et al* 2016 The mechanism of toluene-assisted crystallization of organic–inorganic perovskites for highly efficient solar cells *J. Mater. Chem.* A **4** 4464–71
[75] Maiti A *et al* 2020 Defects and their passivation in hybrid halide perovskites toward solar cell applications *Sol. RRL* **4** 2000505
[76] Agiorgousis M L *et al* 2014 Strong covalency-induced recombination centers in perovskite dolar vell material $CH_3NH_3PbI_3$ *J. Am. Chem. Soc.* **136** 14570–5
[77] Yin W-J, Shi T and Yan Y 2014 Unusual defect physics in $CH_3NH_3PbI_3$ perovskite solar cell absorber *Appl. Phys. Lett.* **104** 063903
[78] Bischak C G *et al* 2017 Origin of reversible photoinduced phase separation in hybrid perovskites *Nano Lett.* **17** 1028–33
[79] Carrillo J *et al* 2016 Ionic reactivity at contacts and aging of methylammonium lead triiodide perovskite solar cells *Adv. Energy Mater.* **6** 1502246
[80] Abdi-Jalebi M *et al* 2018 Maximizing and stabilizing luminescence from halide perovskites with potassium passivation *Nature* **555** 497–501
[81] Shao Y *et al* 2014 Origin and elimination of photocurrent hysteresis by fullerene passivation in $CH_3NH_3PbI_3$ planar heterojunction solar cells *Nat. Commun.* **5** 5784
[82] Aberle A G 2001 Overview on SiN surface passivation of crystalline silicon solar cells *Sol. Energy Mater. Sol. Cells* **65** 239–48
[83] Son D-Y *et al* 2016 Self-formed grain boundary healing layer for highly efficient $CH_3NH_3PbI_3$ perovskite solar cells *Nat. Energy* **1** 16081
[84] Bi C *et al* 2017 Spontaneous passivation of hybrid perovskite by sodium ions from glass substrates: mysterious enhancement of device efficiency revealed *ACS Energy Lett.* **2** 1400–6
[85] Hoex B *et al* 2008 Silicon surface passivation by atomic layer deposited Al_2O_3 *J. Appl. Phys.* **104** 044903
[86] Aberle A G 2000 Surface passivation of crystalline silicon solar cells: a review *Prog. Photovolt. Res. Appl.* **8** 473–87

[87] Bi D *et al* 2016 Efficient luminescent solar cells based on tailored mixed-cation perovskites *Sci. Adv.* **2** e1501170

[88] Chen Q *et al* 2014 Controllable self-induced passivation of hybrid lead iodide perovskites toward high performance solar cells *Nano Lett.* **14** 4158–63

[89] Saidaminov M *et al* 2018 Suppression of atomic vacancies via incorporation of isovalent small ions to increase the stability of halide perovskite solar cells in ambient air *Nat. Energy* **3** 648–54

[90] Man T *et al* 2023 *In situ* passivation of two-dimensional perovskites by external electric field *Adv. Opt. Mater.* **11** 2300969

[91] Yang W S *et al* 2015 High-performance photovoltaic perovskite layers fabricated through intramolecular exchange *Science* **348** 1234–7

[92] Shao Y, Yuan Y and Huang J 2016 Correlation of energy disorder and open-circuit voltage in hybrid perovskite solar cells *Nat. Energy* **1** 15001

[93] Kim S G and Zhu K 2023 Chemical design of organic interface modifiers for highly efficient and stable perovskite solar cells *Adv. Energy Mater.* **13** 2300603

[94] Dunfield S P *et al* 2020 From defects to degradation: a mechanistic understanding of degradation in perovskite solar cell devices and modules *Adv. Energy Mater.* **10** 1904054

[95] Caliò L *et al* 2018 A generic route of hydrophobic doping in hole transporting material to increase longevity of perovskite solar cells *Joule* **2** 1800–15

[96] Luo S and Daoud W A 2015 Recent progress in organic–inorganic halide perovskite solar cells: mechanisms and material design *J. Mater. Chem.* A **3** 8992–9010

[97] Habisreutinger S N *et al* 2014 Carbon nanotube/polymer composites as a highly stable hole collection layer in perovskite solar cells *Nano Lett.* **14** 5561–8

[98] Zhang F *et al* 2017 Isomer-pure Bis-PCBM-assisted crystal engineering of perovskite solar cells showing excellent efficiency and stability *Adv. Mater.* **29** 1606806

[99] Wang S *et al* 2019 Targeted therapy for interfacial engineering toward stable and efficient perovskite solar cells *Adv. Mater.* **31** 1903691

[100] Boldyreva A G *et al* 2020 Unraveling the impact of hole transport materials on photostability of perovskite films and p–i–n solar cells *ACS Appl. Mater. Interfaces* **12** 19161–73

[101] Chauhan A K and Kumar P 2019 Photo-stability of perovskite solar cells with Cu electrode *J. Mater. Sci. Mater. Electron.* **30** 9582–92

[102] Chauhan A and Kumar P 2017 Degradation in perovskite solar cells stored in different environmental conditions *J. Phys.* D: Appl. Phys. **50** 325105

[103] Du T *et al* 2023 Efficient, stable, and fully printed carbon-electrode perovskite solar cells enabled by hole-transporting bilayers *Joule* **7** 1920–37

[104] Que M *et al* 2021 Carbon-based electrodes for perovskite solar cells *Mater. Adv.* **2** 5560–79

[105] Grancini G *et al* 2017 One-year stable perovskite solar cells by 2D/3D interface engineering *Nat. Commun.* **8** 15684

[106] Tong J *et al* 2025 22.1% carbon-electrode perovskite solar cells by spontaneous passivation and self-assembly of hole-transport bilayer *ACS Nano* **19** 12960–70

[107] Zhang H *et al* 2022 High-efficiency (> 20%) planar carbon-based perovskite solar cells through device configuration engineering *J. Colloid Interface Sci.* **608** 3151–8

[108] Zhu C *et al* 2019 Strain engineering in perovskite solar cells and its impacts on carrier dynamics *Nat. Commun.* **10** 815

[109] Chen Y *et al* 2020 Strain engineering and epitaxial stabilization of halide perovskites *Nature* **577** 209–15

[110] Zhao J *et al* 2017 Strained hybrid perovskite thin films and their impact on the intrinsic stability of perovskite solar cells *Sci. Adv.* **3** eaao5616

[111] Steele J A *et al* 2019 Thermal unequilibrium of strained black $CsPbI_3$ thin films *Science* **365** 679–84

[112] Seong S, Liu Y and Gong X 2022 Mechanical study of perovskite solar cells: opportunities and challenges for wearable power source *Opt. Mater. Express* **12** 772–87

[113] Dai Z *et al* 2020 Effect of grain size on the fracture behavior of organic–inorganic halide perovskite thin films for solar cells *Scr. Mater.* **185** 47–50

[114] Yadavalli S K *et al* 2020 Facile healing of cracks in organic–inorganic halide perovskite thin films *Acta Mater.* **187** 112–21

[115] Lee G *et al* 2019 Ultra-flexible perovskite solar cells with crumpling durability: toward a wearable power source *Energy Environ. Sci.* **12** 3182–91

[116] Li J *et al* 2020 Biological impact of lead from halide perovskites reveals the risk of introducing a safe threshold *Nat. Commun.* **11** 310

[117] Wagner L *et al* 2024 The resource demands of multi-terawatt-scale perovskite tandem photovoltaics *Joule* **8** 1142–60

[118] Xiao X *et al* 2025 Aqueous-based recycling of perovskite photovoltaics *Nature* **638** 670–5

[119] Chen B *et al* 2021 Recycling lead and transparent conductors from perovskite solar modules *Nat. Commun.* **12** 5859

[120] Hagelüken C and Goldmann D 2022 Recycling and circular economy—towards a closed loop for metals in emerging clean technologies *Miner. Econ.* **35** 539–62

[121] Heath G A *et al* 2020 Research and development priorities for silicon photovoltaic module recycling to support a circular economy *Nat. Energy* **5** 502–10

[122] Kadro J M *et al* 2016 Proof-of-concept for facile perovskite solar cell recycling *Energy Environ. Sci.* **9** 3172–9

[123] Bogachuk D *et al* 2024 Remanufacturing perovskite solar cells and modules—a holistic case study *ACS Sustain. Resource Manag.* **1** 417–26

[124] Corsini F and Griffini G 2020 Recent progress in encapsulation strategies to enhance the stability of organometal halide perovskite solar cells *J. Phys. Energy* **2** 031002

[125] Vidal R *et al* 2021 Assessing health and environmental impacts of solvents for producing perovskite solar cells *Nat. Sustain.* **4** 277–85

[126] Akulenko E S *et al* 2023 Eco-design for perovskite solar cells to address future waste challenges and recover valuable materials *Heliyon* **9** e13584

[127] Tian X, Stranks S D and You F 2021 Life cycle assessment of recycling strategies for perovskite photovoltaic modules *Nat. Sustain.* **4** 821–9

[128] Ravilla A *et al* 2024 Life cycle assessment of low-dimensional materials for perovskite photovoltaic cells *Energy Adv.* **3** 800–11

[129] Leccisi E and Fthenakis V 2021 Life cycle energy demand and carbon emissions of scalable single-junction and tandem perovskite PV *Prog. Photovolt. Res. Appl.* **29** 1078–92

[130] Fthenakis V and Leccisi E 2021 Updated sustainability status of crystalline silicon-based photovoltaic systems: life-cycle energy and environmental impact reduction trends *Prog. Photovolt. Res. Appl.* **29** 1068–77

[131] Deng Y *et al* 2018 Surfactant-controlled ink drying enables high-speed deposition of perovskite films for efficient photovoltaic modules *Nat. Energy* **3** 560–6

[132] Chen H-W *et al* 2015 A switchable high-sensitivity photodetecting and photovoltaic device with perovskite absorber *J. Phys. Chem. Lett.* **6** 1773–9

[133] Chen X *et al* 2016 Large-area, high-quality organic–inorganic hybrid perovskite thin films via a controlled vapor–solid reaction *J. Mater. Chem.* A **4** 9124–32

[134] Leyden M R, Jiang Y and Qi Y 2016 Chemical vapor deposition grown formamidinium perovskite solar modules with high steady state power and thermal stability *J. Mater. Chem.* A **4** 13125–32

[135] Chang C-Y *et al* 2016 Achieving high efficiency and improved stability in large-area ITO-free perovskite solar cells with thiol-functionalized self-assembled monolayers *J. Mater. Chem.* A **4** 7903–13

[136] Ye F *et al* 2017 Low-temperature soft-cover deposition of uniform large-scale perovskite films for high-performance solar cells *Adv. Mater.* **29** 1701440

[137] Di Giacomo F *et al* 2018 Up-scalable sheet-to-sheet production of high efficiency perovskite module and solar cells on 6-in. substrate using slot die coating *Sol. Energy Mater. Sol. Cells* **181** 53–9

[138] Chang N L *et al* 2018 Manufacturing cost and market potential analysis of demonstrated roll-to-roll perovskite photovoltaic cell processes *Sol. Energy Mater. Sol. Cells* **174** 314–24

[139] Kim D H *et al* 2018 Outlook and challenges of perovskite solar cells toward terawatt-scale photovoltaic module technology *Joule* **2** 1437–51

[140] Rong Y *et al* 2018 Toward industrial-scale production of perovskite solar cells: screen printing, slot-die coating, and emerging techniques *J. Phys. Chem. Lett.* **9** 2707–13

[141] Park M *et al* 2019 Highly reproducible large-area perovskite solar cell fabrication via continuous megasonic spray coating of $CH_3NH_3PbI_3$ *Small* **15** 1804005

[142] Heo J H *et al* 2016 Highly efficient $CH_3NH_3PbI_{3-x}Cl_x$ mixed halide perovskite solar cells prepared by re-dissolution and crystal grain growth via spray coating *J. Mater. Chem.* A **4** 17636–42

[143] Lee D *et al* 2018 Slot-die coated perovskite films using mixed lead precursors for highly reproducible and large-area solar cells *ACS Appl. Mater. Interfaces* **10** 16133–9

[144] Gujar T P and Thelakkat M 2016 Highly reproducible and efficient perovskite solar cells with extraordinary stability from robust $CH_3NH_3PbI_3$: towards large-area devices *Energy Technol.* **4** 449–57

[145] Leyden M R *et al* 2015 Large formamidinium lead trihalide perovskite solar cells using chemical vapor deposition with high reproducibility and tunable chlorine concentrations *J. Mater. Chem.* A **3** 16097–103

[146] Borchert J *et al* 2019 Impurity tracking enables enhanced control and reproducibility of hybrid perovskite vapor deposition *ACS Appl. Mater. Interfaces* **11** 28851–7

[147] Yan J *et al* 2022 Progress and challenges on scaling up of perovskite solar cell technology *Sustain. Energy Fuels* **6** 243–66

[148] Hu Y *et al* 2019 Standardizing perovskite solar modules beyond cells *Joule* **3** 2076–85

[149] Galagan Y *et al* 2016 Towards the scaling up of perovskite solar cells and modules *J. Mater. Chem.* A **4** 5700–5

[150] Li Z *et al* 2018 Scalable fabrication of perovskite solar cells *Nat. Rev. Mater.* **3** 1–20

[151] Rong Y *et al* 2018 Challenges for commercializing perovskite solar cells *Science* **361** eaat8235

[152] Cui P *et al* 2019 Planar p–n homojunction perovskite solar cells with efficiency exceeding 21.3% *Nat. Energy* **4** 150–9

[153] Liu M, Johnston M B and Snaith H J 2013 Efficient planar heterojunction perovskite solar cells by vapour deposition *Nature* **501** 395–8

[154] Hong L *et al* 2017 Improvement and regeneration of perovskite solar cells via methylamine gas post-treatment *Adv. Funct. Mater.* **27** 1703060

[155] Mei A *et al* 2014 A hole-conductor–free, fully printable mesoscopic perovskite solar cell with high stability *Science* **345** 295–8

[156] Hu Y *et al* 2017 Stable large-area (10 × 10 cm^2) printable mesoscopic perovskite module exceeding 10% efficiency *Sol. RRL* **1** 1600019

[157] Chiang C-H, Lin J-W and Wu C-G 2016 One-step fabrication of a mixed-halide perovskite film for a high-efficiency inverted solar cell and module *J. Mater. Chem.* A **4** 13525–33

[158] Singh M *et al* 2018 A novel ball milling technique for room temperature processing of TiO_2 nanoparticles employed as the electron transport layer in perovskite solar cells and modules *J. Mater. Chem.* A **6** 7114–22

[159] Chen C *et al* 2015 Radio frequency magnetron sputtering deposition of TiO_2 thin films and their perovskite solar cell applications *Sci. Rep.* **5** 17684

[160] Sakhatskyi K *et al* 2025 A decade of lead halide perovskites for direct-conversion x-ray and gamma detection: technology readiness level and challenges *Adv. Mater.* **37** 2418465

[161] Development Roadmap PEPPERONI https://pepperoni-project.eu/development/

[162] Mattiello S *et al* 2022 Sustainable, efficient, and scalable preparation of pure and performing spiro-OMeTAD for perovskite solar cells *ACS Sustain. Chem. Eng.* **10** 4750–7

[163] Tepliakova M M *et al* 2020 Suzuki polycondensation for the synthesis of polytriarylamines: a method to improve hole-transport material performance in perovskite solar cells *Tetrahedron Lett.* **61** 152317

[164] Seyisi T *et al* 2025 Major challenges for commercialization of perovskite solar cells: a critical review *Energy Rep.* **13** 1400–15

[165] Heath G A *et al* 2022 A critical review of the circular economy for lithium-ion batteries and photovoltaic modules—status, challenges, and opportunities *J. Air Waste Manage. Assoc.* **72** 478–539

[166] Samantaray M R *et al* 2024 Scalable fabrication methods of large-area (n–i–p) perovskite solar panels *Sol. RRL* **8** 2400235

[167] Larini V *et al* 2025 Circular management of perovskite solar cells using green solvents: from recycling and reuse of critical components to life cycle assessment *EES Solar* **1** 378–90

[168] Tian X *et al* 2025 Perspectives for sustainability analysis of scalable perovskite photovoltaics *Energy Environ. Sci.* **18** 194–213

IOP Publishing

Perovskites

Fundamentals, properties, preparation and applications

N M Ravindra, Priyanka Singh, Leqi Lin and Pankaj Kumar

Chapter 9

Future directions in perovskite materials and technologies

This chapter outlines the future directions for advancing perovskite technologies for stable, and sustainable applications. It highlights the next-generation perovskite materials such as lead-free, 2D, high-entropy, and perovskite-inspired compounds, focusing on improving the device performance with reduced toxicity. Stability is addressed through compositional tuning and advanced passivation strategies. Additionally, the role of digital and predictive tools such as machine learning (ML) and density functional theory (DFT), respectively, are emphasized for accelerating materials discovery and the design of materials. Furthermore, the optical modeling of perovskite materials is described utilizing bandgap–refractive index (E_g–n) relationships.

9.1 Introduction

Perovskite materials, defined by the ABX_3 crystal structure, have gained prominence as advanced functional materials owing to their remarkable optoelectronic properties, structural flexibility, and facile chemical tunability [1]. In this structure of the perovskite, a large A-site cation (such as Cs^+, MA^+, or FA^+) occupies the interstitial spaces between corner-sharing BX_6 octahedra, whereas the smaller B-site metal cation (e.g. Pb^{2+}or Sn^{2+}) is octahedrally coordinated by six halide X-site anions (e.g. Cl^-, Br^-, or I^-), forming a robust and highly adaptable three-dimensional (3D) framework. Despite the remarkable rise of perovskites in photovoltaic (PV) technology, the scope of applications of perovskites now extends far beyond the utilization of solar energy. Following the broad range of applications, discussed in chapter 7, the intrinsic advantages of perovskites, such as high absorption coefficients, long carrier-diffusion lengths, low exciton binding energies, and compatibility with low-temperature, solution-based processing, make them ideal for integration into wide range of optoelectronic devices [2].

Despite the considerable advancements, significant challenges remain in achieving long-term operational stability, and scalability for mass production of PSCs (outlined in chapter 8). Degradation in perovskite layers directly impact the critical performance parameters of PSCs, such as open-circuit voltage (V_{oc}), short-circuit current density (J_{sc}), fill factor (FF), and power conversion efficiency (PCE). Strategic innovations in perovskite research are paving the way for more stable, efficient, and environmentally friendly materials and devices. This chapter focuses on the exploration of emerging and the next-generation perovskite materials that aim to overcome their toxicity and instability challenges. The device engineering, along with advanced additive and passivation strategies, have significantly enhanced the device lifetimes and operational robustness. Furthermore, the integration of digital design tools, such as ML and DFT, is accelerating the discovery and optimization of perovskite materials by enabling high-throughput and predictive screening [3].

9.2 Next-generation perovskites

Commonly used hybrid halide perovskites such as methylammonium lead iodide ($MAPbI_3$) have achieved remarkable success in optoelectronic performance; yet their long-term operational stability and environmental safety, especially due to the presence of lead, remain serious concerns. Future innovations focus on compositional engineering, including lead-free and halide-free formulations, hybrid organic–inorganic architectures, and vacancy-ordered or double perovskites. Notably, 2D and quasi-2D perovskites have emerged as promising candidates due to their superior environmental stability and thermal resilience, which hold promise for enhancing durability and enabling broader integration of perovskites in the next-generation optoelectronic technologies. Next-generation perovskites mark a significant transition to more robust, adaptable, and eco-friendly materials for future technological applications [2, 4–6].

9.2.1 Lead-free perovskites

While chapter 7 provides an overview of the development of lead-based PSCs, this section focuses on emerging lead-free alternatives for PV applications, driven by significant research efforts to address the environmental and health hazards associated with lead-containing materials. A range of low-toxic cations have been proposed as potential replacements for Pb in halide perovskites, including Ag [7], Bi [8], Sb [9], Ti [10], Ge [11], and Sn [12]. Among these candidates, Sn-based halide perovskites have demonstrated the highest PCEs, making them the most extensively studied alternatives for PSCs. Notably, compounds such as $MASnI_3$ have shown promising PV performance, that has been attributed to their narrow bandgaps and facilitate high J_{sc} ($\sim$20–25 mA cm^{-2}) [13, 14]. However, their practical application is severely hindered by the instability caused by the rapid oxidation of Sn^{2+} to Sn^{4+}, leading to self-doping, increased charge carrier recombination, and consequently low V_{oc} ($\sim$0.5 V) [13, 15]. Various approaches have been used to address these challenges. The incorporation of ethylenediammonium (en) cations has enabled the formation of three-dimensional (3D) hollow perovskite structures with tunable bandgaps (1.3–1.9 eV), resulting in enhanced V_{oc} and PCEs. Devices using both ethylammonium (en) and guanidinium (GA) cations have achieved efficiencies up to 9.6% [13, 14, 16]. In addition, 2D Sn-based perovskites, employing large organic spacers such as phenylethylammonium (PEA) and butylammonium (BA), e.g. $(PEA)_2(MA)_{n-1}Sn_nI_{3n+1}$ and $(BA)_2(MA)_{n-1}Sn_nI_{3n+1}$, have demonstrated improved environmental stability and PCEs reaching 9.4% in inverted architectures [15, 17]. Figure 9.1 illustrates the important structural modifications and performance trends of these Sn-based perovskites, including 3D/2D hybrids that show enhanced stability and device efficiency. 3D/2D perovskites are discussed in the subsequent section.

Beyond Sn, alternative chemistries are being actively investigated to overcome the intrinsic limitations of Sn-based perovskites. These include Bi and Sb based double perovskites (e.g. $Cs_2AgBiBr_6$) and vacancy-ordered halide structures, which offer enhanced chemical robustness and also eliminate toxic elements [18, 19]. Although these materials typically yield lower PCEs compared to traditional lead-based PSCs, their superior environmental safety and long-term stability make them promising candidates for the next-generation, sustainable optoelectronic devices. Another notable perovskite structure is Cs_2SnX_6 (X = I, Br), which is chemically stable due to its Sn^{4+} oxidation state. Despite possessing a direct bandgap in the range of 1.3–1.6 eV, its non-3D structure results in poor charge transport, limiting its PV performance [20]. Continued efforts to enhance film morphology and charge carrier diffusion are expected to improve its applicability in solar cells. If future progress is made in extending the carrier-diffusion lengths, such as making high-quality films with big grains and less defects, it would be possible to increase the PCEs of perovskite solar cells.

Ge-based perovskites share some similarities with Pb-based ones but are less comparable than Sn analogs [21]. In the $AGeI_3$ family, $CsGeI_3$ (bandgap ~1.6 eV) is the most suitable candidate for solar applications, while $MAGeI_3$ and $FAGeI_3$ have wider bandgaps (~1.9 and 2.2 eV). However, their efficiencies remain lower than that of Sn-based perovskites [13]. Sn-Ge alloyed perovskites, such as $CsSn_{0.5}Ge_{0.5}I_3$, show better performance, with a 1.5 eV bandgap and 7.11% PCE (figure 9.1(b)) [8]. These devices retain 92% of their initial efficiency after 500 h under 1 sun exposure at 45 °C in N_2 environment and 91% of their initial efficiency after 100 h in air (figure 9.1(c)), aided by a protective GeO_2 surface layer.

In comparison, $MAPbI_3$ continues to set performance benchmarks, with certified PCEs exceeding 25% [22]. However, toxicity and long-term instability remain critical barriers for commercialization [8, 23–25]. As shown in figure 9.1, a side-by-side comparison of device performance and their stability highlights the trade-offs between high-efficiency lead-based PSCs and safer, but less efficient, lead-free alternatives.

9.2.2 Two-dimensional (2D) and quasi-2D perovskites

2D and quasi-2D halide perovskite materials have emerged as promising alternatives to conventional 3D structures, particularly due to their superior environmental

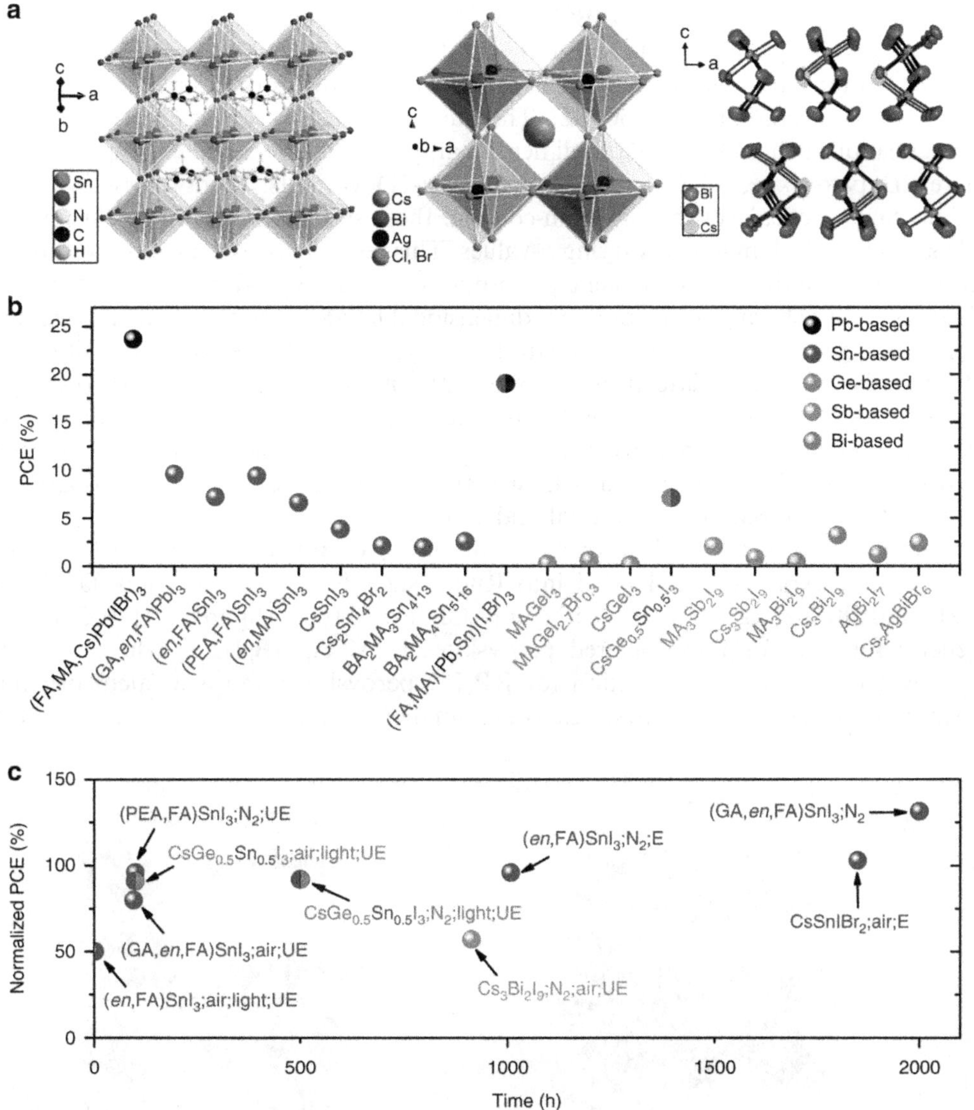

Figure 9.1. Comparison of crystal structure, PCE, and device stability across various perovskite materials. (a) Crystals structures of $FASnI_3$ (left), $Cs_2AgBi(ClBr)_6$ (middle), and $Cs_3Bi_2I_9$ (right) illustrating the diversity in lattice arrangements and dimensionality. (b) Reported PCEs of solar cells using Pb, Sn, Ge, Sb, and Bi-based perovskite absorbers [8, 14, 15, 27]. (c) Stability profiles of Sn, Ge, and Bi-based PSCs under different environmental conditions such as encapsulated (E), unencapsulated (UE), ambient atmosphere, inert nitrogen (N_2) glovebox, and light exposure [8, 14, 15, 27]. ((a) The structure of $Cs_3Bi_2I_9$ (right) reproduced with permission from [26]. Copyright 2017 American Chemical Society. (b) and (c) Reproduced from [13]. CC BY 4.0.)

stability [28]. This enhanced robustness offers notable advantages in applications such as in PV and light-emitting diodes (LEDs) [29–31]. The low-dimensional materials are derived from 3D perovskite frameworks that are typically composed of corner-sharing $[BX_6]$ octahedra ($B^{2+} = Pb^{2+}$, $X^- =$ halide) by slicing along specific crystallographic planes, most commonly in (100), (110), or (111) orientations. Bulky organic cations are then introduced between the layers to form a layered structure, as depicted in figure 9.2 [28, 30, 32]. The number of octahedral layers (n) between spacer cations defines the material dimensionality such as $n = 1$ and corresponds to a pure 2D perovskite, while $n > 1$ forms quasi-2D variants. During practical film fabrication, particularly through spin-coating, these systems frequently yield mixed-phase films with domains of varying n-values. This requires precise process control to ensure phase purity and consistent performance. While typical perovskite structure features corner-sharing octahedra, low-dimensional halide perovskites can also exhibit edge-sharing or face-sharing octahedral arrangements. These variants, although structurally related, deviate from the strict definition of a perovskite. In order to account for this diversity, various terms such as perovskite-like, halometalate, and perovskitoid have been proposed. A more systematic approach is to use the label perovskite-related, supplemented with structural qualifiers (e.g. '2D corner-sharing') to reflect the connectivity of material and geometry [32].

Among low-dimensional perovskites, the (100) oriented layered structures are most widely explored, subdivided into Ruddlesden–Popper (RP), Dion–Jacobson (DJ), and alternating cation in interlayer (ACI or AC) types (figure 9.2) [29–31]. The general formula for these layered perovskites is $A'_mA_{n-1}B_nI_{3n+1}$, where A' is a monovalent organic spacer cation for RP 2D perovskites and a divalent organic spacer cation for DJ 2D perovskites, A is a smaller organic or inorganic cation, B is

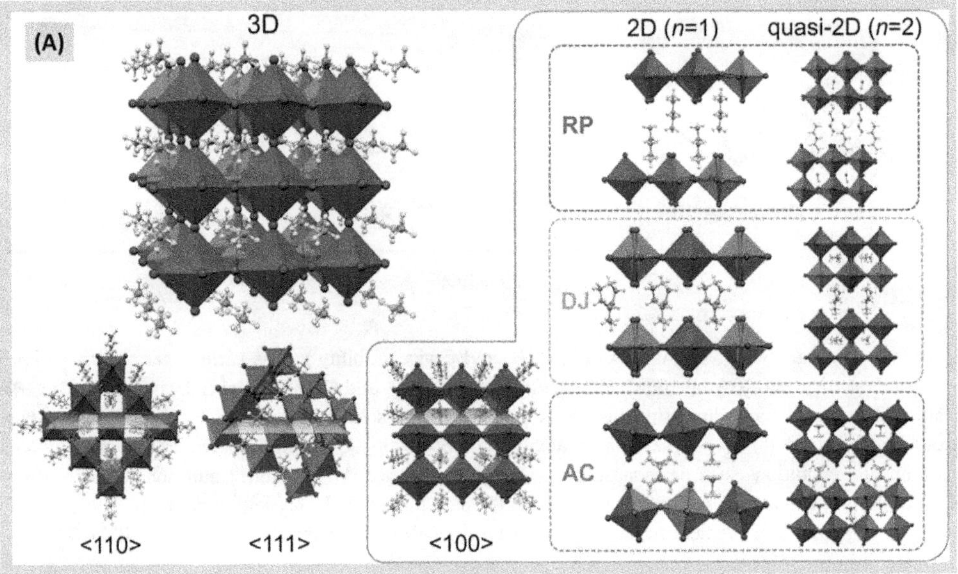

Figure 9.2. Schematic representation of structural configurations of 2D and quasi-2D halide perovskites formed by inserting organic spacer cations between layers of corner-sharing lead halide octahedra. $n = 1$ corresponds to pure 2D perovskites whereas $n > 1$ corresponds to quasi-2D structures along the crystallographic directions such as (100), (110), and (111). (Reproduced from [28]. CC BY 4.0.)

a metal cation and I is a halide anion. Here m represents the number of spacer cation and n is the number of octahedral layers in the 2D perovskite. RP perovskites feature bilayered spacers and (1/2, 1/2) octahedral shifts; DJ perovskites use divalent spacers and show aligned (0,0) octahedral layers [29, 31]; ACI perovskites exhibit a (1/2,0) shift and typically involve GA with small interlayer cations. These classifications originate from oxide perovskite chemistry but are often only loosely applicable to halide analogs, which exhibit more structural flexibility. Therefore, there is an ongoing re-evaluation of these classification schemes to better suit halide-based materials [32, 33].

Structural differences significantly affect physical properties. For example, RP phases offer facile exfoliation and wider interlayer spacing but may limit charge transport. In contrast, DJ phases present tighter layer alignment and better electronic coupling across layers. The choice of spacer cation plays a critical role; cation candidates such as butylammonium (BA^+), octylammonium (OA^+), phenethylammonium (PEA^+), thiophenemethylammonium ($ThMA^+$) and 5-aminovaleric acid (5-AVA) are some of the examples of monovalant spacer cations widely used in RP structures, whereas 1,4-butanediammonium (BDA), 4-aminomethylpiperidine (4-AMP), 3-(aminomethyl)piperidinium) ($3AMP^{2+}$), and propane-1,3-diammonium (PDA^{2+}) are some of the examples of divalent spacer cations for DJ systems [30, 31]. Figure 9.2 demonstrates the crystal structures of 2D (RP, DJ and AC systems) and 3D perovskites.

Compared to 3D perovskites, 2D variants offer better moisture resistance, improved photostability, and superior thermal endurance. However, the insulating nature of organic spacers can impede charge carrier transport, leading to reduced PCEs in purely 2D PSCs. To overcome this, hybrid architectures combining 2D and 3D phases are frequently employed, either as mixed-phase light-absorbing layers or by using 2D perovskites as protective capping layers on 3D films. This approach balances efficiency and long-term device stability [29–31]. A concise comparison of important material properties is presented in table 9.1, highlighting the advantages and limitations of 2D and 3D perovskites.

Despite the growing interest in 2D and quasi-2D perovskites, several challenges persist. The broad range of possible compositions remains only partially explored, and the complex relationship between spacer chemistry, film formation, and optoelectronic properties is not yet fully understood. Moreover, current research is heavily focused on iodide-based compounds, with relatively limited attention given to bromide-containing perovskites [34]. Fabrication methods of 2D perovskites are also very critical. For example, an excess of 2D precursor, instead of improving the device stability and efficiency, can disrupt the beneficial balance of 3D/2D hybrid-perovskite [35]. Therefore, a refined understanding of material design, processing parameters, and device architecture is essential for realizing the full potential of 3D/2D hybrid-perovskite [28].

9.2.3 High-entropy perovskites

High-entropy perovskites (HEPs), inspired by the concept of high-entropy alloys, represent a novel class of materials where five or more different elements are

Table 9.1. Comparison of fundamental properties of 2D and 3D halide perovskites [28–35].

S. No.	2D perovskites	3D perovskites
1.	Lower power conversion efficiencies	Higher power conversion efficiencies
2.	Larger bandgap, insufficient light absorption	Smaller bandgap, broad absorption spectrum and sufficient light absorption
3.	Lower charge carrier mobilities, inefficient charge extraction	Higher charge carrier mobilities, efficient charge extraction
4.	Lower dielectric constant, higher exciton binding energy	Higher dielectric constant, lower exciton binding energy
5.	Better photo and thermal stability	Lower photo and thermal stability
6.	Examples: BA_2PbI_4, $BA_2MA_3Pb_4I_{13}$, $BA_2(FA,MA)_3Pb_4I_{13}$, $(4FPEA)_2MA_4Pb_5I_{16}$, $(ThMA)_2MA_2Pb_3I_{10}$, $(3AMP)MA_2Pb_3I_{10}$, $(3AMP)(MA)_2Pb_3I_{10}$, $(PDA)(MA)_3Pb_4I_{13}$, $(GA)(MA)_3Pb_3I_{10}$, etc	Examples: $MAPbI_3$, $MAPbI_{3-x}Cl_x$, $MAPbBr_3$, $FAPbI_3$, $FA_xMA_{1-x}PbI_3$, $CsPbI_3$ etc

incorporated into the A and/or B sites of the ABX_3 perovskite structure (figure 9.3) [36, 37]. This multi-elemental configuration significantly increases the configurational entropy, which in turn enhances thermodynamic stability by suppressing phase segregation and structural decomposition under operational conditions. As a result, HEPs have garnered growing attention for their potential to improve the long-term reliability of perovskite-based devices. The compositional diversity of HEPs leads to several advantageous properties, such as improved defect tolerance, longer carrier lifetimes, and enhanced mechanical and structural robustness. These attributes make them well-suited for demanding applications in energy conversion and energy storage technologies, such as thermoelectric, PV, lithium-ion batteries, solid oxide fuel cells (SOFCs), electrocatalysis etc [36, 37]. Additionally, targeted doping with cations such as rubidium (Rb^+), potassium (K^+), and various lanthanides enables fine-tuning of critical functional parameters, such as bandgap energy, ionic conductivity, and crystal orientation. Through modulation of the crystal field environment, such dopants also enhance optical absorption and electronic transport, expanding the potential of HEPs in optoelectronic and solid-state devices.

Recent experimental work has validated the stability of these complex systems. For example, Jiang *et al* [38] reported the synthesis of various HEP oxides incorporating multiple cations, where six out of thirteen compositions-including $Sr(Zr_{0.2}Sn_{0.2}Ti_{0.2}Hf_{0.2}Mn_{0.2})O_3$, $Sr(Zr_{0.2}Sn_{0.2}Ti_{0.2}Hf_{0.2}Nb_{0.2})O_3$, $Ba(Zr_{0.2}Sn_{0.2}Ti_{0.2}Hf_{0.2}Ce_{0.2})O_3$, $Ba(Zr_{0.2}Sn_{0.2}Ti_{0.2}Hf_{0.2}Y_{0.2})O_3$, $Ba(Zr_{0.2}Sn_{0.2}Ti_{0.2}Hf_{0.2}Nb_{0.2})O_3$ and $(Sr_{0.5}Ba_{0.5})(Zr_{0.2}Sn_{0.2}Ti_{0.2}Hf_{0.2}Nb_{0.2})O_3$, formed stable, single-phase cubic structures [38]. Interestingly, the formation and thermal stability of these perovskites were found to be governed more by Goldschmidt's tolerance factor (discussed in section 9.4.1) than by cation-size mismatches, emphasizing the importance of geometric compatibility in

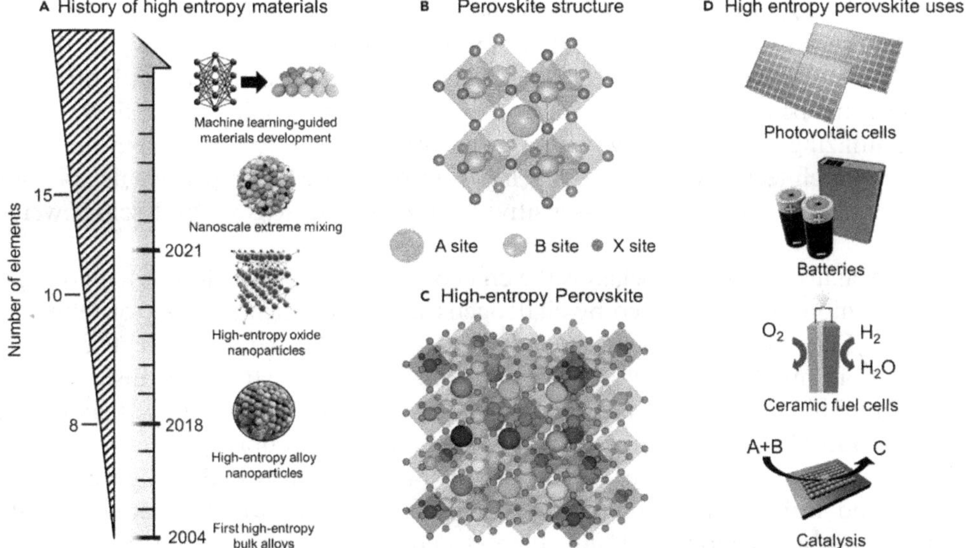

Figure 9.3. Schematic representation of the development and application of HEPs. (A) History of HEPs. (B) Schematic diagram of the perovskite structure. (C) Schematic diagram of the HEP structure. (D) Energy applications of HEP materials. (Reproduced with permission from [41]. Copyright 2023 Elsevier.)

designing robust HEP systems. HEPs have demonstrated great promise for both thermoelectric and PV applications. For instance, $Sr(Ti_{0.2}Fe_{0.2}Mo_{0.2}Nb_{0.2}Cr_{0.2})O_3$, a B-site HEP oxide incorporating five transition metals, was developed as an eco-friendly material for high-temperature thermoelectric applications [39]. The high entropy in this structure enhanced phonon scattering, reducing the lattice thermal conductivity to $0.7\,W\,m^{-1}\,K$ at 1100 K and showcasing its viability for thermal management. In the domain of PV, another study [40] introduced HEPs formed by combining various A-site organic cations. Integrated into inverted PSCs, these materials achieved a certified efficiency of 25.5% and retained over 98% of their initial performance after 1000 h of continuous illumination—demonstrating a scalable and error-tolerant strategy for the development of stable, high-efficiency devices that are suitable for industrial manufacturing [40].

Figure 9.3 illustrates the evolution and potential of HEPs across multiple energy domains. Interest in high-entropy materials has grown steadily since 2004 (figure 9.3(A)). Figure 9.3(B) presents the conventional ABX_3 perovskite structure, while figure 9.3(C) contrasts this with a high-entropy variant that incorporates multiple elements at each crystallographic site. Figure 9.3(D) highlights the broad application scope of HEPs-including PV, batteries, SOFCs, and catalysis, demonstrating their importance in the next-generation energy technologies [41].

9.2.4 Emerging perovskite-inspired materials

The field of perovskite materials is evolving beyond traditional PV applications, expanding into diverse functional domains such as spintronics, neuromorphic computing, piezoelectric sensing, and energy conversion [2, 42]. These emerging perovskite-inspired systems leverage structural and chemical versatility to deliver tuneable multifunctional properties. Hybrid halide perovskites, doped with magnetic ions [43], are being explored for data storage and spintronic applications, while devices utilizing memristive switching behavior show promise for neuromorphic computing architectures. On the other hand, ferroelectric oxide perovskites [44, 45] are under development for high-sensitivity piezoelectric sensors and self-powered systems.

A particularly notable subclass is the chalcogenide perovskites [46], where oxygen or halide anions are substituted by chalcogens sulfur (S), selenium (Se), or tellurium (Te) [47]. Due to the larger ionic radii and different bonding characteristics of chalcogen atoms, these materials often exhibit significant structural distortions—frequently failing the octahedral factor [48, 49] and only marginally satisfying the Goldschmidt tolerance factor. As a result, they adopt non-cubic symmetries such as distorted orthorhombic or hexagonal phases [47]. As illustrated in figure 9.4(A), $BaZrS_3$ and $BaHfS_3$ maintain distorted perovskite frameworks, whereas $BaTiS_3$ crystallizes in a hexagonal $BaNiO_3$-type structure featuring one-dimensional TiS_6 chains aligned along the c-axis [50]. These structural variations substantially impact their optoelectronic and transport properties. With their high thermal and chemical stability, low toxicity, and unique electronic structures, chalcogenide perovskites are promising candidates for sustainable photonic and high-temperature energy applications [48, 49].

Another important family is the oxide zirconate perovskites $AZrO_3$ (A = Ca, Sr, Ba) which offer superior stability, high proton conductivity, and strong dielectric properties. These features make zirconates attractive for use in SOFCs, high-temperature dielectrics, and radiation-tolerant electronics. Their structural robustness stems from tolerance factors that comfortably fall within stable limits, leading to minimal octahedral tilting and strong metal–oxygen bonds [51]. Recent first-principles studies, based on DFT, have deepened understanding of these systems [52]. A notable investigation performed using the Vienna Ab Initio Simulation Package (VASP) employed both the generalized gradient approximation Perdew–Burke–Ernzerhof (GGA-PBE) and Heyd–Scuseria–Ernzerhof (HSE06) functionals to evaluate the structural, electronic, elastic, and optical properties of cubic-phase $BaZrO_3$, $SrZrO_3$ and $CaZrO_3$ (zirconates) [52]. This study revealed that while the PBE functional severely underestimates bandgaps due to self-interaction errors, the HSE06 hybrid functional offers improved predictive accuracy. It also showed that spin–orbit coupling has negligible impact on the electronic structures. All three materials exhibit indirect wide bandgaps with electronic transitions occurring between the R and Γ points in the Brillouin zone (figure 9.4(B)). The partial and total density of states (TDOS) highlight the contributions of both A and B site orbitals, demonstrating the influence on electronic properties by A-site substitution (figure 9.4(C)). Mechanically, $CaZrO_3$ displays a degree of ductility, whereas $SrZrO_3$ and $BaZrO_3$ are more brittle-offering a pathway for mechanical property tuning through cation engineering [52]. Additionally, phonon dispersion analyses confirm the dynamic stability of these materials, and low defect formation energies for

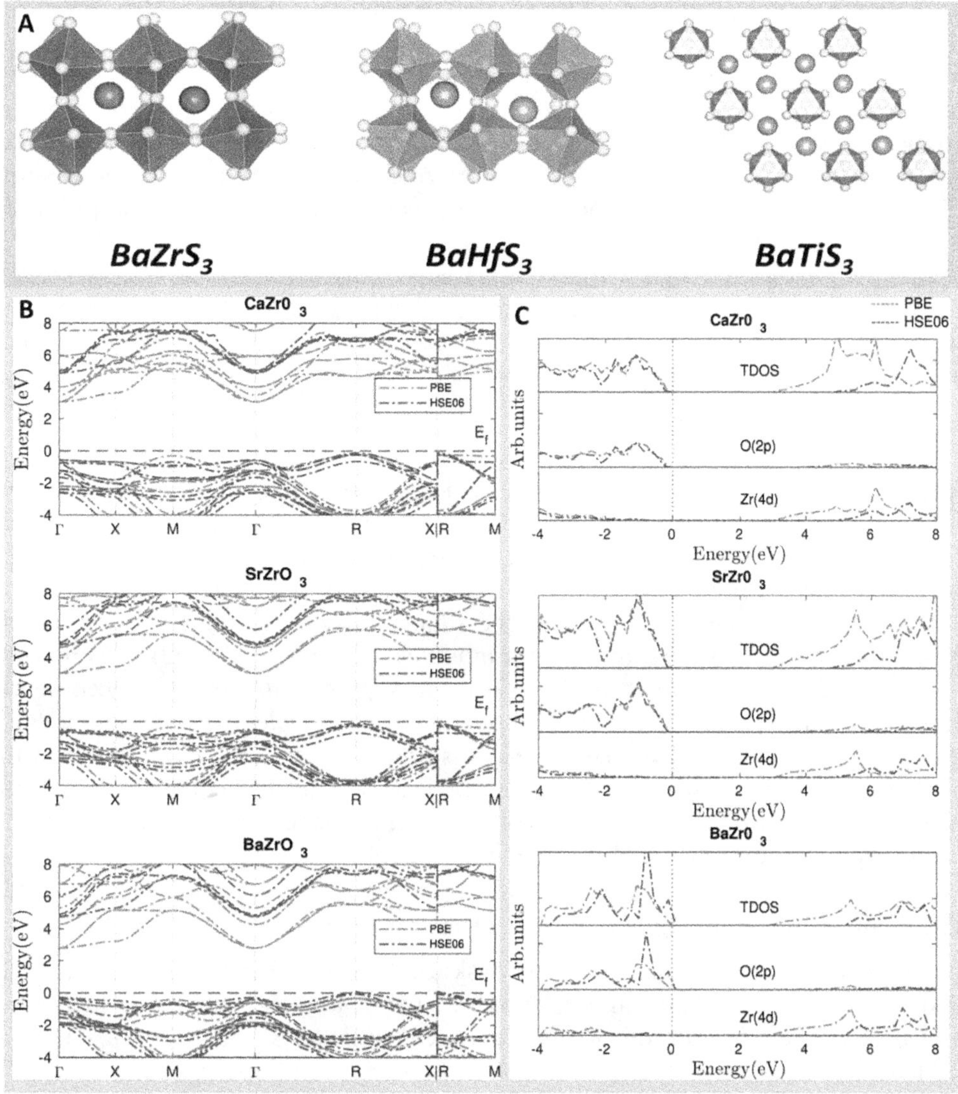

Figure 9.4. (A) Crystal structures of $BaZrS_3$, $BaHfS_3$, and $BaTiS_3$. $BaZrS_3$ and $BaHfS_3$ crystallize in distorted perovskite structures, whereas $BaTiS_3$ adopts a hexagonal $BaNiO_3$-type structure. (B) The electronic band structures and (C) TDOS of $CaZrO_3$, $SrZrO_3$ and $BaZrO_3$ (zirconate) perovskites. ((A) Reproduced from [47]. CC BY 3.0. (B) and (C) Reproduced with permission from [52]. Copyright 2021 The Author(s). Published by IOP Publishing Ltd. CC BY 4.0.)

oxygen vacancies suggest exceptional tolerance to thermal and oxidative stress [51]. Emerging doping strategies—such as lanthanide or transition metal substitution—are being used to reduce the bandgap and enhance optoelectronic response while preserving the intrinsic stability of the zirconate lattice [53].

Emerging material classes such as chalcogenide perovskites, zirconates, and functional oxide derivatives highlight the role of computational design in advancing perovskite-inspired materials beyond energy harvesting. The multifunctionality, tunable chemistry, and structural flexibility of these materials position them as strong candidates for the next-generation optoelectronic, ionic, and sensing applications. Table 9.2 provides an overview of emerging and next-generation perovskites, highlighting their compositions, distinctive properties, and applications across PV, optoelectronics, catalysis, and energy storage.

Table 9.3 shows the PCEs, reported for solar cells based on different emerging perovskite materials, including lead-free, 2D/quasi-2D, doped, high-entropy, and perovskite-inspired compositions. Theoretical PCEs distinguish computational predictions from experimental results.

9.3 Stability engineering

The long-term stability of perovskite devices is essential for their commercial success, particularly in PV and optoelectronics. Critical challenges include degradation caused by heat, light, moisture, oxygen, and electrical stress. These issues, along with their underlying mechanisms and mitigation strategies, are discussed in chapter 8. To enhance stability, researchers have developed a range of engineering strategies, including A-, B-, and X-site modifications within the ABX_3 perovskite structure and surface passivation techniques to suppress the defect-induced degradation. Compositional engineering also plays a vital role by replacing volatile organic cations with more stable alternatives and using halide mixing to adjust the bandgap and enhance moisture resistance. These approaches collectively improve the structural, thermal, chemical, and operational resilience of perovskite materials, advancing their durability and commercial potential.

9.3.1 Compositional tuning and the role of tolerance factor

The structural stability of ABX_3 perovskites (figure 9.3(B)) can be effectively predicted using the Goldschmidt tolerance factor (t) and the octahedral factor (μ), both based on the ionic radii of the constituent ions A, B and X [74–77] and are defined as

$$t = \frac{(r_A + r_X)}{\sqrt{2}\,(r_B + r_X)} \tag{9.1}$$

$$\mu = \frac{r_B}{r_X}, \tag{9.2}$$

where r_A, r_B, and r_X are the ionic radii of the A-site cation, B-site metal cation, and X-site anion, respectively. The tolerance factor serves as a simple yet powerful indicator of whether a given ABX_3 composition will form a stable perovskite structure or not. Cubic perovskites are generally formed when $0.91 \leqslant t \leqslant 1.00$ and $\mu \geqslant 0.41$, while distorted orthorhombic structures appear for $0.71 \leqslant t < 0.91$ with $\mu \geqslant 0.41$. Non-perovskite phases generally emerge when $t < 0.71$, $t > 1.00$, or μ

Table 9.2. Overview of emerging and next-generation perovskites.

Type	Composition/modification	Distinctive properties	Potential applications	Sources
2D/quasi-2D perovskites	Ruddlesden–Popper (RP), Dion–Jacobson (DJ) and alternate cations in interlayers (ACI) phases	Layered structure, moisture resistance, tuneable optoelectronics	Solar cells, LEDs, photodetectors	[28]
Mixed cation/anion perovskites	FA–Cs, MA–FA–Cs, Cl–Br–I mixed	Bandgap tunability, improved thermal and phase stability	Tandem solar cells, x-ray detectors	[54]
Lead-free perovskites	Sn, Bi, Cu, Ge based	Low toxicity, environmental compatibility	Solar cells, sensors	[8, 15, 23]
All-inorganic perovskites	$CsBX_3$ (B = Pb, Sn, X = I, Br, Cl)	High thermal stability, strong light absorption	Lasers, solar cells, LEDs	[55]
Double perovskites	$A_2BB'X_6$ (e.g. $Cs_2AgBiBr_6$)	Non-toxic composition, structural robustness	Radiation detection, photonics	[56]
Halide-doped perovskites	Br/Cl doping in $MAPbI_3$ or $CsPbI_3$	Bandgap modulation, enhanced phase stability	LEDs, solar cells	[57, 58]
Quantum dot perovskites	$CsPbX_3$ QDs (nanocrystals)	Quantum confinement, high photoluminescence	Displays, bioimaging, optoelectronics	[59]
Perovskite–polymer composites	Hybrid structures with flexible polymers	Mechanical flexibility, improved durability	Flexible electronics, wearables	[60]
High-entropy perovskites (HEPs)	Multi-cations at A/B sites (e.g. $(La_{0.2}Nd_{0.2}Sm_{0.2}Y_{0.2}Gd_{0.2})BO_3$)	High configurational entropy, exceptional thermal/chemical stability	Fuel cells, catalysis, energy storage, solar cells	[41]
Zirconate-based perovskites	ABO_3 where B = Zr (e.g. $BaZrO_3$, $SrZrO_3$)	Chemical robustness, high proton conductivity	SOFCs, sensors, water electrolysis	[52]
Chalcogenide perovskites	ABX_3 (X = S, Se; e.g. $BaZrS_3$, $SrHfS_3$)	Narrow bandgap, strong light absorption, lead-free	Solar cells, thermoelectrics	[47]
Ferroelectric oxide perovskites	ABO_3-type (e.g. $BaTiO_3$, $PbTiO_3$, $KNbO_3$)	Spontaneous polarization, switchable electric dipoles, piezoelectric and pyroelectric effects	Non-volatile memory (FeRAM), actuators, sensors, solar cells	[45]

< 0.41 [74–78]. These thresholds are indicative and can vary depending on the ionic radius definitions and material compositions [79–81], prompting refinements to these geometric criteria [79, 81, 82]. Nonetheless, the combined use of t and μ remains a valuable screening tool, particularly for evaluating the formability of chalcogenide perovskites [79, 83, 84]. This approach is more reliable than relying solely on the tolerance factor, in particular when aiming to replicate the favorable charge transport and optical properties of hybrid lead halide perovskites, which stem from a continuous network of corner-sharing PbX_6 octahedra [85]. A stable octahedral configuration usually corresponds to $0.41 \leqslant \mu \leqslant 0.73$ and $0.85 \leqslant t \leqslant 1.05$; values outside these ranges often result in alternative phases such as hexagonal $BaNiO_3$-type or needle-like structures [48, 75, 86, 87].

Hybrid halide perovskites like $MAPbI_3$ often lie near the stability limits defined by tolerance factors, making them prone to phase instability under thermal or moisture stress. To address this, advanced stability models, incorporating octahedral and bond valence factors, are used for predicting robust structures, especially in lead-free and low-dimensional perovskites. These models are increasingly applied in high-throughput and machine learning (ML) driven materials discovery. The tolerance factor remains a key design tool, while performing compositional engineering, such as replacing volatile or reactive ions. It further enhances thermal, chemical, and operational stability. Together, these strategies enable the development of durable, high-performance perovskite optoelectronic devices.

9.3.1.1 A-site engineering

Substitution of the commonly used methylammonium (MA^+) cation in lead halide perovskites, with larger or inorganic alternatives such as formamidinium (FA^+) or cesium (Cs^+), has been shown to significantly enhance thermal and structural stability [88]. FA-based perovskites, such as $FAPbI_3$, demonstrate lower volatility and superior phase stability compared to $MAPbI_3$. However, $FAPbI_3$ tends to transit to the yellow non-perovskite δ-phase at room temperature, which is optoelectronically inactive [89]. This issue can be mitigated by partially substituting FA^+ with Cs^+, thereby stabilizing the desired black α-phase [89]. Furthermore, stabilization and performance enhancement have also been achieved with triple-cation systems incorporating FA^+, MA^+, and Cs^+. These compositions benefit from entropy-driven stabilization, which improves both phase reproducibility and environmental durability [88].

A landmark study by Eperon *et al* demonstrated the tunability of the bandgap in ABX_3 perovskites by altering the A-site cation size [90]. As illustrated in figure 9.5(A), increasing the ionic radius from Cs^+ to FA^+ shifts the absorption edge toward longer wavelengths (i.e. the larger the A-site cation, the smaller the bandgap); the bandgap narrows from 1.73 eV for $CsPbI_3$ to 1.57 eV for $MAPbI_3$ and further to 1.48 eV for $FAPbI_3$. This trend is attributed to the enlarged unit cell volume and associated lattice relaxation with larger cations. Building on this principle, Pellet *et al* [91] incorporated mixed MA/FA cations to enhance light absorption and suppress structural instabilities. These devices showed increased J_{sc} and improved PCE relative to single-cation counterparts, validating the mixed-cation strategy for optoelectronic optimization.

Table 9.3. Reported PCEs for solar cells based on different emerging perovskite materials.

Type	Composition	PCE (%)	References
Lead-free perovskites	$CsSnI_3$	24.4[t]	[61]
	$MASnI_3$	28.0[t]	[62]
	$Cs_2AgBiBr_6$	6.37	[63]
2D and quasi-2D perovskites	$(CDMA)(MA)_{n-1}Pb_nI_{3n+1}$	19.11	[64]
	2D–3D perovskite (FA-MA mixed)	20.12	[65]
Doped/alloyed perovskites	$Cs_xFA_{1-x}PbX_3$	24.0	[66]
	Rb/Cl-doped FA-based $MA_{0.85}FA_{0.15}PbI_3$	22.54	[67]
	$CH_3NH_3Pb_{0.90}Ba_{0.10}I_{3-x}Cl_x$	9.7	[68]
High-entropy hybrid perovskites	4/6/8/10/12C HEHP (five equimolar A-site organic cations in $APbI_3$)	25.7	[40, 69]
Perovskite-inspired materials	Cu_2AgBiI_6	5.52	[70]
	$AgBiS_2$	9.32	[71]
Novel architectures/ predictions	Sn-Ge co-doped $MASnI_3$ $CsSn_{1-x}Ge_xI_3$	23.55[t]	[72]
	Mesoporous TiO_2 with graphene (m-TiO_2 + G)/ perovskite/graphene oxide (GO)	18.2	[73]

Note: 't' represents the theoretical PCEs obtained through simulation tools such as DFT and SCAPS-1D.

Complementing these experimental findings, Filip *et al* [92] provided theoretical insights by predicting a wide bandgap range (0.99–1.79 eV) across various hypothetical A-site cation substitutions. Although certain compositions, such as $RbPbI_3$, are unsuitable as standalone absorbers due to excessively wide bandgaps, small additions of Rb^+ have been shown to synergize with CsFA-based perovskites, enhancing both stability and PCE [93].

As discussed in section 9.3.5, zirconate perovskites ($CaZrO_3$, $BaZrO_3$ and $SrZrO_3$), although characterized by wider bandgaps and insulating behavior, offer a compelling alternative due to their exceptional thermal and chemical resilience. These materials represent a foundational platform for the development of perovskite systems prioritizing long-term stability.

9.3.1.2 B-site engineering

Although Pb remains the dominant B-site cation in high-performance PSCs due to its exceptional optoelectronic characteristics, increasing concerns about its toxicity have spurred significant research into safer alternatives, as previously discussed in section 9.3.1 (figure 9.5(B) and figure 9.1(b) and (c)) [12, 88, 95, 96]. Among these, Sn^{2+} and germanium (Ge^{2+}), both group 14 elements, have garnered significant attention as illustrated in figures 9.1(b) and 9.1(c). Sn-based perovskites, such as $FASnI_3$, offer narrower bandgaps and comparable charge transport properties, but suffer from severe instability due to the spontaneous oxidation. Strategies such as

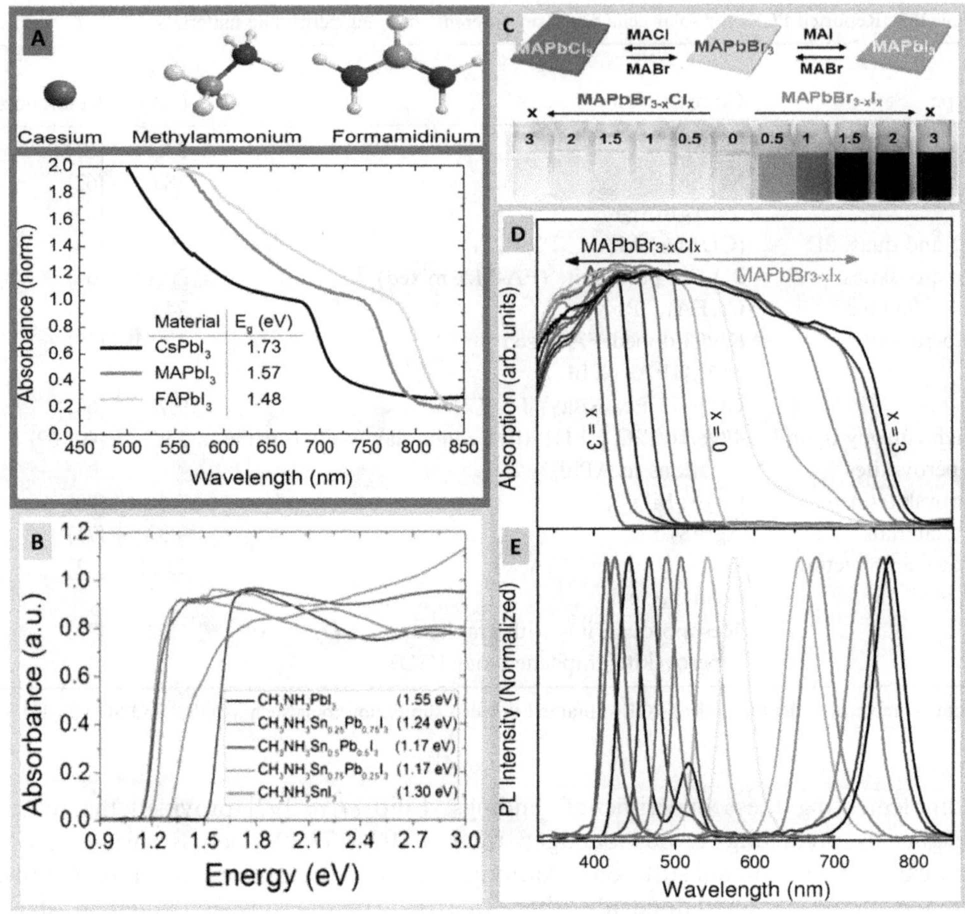

Figure 9.5. Bandgap tunability in metal halide perovskites. (A) Structures of the three A-site cations and UV–vis absorption spectra for the $APbI_3$ perovskites, where A is either Cs, MA, or FA. (B) UV–vis absorption spectra of the mixed tin–lead perovskites ($CH_3NH_3Sn_{1-x}Pb_xI_3$). (C) Photograph of $MAPbBr_{3-x}Cl_x$ and $MAPbBr_{3-x}I_x$ colloidal solutions. (D) UV–vis absorption and (E) photoluminescence (PL) spectra of $MAPbBr_{3-x}Cl_x$ and $MAPbBr_{3-x}I_x$ nanocrystal films where $x = 0$, 0.5, 1, 1.5, 2, and 3. ((A) Adapted with permission from [90]. Copyright 2014 Royal Society of Chemistry. (B) Reproduced with permission from [94]. Copyright 2014 American Chemical Society. (C), (D) and (E) Reproduced with permission from [58]. Copyright 2015 American Chemical Society.)

alloying with Ge and incorporating reducing agents, such as SnF_2, have been explored to suppress this degradation.

A significant advancement came in 2013, when Stoumpos *et al* demonstrated that Pb–Sn mixed-halide perovskites could be synthesized across the entire compositional range, enabling bandgap tunability between 1.1 and 1.7 eV [97]. This laid the groundwork for subsequent studies by Ogimi *et al* [98] and Hao *et al* in 2014 [12], who independently fabricated $MASn_{1-x}Pb_xI_3$ devices and achieved PCEs of ~7% at $x = 0.5$ [12, 98]. These compositions exhibited an absorption onset at 1060 nm, corresponding to a bandgap of $\sim$1.17 eV, approaching the ideal for single-junction solar cells [99]. Interestingly, the bandgap evolution in $MASn_{1-x}Pb_xI_3$ deviates from Vegard's law, instead showing a 'bowing' effect with the minimum bandgap at $x = 0.5$. This phenomenon is explained by first-principles calculations which highlight the competing influences of SOC and lattice distortion during phase transitions [100]. Due to these favorable low-bandgap properties, Pb–Sn mixed perovskites have emerged as ideal candidates for the rear sub-cell in all-perovskite tandem solar cells (TSCs) [88]. Notably, the PCE of Pb–Sn based TSCs surged from just 4.2% in 2014 [98] to an impressive 21.1% by 2019 [101], highlighting their immense potential in scalable, high-efficiency PV devices.

9.3.1.3 X-site engineering

Bandgap tuning in perovskites can also be achieved by modifying the halide composition (X-site), markedly through partial substitution of iodide (I^-) with bromide (Br^-) or chloride (Cl^-) [88]. In $MAPbX_3$ based perovskites, this enables a bandgap shift from 1.55 eV ($MAPbI_3$) to 3.11 eV ($MAPbCl_3$), attributed to the increasing electronegativity of the halogens (Cl > Br > I), which alters the valence band position [58, 102] (figure 9.5(C), (D) and (E)). Mixed-halide perovskites, such as $MAPbI_{3-x}Br_x$, exhibit continuous bandgap tuning up to $\sim$2.3 eV without a significant bowing effect, as demonstrated by Noh *et al* [103]. While this makes them suitable for high V_{oc} devices and tandem applications [104, 105], high Br-content (>30%) compositions are prone to light-induced phase segregation, which degrades performance and stability [106].

Chloride, despite its limited solubility in iodide-based perovskites [107], has been shown to enhance carrier lifetimes and diffusion lengths even at low concentrations [108]. To mitigate halide segregation, additives such as potassium iodide (KI), rubidium iodide (RbI) have been effectively employed, helping stabilize mixed-halide phases under illumination [109]. Thus, halide engineering, complemented by additive strategies, offers a valuable route for balancing tunability and operational stability in perovskite materials.

9.3.2 Approaches to stability enhancement

Stability enhancement in PSCs is critically supported by two complementary strategies: additive engineering and defect passivation. These approaches work synergistically to suppress non-radiative recombination and minimize structural and interfacial defects, thereby improving film morphology, crystallinity, and device reproducibility. Additive engineering primarily influences the crystallization process during film formation, reducing defect density at the source. In contrast, defect passivation addresses existing imperfections, particularly at grain boundaries and interfaces, by chemically neutralizing trap states. Together, these techniques form a comprehensive framework for improving the efficiency, stability, and operational lifespan of PSCs. The nature and impact of defects in perovskites are discussed in chapter 8; the strategies outlined here provide solutions to those issues.

9.3.2.1 *Additive engineering*

Additive engineering involves the deliberate incorporation of small quantities of functional molecules or ions into the perovskite precursor solution to regulate crystallization dynamics, suppress defect formation, and enhance film morphology. Additives improve both the initial performance and long-term durability of PSCs [110]. As shown in figure 9.6, additive-containing PSCs consistently exhibit superior initial efficiencies across various additive types (illustrated in figure 9.6(A)) due to improved crystallinity and defect passivation. More importantly, these devices demonstrate greater operational stability (figure 9.6(B)), exhibiting lower performance degradation over time compared to additive-free devices [111].

Among inorganic additives, KI, RbI, and CsI have been studied extensively for their capacity to occupy interstitial or substitutional sites within the perovskite lattice [112, 113]. These additives effectively passivate deep-level traps, increase grain size, and stabilize the perovskite phase under thermal and environmental

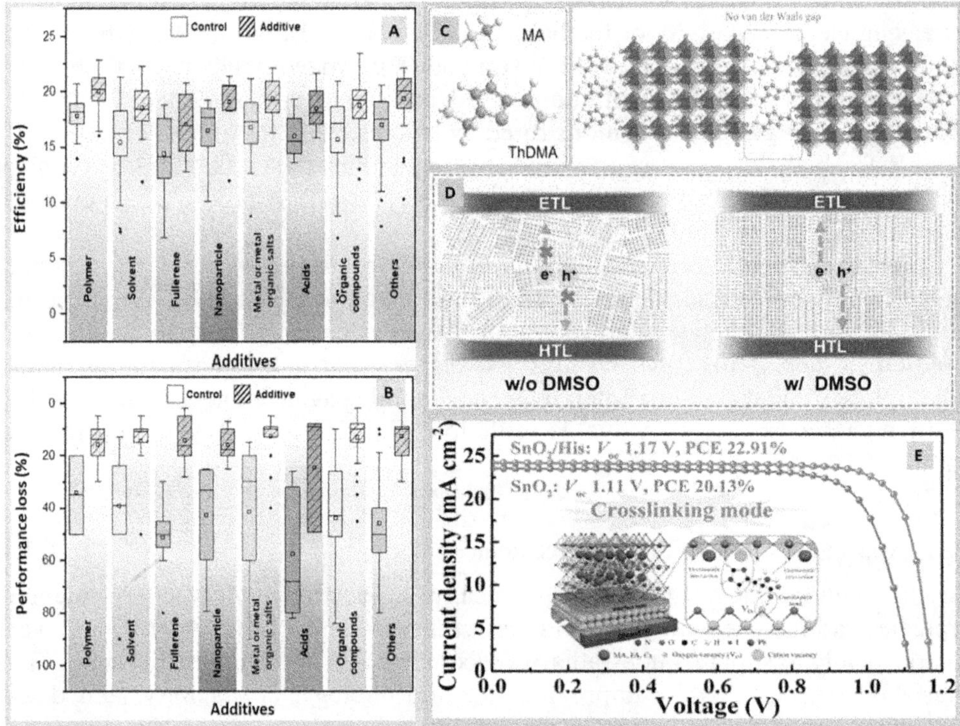

Figure 9.6. (A) Efficiency and (B) performance loss (reduction in efficiency after stability testing) of PSCs incorporating various types of additives. (C) 3D chemical structures of MA, ThDMA and schematic representation of the crystal structure of $(ThDMA)(MA)_4Pb_5I_{16}$. (D) Schematic illustration of the film morphology and corresponding charge transport pathways with and without DMSO treatment. (E) *J–V* curves of PSCs using SnO_2 and histidine-modified SnO_2 (SnO_2/His) electron transport layers. ((A) and (B) Reproduced with permission from [111]. Copyright 2021 Elsevier. (C) and (D) Reproduced with permission from [120]. Copyright 2020 American Chemical Society. (E) Reproduced with permission from [121]. Copyright 2022 American Chemical Society.)

stresses. Furthermore, they contribute to local charge redistribution, facilitating more efficient carrier transport and improving resistance to moisture ingress. Organic additives, including small ammonium salts and chelating molecules such as ethylenediamine (EDA), are effective in mitigating halide segregation in mixed-halide systems and suppressing phase separation. Meanwhile, polymeric additives such as polyethylene glycol (PEG) and polyethylene oxide (PEO) enhance film flexibility, reduce interfacial strain, and passivate grain boundaries, contributing to mechanical and chemical robustness [114, 115].

Lewis bases, such as thiocyanate (SCN^-) and dimethyl sulfoxide (DMSO), serve dual roles by coordinating with undercoordinated Pb^{2+} ions and modulating solvent–solute interactions [116, 117]. This dual interaction promotes more uniform nucleation and defect-suppressed crystallization, which translates to better opto-electronic properties and device stability.

Recent advancements in perovskites have also incorporated ML to accelerate the identification and optimization of effective additive formulations [118, 119]. By enabling high-throughput screening (HTS) of additive candidates, ML has emerged as a powerful tool to enhance both PCE and environmental stability. These approaches, which leverage digital and predictive design tools, are further explored in section 9.5. For example, figure 9.6(C) depicts the molecular structures of MA and the spacer cation 2,5-thiophenedimethylammonium (ThDMA), along with the crystal structure of the DJ perovskite $(ThDMA)(MA)_4Pb_5I_{16}$ [120]. In this configuration, ThDMA cation separates adjacent $[PbI_6]^{4-}$ octahedral planes without introducing a van der Waals gap, unlike in RP structures. This structural arrangement results in reduced interlayer spacing and enhanced lattice integrity. Additives also significantly impact crystal orientation and film morphology. As shown in figure 9.6(D), the control perovskite film exhibits small, disordered grains with poor vertical alignment, which hinders efficient charge transport. Introducing DMSO into the precursor, crystallization kinetics is favorably altered, producing larger, vertically aligned grains that facilitate improved carrier extraction and superior device performance [120]. In addition to bulk crystallization control, interface-targeted additives play a vital role in enhancing PSC performance. For planar SnO_2 based devices (figure 9.6(E)), histidine (His) has been demonstrated as an effective multifunctional agent that chemically bridges SnO_2 and perovskite layers [121]. Through interactions with oxygen vacancies, cationic sites, and iodide ions, His reduces interfacial trap density and non-radiative recombination. This results in improved film quality, enhanced charge extraction, and better interfacial contact. Consequently, His-modified devices achieve a higher PCE of 22.91% (up from 20.13%) and V_{oc} of 1.17 V, with reduced hysteresis and superior humidity and thermal stability (figure 9.6(E)).

9.3.2.2 Defects passivation

Chemical route for defects passivation targets surface and interfacial defects such as undercoordinated Pb^{2+}/I^- ions, iodide vacancies, Pb clusters, and A-site cation deficiencies, that serve as non-radiative recombination centers and limit device efficiency. These defects are especially prevalent at grain boundaries and surfaces in polycrystalline films, where they disrupt charge extraction and accelerate degradation. To mitigate these challenges, researchers have developed molecular passivation strategies that employ small molecules or ions to chemically interact with defect sites. For example, as illustrated in figure 9.6(D), the incorporation of Lewis base molecules such as SCN^- and DMSO can effectively passivate undercoordinated Pb^{2+} sites via coordinate bonding [122, 123]. This coordinate bonding reduces the density of deep-level trap states and simultaneously modulates the crystallization kinetics, resulting in films with larger grain sizes, improved orientation, and enhanced photoluminescence quantum yields. These improvements directly contribute to higher V_{oc} and overall better device performance. Another widely explored strategy involves the controlled use of excess PbI_2, which tends to segregate at grain boundaries and interfaces during film formation. This residual PbI_2 serves as a self-passivating layer, mitigating trap-assisted recombination and improving both carrier lifetime and device stability, especially under light exposure and thermal stress [93, 124].

While conventional passivating agents are effective, recent innovations have turned toward the use of aromatic sulfonate-based ligands, which offer more targeted and robust passivation at both surfaces and grain boundaries. Although surface treatments reduce non-radiative losses at the perovskite/transport layer interface [125], many traditional ligands bind only to isolated active sites and often aligning perpendicularly to the perovskite surface. This alignment can limit packing density and introduce resistive barriers to charge transport. To address this limitation, Chen *et al* [126] introduced a novel class of benzenesulfonate (BZS) ligands that offer dual benefits: they can bridge neighboring Pb^{2+} defect sites and align parallel to the perovskite surface, thereby improving passivation coverage and reducing interfacial energy barriers. Devices without BZS treatment achieved a PCE of approximately 24%, whereas those treated with 4-chlorobenzenesulfonate (4Cl-BZS) demonstrated a significantly enhanced PCE of 26.3%. Furthermore, the 4Cl-BZS-treated devices retained 95% of their initial efficiency after 1200 h of continuous 1 sun illumination under 50% relative humidity at a device temperature of 65 °C. Together, strategies involving Lewis base coordination, controlled PbI_2 excess, and sulfonate ligand surface engineering, exemplify the essential role of chemical passivation in PSCs development. These methods significantly enhance both the efficiency and durability of devices by stabilizing defect-prone regions, in particular at critical interfaces under light and thermal stress [93, 122–126].

9.3.2.3 Emerging passivation techniques

Following the conventional chemical strategies, several advanced passivation methods have emerged to further minimize interfacial losses and enhance long-term operational stability in PSCs. These include physical, energetic, and field-effect passivation approaches, which complement chemical treatments by targeting distinct degradation pathways [127–130].

Physical passivation involves depositing ultrathin barrier layers such as plasma polymers, 2D materials, or metal oxides on the perovskite surface to block moisture, oxygen, and UV-induced degradation [131]. These layers also reduce ion migration, contributing to improved environmental resilience and shelf-life. For instance, polymer encapsulants and atomic layer deposition (ALD) of oxides have shown significant improvements in device durability under damp heat and light soaking.

Energetic passivation targets energy level mismatches at critical interfaces. By introducing buffer layers, self-assembled monolayers (SAMs), or tailored dipole modifiers, it is possible to align the perovskite energy levels with adjacent transport layers, facilitating carrier extraction and reducing recombination losses [132]. These strategies also contribute to improve V_{oc} and FF.

Field-effect passivation employs materials that induce local electric fields, such as ferroelectric compounds or polar molecules, to repel minority carriers from trap-rich regions [129, 133]. These built-in fields reduce surface recombination and enhance charge separation. Recent studies show that histidine and other zwitterionic compounds form oriented dipole layers that simultaneously passivate oxygen and cation vacancies, while improving charge transport and device uniformity.

Collectively, these emerging strategies establish a comprehensive passivation framework. The integration of additive engineering with advanced chemical, physical, or electrostatic passivation enables devices with enhanced efficiency retention and long-term stability under real-world conditions. This synergy between defect mitigation and interfacial design is critical to advancing the commercialization of perovskite photovoltaics.

9.4 Digital and predictive design tools

The development of perovskite materials is rapidly advancing with the adoption of digital and predictive tools that replace traditional trial-and-error approaches with data-driven, efficient strategies [134]. These tools accelerate the discovery and optimization of perovskites by revealing complex structures, optical and electrical properties, and performance relationships [135].

9.4.1 High-throughput and machine learning approaches

HTS, empowered by DFT, enables the simulation of thousands of perovskite compositions to assess the fundamental properties such as bandgap, formation energy, phase stability, and ion migration barriers. HTS has led to the discovery of novel lead-free perovskites and optimized cation/anion configurations. Notably, Liu *et al* [136] analysed 891 ABO_3 structures using ML, identifying 331 perovskite phases, with 174 exhibiting a formation probability $\geqslant$85% (figure 9.7(A)). Further screening yielded 37 thermodynamically stable and 13 metastable perovskites (0–70 meV atom E_{hull}). Li *et al* [137] introduced a transfer learning approach, predicting formation energies and screening 764 potential ABX_3 perovskites, 98 of which were DFT-validated.

ML and artificial intelligence (AI) are central to the next-generation predictive modeling. Supervised models-including neural networks, support vector machines (SVMs), and decision trees have been used to forecast the stability and bandgaps from compositional data. For example, Pilania *et al* [138] used an SVM classifier to

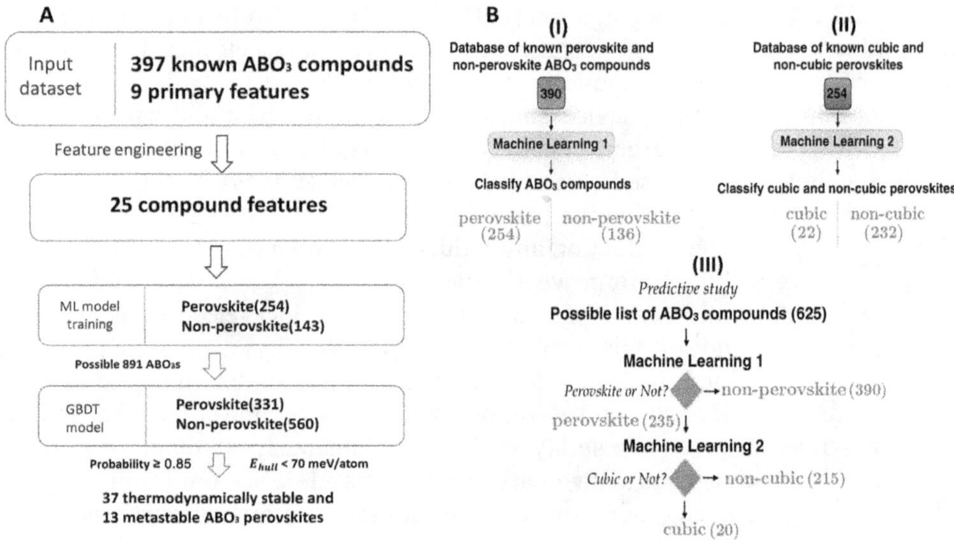

Figure 9.7. Workflows of ML in ABO_3 perovskites. (A) Workflow illustrating the prediction process for assessing the stability and metastability of ABO_3 perovskites. (B) Schematic workflow for evaluating the cubic-phase stability of ABO_3 perovskites, based on computational modeling. ((A) Adapted with permission from [136]. Copyright 2020 Elsevier. (B) Reproduced with permission from [3]. Copyright 2018 American Physical Society.)

assess halide perovskite formability based on structural descriptors such as tolerance and octahedral factors, successfully proposing 40 new ABX_3 materials. Similarly, Balachandran *et al* [3] achieved over 90% accuracy using decision tree classifiers, identifying 235 perovskites from a pool of 625 candidates with 87 selected for further investigation (figure 9.7(B)).

Materials informatics platforms such as the Materials Project, NOMAD, and OQMD offer large datasets to support AI model training and trend mining [139]. These informatic frameworks aid in hypothesis generation, statistical validation, and cross-material comparison across vast chemical spaces. Inverse design strategies, in which desired material properties are predefined and optimal compositions or structures are predicted computationally, are rapidly gaining attraction in materials research [140]. These approaches are significantly empowered by digital twins and Multiphysics simulations, including finite element modeling (FEM) and drift-diffusion techniques, which enable predictive insights into thermal, mechanical, and degradation behaviors under realistic operating conditions [141, 142].

In parallel, closed-loop experimentation, which integrates automated synthesis platforms with real-time feedback from advanced characterization tools-facilitates iterative material optimization. In this framework, ML algorithms intelligently direct the experimental process, while robotic systems perform precise fabrication and property evaluation, thereby enhancing throughput, consistency, and reproducibility.

9.4.2 Modeling of bandgap–refractive index relationships

Following the HTS and DFT methodologies discussed in section 9.4.1, predictive modeling of the optical properties plays a vital role in tailoring perovskite materials for optoelectronic applications. Understanding the link between electronic structure and optical response is critical for rational material selection and device design. A fundamental correlation exists between the optical bandgap (E_g) and the refractive index (n), to estimate optical parameters from compositional data through E_g–n semi-empirical formulas [143].

Despite the growing interest in perovskites, a comprehensive analysis of the E_g–n relationship, specifically for all-inorganic perovskites, has been largely missing in the literature. One of the first detailed examinations of the E_g–n correlation for perovskites was presented by Lamichhane and Ravindra [144]. Establishing reliable E_g–n models is essential not only for assessing the potential of perovskites in electronic and photonic applications but also for populating predictive materials databases used in emerging frameworks such as 'propnet' [145]. As materials informatics continues to advance, such structure–property correlations will be pivotal in accelerating the design of the next-generation perovskite-based technologies.

Moss [146, 147], in 1950, proposed a foundational correlation between the refractive index n and the energy gap E_g, given by

$$n^4 E_g = 95\ \text{eV}. \tag{9.3}$$

However, to address the structural limitations of the Moss relation, Penn [148] proposed an improved model in 1962 by modifying Callaway's approximation of the dielectric constant through the inclusion of the Umklapp process for isotropic semiconductors. Subsequently, Gupta and Ravindra [149] found that the difference between the Penn gap and the E_g remains nearly constant, and presented a linear relationship between n and E_g [150], given by the following equation:

$$n = 4.084 - 0.62 E_g\ (\text{eV}). \tag{9.4}$$

However, since equation (9.4) imposes an upper limit on the refractive index, several empirical models have been proposed to address both the structural constraints of the Moss relation and the refractive index limitations of the Ravindra relation [151–153]. Among these, Herve and Vandamme [154], introduced a widely referenced relation based on oscillatory theory:

$$n = \sqrt{1 + \left(\frac{A}{E_g + B}\right)^2}. \tag{9.5}$$

In this model, $A = 13.6$ eV and $B = 3.4$ eV. While the relation performs well and aligns closely with the experimental data for many optoelectronic materials, it exhibits notable limitations when applied to compounds in the IV–VI group [153].

Accurate optical modeling is essential for optimizing perovskite-based devices; however, their optical behavior remains challenging due to their compositional diversity. The E_g–n relationship is particularly critical for simulating performance in multilayer photonic systems, such as tandem solar cells, where refractive index contrast governs light trapping, interference, and current matching [155]. Notably, changing the bandgap through halide substitution, cation mixing, or dimensional control, as discussed in section 9.4, often causes nonlinear changes in the refractive index, making traditional models based on a simple inverse relationship less reliable.

Considering the limitations of these models, Lamichhane and Ravindra [144] investigated the correlation between the E_g and n in perovskite materials. Using the Wemple–DiDomenico single oscillator approximation [156] and incorporating the unique structural characteristics of perovskites, a simplified model for both oxide and halide perovskite systems was proposed. The performance of this model was subsequently compared with the Moss equation (9.3), Ravindra equation (9.4), and Herve–Vandamme equation (9.5) models, as well as with the experimental refractive index data reported in the literature.

The proposed model is suitable for lead halide perovskites such as $MAPbX_3$ where photon–electron interactions and band structure intricacies dominate optical behavior. Additionally, it addresses the structural dependencies more effectively than traditional methods, offering improved predictive accuracy for complex perovskite systems. The corresponding equations for oxide and halide perovskite systems are presented below:

$$n = \sqrt{1 + \frac{(r_A + r_B)}{r_x}\frac{11.84\ \text{eV}}{E_g}} \tag{9.6a}$$

$$n = \sqrt{1 + \frac{(r_A + r_B)}{r_x}\frac{5.92\ \text{eV}}{E_g}}. \tag{9.6b}$$

As shown in figure 9.8, the proposed model accurately predicts the refractive index of oxide perovskites (figure 9.8(A)) and halide perovskite (figure 9.8(B)) and aligns with the well-established models such as Wemple–DiDomenico, Moss, Ravindra, and Herve–Vandamme. The equations (9.6a) and (9.6b) are specific to perovskites and include both the energy gap and structural parameters, r_A, r_B and r_X. It is important to note in figure 9.8 that Ravindra relation predicts negative n values for energy gaps exceeding 6.6 eV. This could be attributed to non-parallelism in the trajectories of valence and conduction bands along the symmetry directions in perovskites [157–159].

To enable reliable optical modeling, standardized composition-specific datasets and advanced *in situ* techniques such as spectroscopic ellipsometry are essential for accurately extracting optical constants. Integrating validated E_g–n relationships into digital design tools such as HTS-DFT and AI models can greatly accelerate the development of optimized optoelectronic and photonic devices. The complexity of perovskites, due to changes in composition, structure, temperature, and phase, makes it important to use advanced models based on first-principles and machine learning. To fully unlock their potential in future technologies, it is essential to improve the E_g–n relationship using accurate experiments and digital tools.

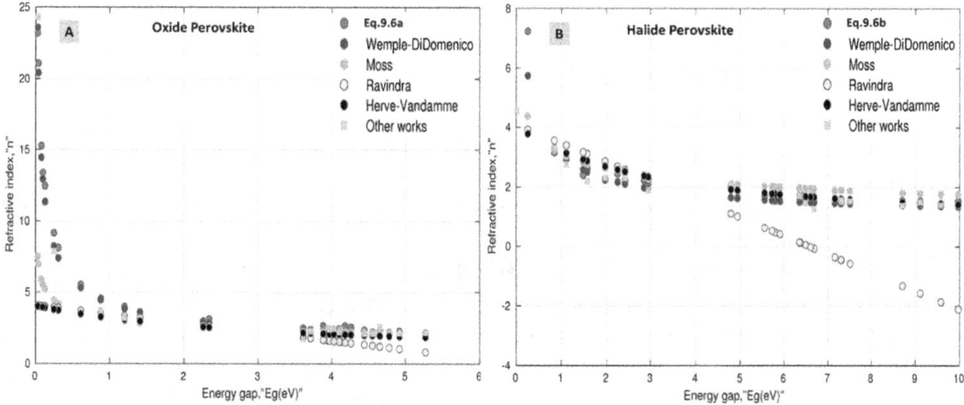

Figure 9.8. Simulated behavior of various models for (A) oxide and (B) halide perovskites. (Adapted from [144]. CC BY 4.0.)

9.5 Conclusions

This chapter highlights the advances in the next-generation perovskites and stability engineering. Lead-free, 2D/quasi-2D, and high-entropy materials offer safer, more stable alternatives, while compositional tuning and passivation strategies improve durability. Digital design tools such as ML, DFT and optical modeling further accelerate innovation, guiding the development of efficient, stable, and sustainable perovskite technologies.

References

[1] Jung H S and Park N G 2015 Perovskite solar cells: from materials to devices *Small* **11** 10–25

[2] Zhang L *et al* 2023 Advances in the application of perovskite materials *Nano-Micro Lett.* **15** 177

[3] Balachandran P V *et al* 2018 Predictions of new ABO_3 perovskite compounds by combining machine learning and density functional theory *Phys. Rev. Mater.* **2** 043802

[4] Korde V B *et al* 2024 Review: perovskite nanostructures materials versatile platform for advance biosensor applications *Sens. Actuators Rep.* **7** 100201

[5] Ullah S *et al* 2024 Emerging applications of perovskite oxides in electrochemical reduction of carcinogenic nitrate to ammonia: a recent review *Inorg. Chem. Front.* **11** 7204–37

[6] Pandiyan A *et al* 2023 A comprehensive review on perovskite and its functional composites in smart textiles: progress, challenges, opportunities, and future directions *Prog. Mater Sci.* **140** 101206

[7] Du K z *et al* 2017 Bandgap engineering of lead-free double perovskite $Cs_2AgBiBr_6$ through trivalent metal alloying *Angew. Chem. Int. Ed.* **56** 8158–62

[8] Chen M *et al* 2019 Highly stable and efficient all-inorganic lead-free perovskite solar cells with native-oxide passivation *Nat. Commun.* **10** 16

[9] Zuo C and Ding L 2017 Lead-free perovskite materials $(NH_4)_3Sb_2I_xBr_{9-x}$ *Angew. Chem. Int. Ed.* **56** 6528–32

[10] Chen M *et al* 2018 Cesium titanium (IV) bromide thin films based stable lead-free perovskite solar cells *Joule* **2** 558–70

[11] Krishnamoorthy T *et al* 2015 Lead-free germanium iodide perovskite materials for photovoltaic applications *J. Mater. Chem.* A **3** 23829–32

[12] Hao F *et al* 2014 Lead-free solid-state organic–inorganic halide perovskite solar cells *Nat. Photonics* **8** 489–94

[13] Ke W and Kanatzidis M G 2019 Prospects for low-toxicity lead-free perovskite solar cells *Nat. Commun.* **10** 965

[14] Jokar E *et al* 2019 Robust tin-based perovskite solar cells with hybrid organic cations to attain efficiency approaching 10% *Adv. Mater.* **31** 1804835

[15] Ke W, Stoumpos C C and Kanatzidis M G 2019 'Unleaded' perovskites: status quo and future prospects of tin-based perovskite solar cells *Adv. Mater.* **31** 1803230

[16] Ke W *et al* 2017 Enhanced photovoltaic performance and stability with a new type of hollow 3D perovskite {en}$FASnI_3$ *Sci. Adv.* **3** e1701293

[17] Mao L, Stoumpos C C and Kanatzidis M G 2018 Two-dimensional hybrid halide perovskites: principles and promises *J. Am. Chem. Soc.* **141** 1171–90

[18] Slavney A H *et al* 2016 A bismuth–halide double perovskite with long carrier recombination lifetime for photovoltaic applications *J. Am. Chem. Soc.* **138** 2138–41

[19] Rahane S N *et al* 2024 Lead-free $Cs_2AgBiCl_6$ double perovskite: experimental and theoretical insights into the self-trapping for optoelectronic applications *ACS Phys. Chem. Au* **4** 476–89

[20] Lee H-S *et al* 2018 Enhanced efficiency of crystalline Si solar cells based on kerfless-thin wafers with nanohole arrays *Sci. Rep.* **8** 3504

[21] Stoumpos C C *et al* 2015 Hybrid germanium iodide perovskite semiconductors: active lone pairs, structural distortions, direct and indirect energy gaps, and strong nonlinear optical properties *J. Am. Chem. Soc.* **137** 6804–19

[22] NREL 2025 Best Research-Cell Efficiency Chart *US Department of Energy, Office of Energy Efficiency and Renewable Energy* https://www2.nrel.gov/pv/cell-efficiency

[23] Magdalin A E *et al* 2023 Development of lead-free perovskite solar cells: opportunities, challenges, and future technologies *Results Eng.* **20** 101438

[24] Jung E H *et al* 2019 Efficient, stable and scalable perovskite solar cells using poly(3-hexylthiophene) *Nature* **567** 511–5

[25] Noel N K *et al* 2014 Lead-free organic–inorganic tin halide perovskites for photovoltaic applications *Energy Environ. Sci.* **7** 3061–8

[26] McCall K M *et al* 2017 Strong electron–phonon coupling and self-trapped excitons in the defect halide perovskites $A_3M_2I_9$ ($A = Cs, Rb$; $M = Bi, Sb$) *Chem. Mater.* **29** 4129–45

[27] Giustino F and Snaith H J 2016 Toward lead-free perovskite solar cells *ACS Energy Lett.* **1** 1233–40

[28] Leung T L *et al* 2022 Stability of 2D and quasi-2D perovskite materials and devices *Commun. Mater.* **3** 63

[29] Gangadharan D T and Ma D 2019 Searching for stability at lower dimensions: current trends and future prospects of layered perovskite solar cells *Energy Environ. Sci.* **12** 2860–89

[30] Liu P, Yu S and Xiao S 2021 Research progress on two-dimensional (2D) halide organic–inorganic hybrid perovskites *Sustain. Energy Fuels* **5** 3950–78

[31] Li X, Hoffman J M and Kanatzidis M G 2021 The 2D halide perovskite rulebook: how the spacer influences everything from the structure to optoelectronic device efficiency *Chem. Rev.* **121** 2230–91

[32] Mercier N 2019 Hybrid halide perovskites: discussions on terminology and materials *Angew. Chem. Int. Ed.* **58** 17912–7

[33] Tremblay M-H *et al* 2019 Structures of $(4\text{-Y-}C_6H_4CH_2NH_3)_2PbI_4$ {Y = H, F, Cl, Br, I}: tuning of hybrid organic inorganic perovskite structures from Ruddlesden–Popper to Dion–Jacobson limits *Chem. Mater.* **31** 6145–53

[34] Vasileiadou E S *et al* 2021 Shedding light on the stability and structure–property relationships of two-dimensional hybrid lead bromide perovskites *Chem. Mater.* **33** 5085–107

[35] Lei N *et al* 2020 Whether addition of phenethylammonium ion is always beneficial to stability enhancement of $MAPbI_3$ perovskite film? *Adv. Mater. Interfaces* **7** 2000197

[36] Chen Q *et al* 2024 High entropy enhanced phase and structural stability and high electromagnetic wave absorption in $CsPbBr_3$ perovskite *Mater. Sci. Semicond. Process.* **175** 108295

[37] Jiang S *et al* 2024 Doping strategies for inorganic lead-free halide perovskite solar cells: progress and challenges *Phys. Chem. Chem. Phys.* **26** 4794–811

[38] Jiang S *et al* 2018 A new class of high-entropy perovskite oxides *Scr. Mater.* **142** 116–20

[39] Banerjee R *et al* 2020 High-entropy perovskites: an emergent class of oxide thermoelectrics with ultralow thermal conductivity *ACS Sustain. Chem. Eng.* **8** 17022–32

[40] Tian Y *et al* 2024 High-entropy hybrid perovskites with disordered organic moieties for perovskite solar cells *Nat. Photonics* **18** 960–6

[41] Wang Y *et al* 2023 High-entropy perovskites materials for next-generation energy applications *Joule* **7** 848–54

[42] Vats G *et al* 2023 Optical memory, switching, and neuromorphic functionality in metal halide perovskite materials and devices *Adv. Mater.* **35** 2205459

[43] Jin L *et al* 2022 Magnetic cations doped into a double perovskite semiconductor *J. Mater. Chem.* C **10** 3232–40

[44] Gao W *et al* 2020 A review of flexible perovskite oxide ferroelectric films and their application *J. Materiomics.* **6** 1–16

[45] Li H *et al* 2021 Photoferroelectric perovskite solar cells: principles, advances and insights *Nano Today* **37** 101062

[46] Njema G G and Kibet J K 2025 A review of chalcogenide-based perovskites as the next novel materials: solar cell and optoelectronic applications, catalysis and future perspectives *Next Nanotechnol.* **7** 100102

[47] Agarwal S, Vincent K C and Agrawal R 2025 From synthesis to application: a review of $BaZrS_3$ chalcogenide perovskites *Nanoscale* **17** 4250–300

[48] Sopiha K V *et al* 2022 Chalcogenide perovskites: tantalizing prospects, challenging materials *Adv. Opt. Mater.* **10** 2101704

[49] Sun Y-Y *et al* 2015 Chalcogenide perovskites for photovoltaics *Nano Lett.* **15** 581–5

[50] Perera S *et al* 2016 Chalcogenide perovskites—an emerging class of ionic semiconductors *Nano Energy* **22** 129–35

[51] Celik G and Çabuk S 2013 First-principles study of electronic structure and optical properties of $Sr(Ti,Zr)O_3$ *Cent. Eur. J. Phys.* **11** 387–93

[52] Lamichhane A and Ravindra N M 2021 First-principles study of cubic alkaline-earth metal zirconate perovskites *J. Phys. Commun.* **5** 035006

[53] Rahman M A *et al* 2023 Tailoring the properties of bulk $BaTiO_3$ based perovskites by heteroatom-doping towards multifunctional applications: a review *ECS J. Solid State Sci. Technol.* **12** 103015

[54] Yadegarifard A *et al* 2023 FA/Cs-based mixed Pb–Sn perovskite solar cells: a review of recent advances in stability and efficiency *Nano Energy* **112** 108481

[55] Swarnkar A, Mir W J and Nag A 2018 Can B-Site doping or alloying improve thermal- and phase-stability of all-inorganic $CsPbX_3$ (X = Cl, Br, I) perovskites? *ACS Energy Lett.* **3** 286–9

[56] Greul E *et al* 2017 Highly stable, phase pure $Cs_2AgBiBr_6$ double perovskite thin films for optoelectronic applications *J. Mater. Chem.* A **5** 19972–81

[57] Straus D B and Cava R J 2022 Tuning the band gap in the halide perovskite $CsPbBr_3$ through Sr substitution *ACS Appl. Mater. Interfaces* **14** 34884–90

[58] Jang D M *et al* 2015 Reversible halide exchange reaction of organometal trihalide perovskite colloidal nanocrystals for full-range band gap tuning *Nano Lett.* **15** 5191–9

[59] Castelletto S, Angelis F D and Boretti A 2022 Prospects and challenges of quantum emitters in perovskites nanocrystals *Appl. Mater. Today* **26** 101401

[60] Li Y *et al* 2025 Flexible perovskite light-emitting diodes: recent progress, applications and challenges *NPJ Flex. Electron.* **9** 32

[61] Hossain M K *et al* 2023 Achieving above 24% efficiency with non-toxic $CsSnI_3$ perovskite solar cells by harnessing the potential of the absorber and charge transport layers *RSC Adv.* **13** 23514–37

[62] Panda P *et al* 2025 Toward high-efficiency photovoltaics: $MASnI_3$ and $FASnI_3$ double absorber perovskite solar cells with optimized conversion efficiency of 28% *Physica* B **710** 417232

[63] Zhang Z *et al* 2022 Hydrogenated $Cs_2AgBiBr_6$ for significantly improved efficiency of lead-free inorganic double perovskite solar cell *Nat. Commun.* **13** 3397

[64] Kahana L 2024 Solar cell based on 2D Dion–Jacobson perovskite achieves 19.11% efficiency *PV Magazine—Photovoltaics Markets and Technology* https://pv-magazine.com/2024/07/29/solar-cell-based-on-2d-dion-jacobson-perovskite-achieves-19-11-efficiency/

[65] Liang J *et al* 2022 A finely regulated quantum well structure in quasi-2D Ruddlesden–Popper perovskite solar cells with efficiency exceeding 20% *Energy Environ. Sci.* **15** 296–310

[66] Wang S *et al* 2022 Over 24% efficient MA-free $Cs_xFA_{1-x}PbX_3$ perovskite solar cells *Joule* **6** 1344–56

[67] Sun J *et al* 2023 Synergistic strategy of rubidium chloride regulated SnO_2 and 4-tert-butyl-benzylammonium iodide passivated $MA_xFA_{1-x}PbI_3$ for efficient mixed-cation perovskite solar cells *Chem. Eng. J.* **468** 143722

[68] Wu M-C *et al* 2018 The effect of strontium and barium doping on perovskite-structured energy materials for photovoltaic applications *Appl. Surf. Sci.* **429** 9–15

[69] Bellini E 2024 Solar cell based on high-entropy hybrid perovskite achieves 25.7% efficiency *PV Magazine—Photovoltaics Markets and Technology* https://pv-magazine.com/2024/07/05/solar-cell-based-on-high-entropy-hybrid-perovskite-achieves-25-7-efficiency/

[70] Grandhi G K *et al* 2023 Perovskite-inspired Cu_2AgBiI_6 for mesoscopic indoor photo-voltaics under realistic low-light intensity conditions *Sustain. Energy Fuels* **7** 66–73

[71] Yuan L *et al* 2025 Direct synthesis of semiconductive $AgBiS_2$ NC inks toward high-efficiency, low-cost and environmental-friendly solar cells *Angew. Chem. Int. Ed.* **64** e202416369

[72] Hasan N *et al* 2024 Comprehensive numerical analysis of doping controlled efficiency in lead-free $CsSn_{1-x}Ge_xI_3$ perovskite solar cell *Appl. Phys.* A **131** 34
[73] Agresti A *et al* 2016 Graphene–perovskite solar cells exceed 18% efficiency: a stability study *ChemSusChem.* **9** 2609–19
[74] Tiwari D, Hutter O S and Longo G 2021 Chalcogenide perovskites for photovoltaics: current status and prospects *J. Phys. Energy* **3** 034010
[75] Goldschmidt V M 1926 Die Gesetze der Krystallochemie *Naturwissenschaften* **14** 477–85
[76] Filip M R and Giustino F 2018 The geometric blueprint of perovskites *Proc. Natl. Acad. Sci.* **115** 5397–402
[77] Li C *et al* 2008 Formability of ABX_3 (X = F, Cl, Br, I) halide perovskites *Struct. Sci.* **64** 702–7
[78] Li Z *et al* 2016 Stabilizing perovskite structures by tuning tolerance factor: formation of formamidinium and cesium lead iodide solid-state alloys *Chem. Mater.* **28** 284–92
[79] Brehm J A *et al* 2014 The structural diversity of ABS_3 compounds with d electronic configuration for the B-cation *J. Chem. Phys.* **140** 224703
[80] Zhang H *et al* 2007 Structural stability and formability of ABO_3-type perovskite compounds *Struct. Sci.* **63** 812–8
[81] Travis W *et al* 2016 On the application of the tolerance factor to inorganic and hybrid halide perovskites: a revised system *Chem. Sci.* **7** 4548–56
[82] Bartel C J *et al* 2019 New tolerance factor to predict the stability of perovskite oxides and halides *Sci. Adv.* **5** eaav0693
[83] Ju M G *et al* 2017 Perovskite chalcogenides with optimal bandgap and desired optical absorption for photovoltaic devices *Adv. Energy Mater.* **7** 1700216
[84] Körbel S, Marques M A and Botti S 2016 Stability and electronic properties of new inorganic perovskites from high-throughput *ab initio* calculations *J. Mater. Chem.* C **4** 3157–67
[85] Nishigaki Y *et al* 2020 Extraordinary strong band-edge absorption in distorted chalcogenide perovskites *Sol. RRL* **4** 1900555
[86] Swarnkar A *et al* 2019 Are chalcogenide perovskites an emerging class of semiconductors for optoelectronic properties and solar cell? *Chem. Mater.* **31** 565–75
[87] Li C, Soh K C K and Wu P 2004 Formability of ABO_3 perovskites *J. Alloys Compd.* **372** 40–8
[88] Kothandaraman R K *et al* 2020 Near-infrared-transparent perovskite solar cells and perovskite-based tandem photovoltaics *Small Methods* **4** 2000395
[89] Zhang J *et al* 2022 Maximizing merits of undesirable δ-$FAPbI_3$ by constructing yellow/black heterophase bilayer for efficient and stable perovskite photovoltaics *Adv. Funct. Mater.* **32** 2204642
[90] Eperon G E *et al* 2014 Formamidinium lead trihalide: a broadly tunable perovskite for efficient planar heterojunction solar cells *Energy Environ. Sci.* **7** 982–8
[91] Pellet N *et al* 2014 Mixed-organic-cation perovskite photovoltaics for enhanced solar-light harvesting *Angew. Chem. Int. Ed.* **53** 3151–7
[92] Filip M R *et al* 2014 Steric engineering of metal-halide perovskites with tunable optical band gaps *Nat. Commun.* **5** 5757
[93] Saliba M *et al* 2016 Incorporation of rubidium cations into perovskite solar cells improves photovoltaic performance *Science* **354** 206–9

[94] Hao F *et al* 2014 Anomalous band gap behavior in mixed Sn and Pb perovskites enables broadening of absorption spectrum in solar cells *J. Am. Chem. Soc.* **136** 8094–9

[95] Ke W *et al* 2017 Enhanced photovoltaic performance and stability with a new type of hollow 3D perovskite {en}$FASnI_3$ *Sci. Adv.* **3** e1701293

[96] Yao H *et al* 2020 Strategies for improving the stability of tin-based perovskite ($ASnX_3$) solar cells *Adv. Sci.* **7** 1903540

[97] Stoumpos C C, Malliakas C D and Kanatzidis M G 2013 Semiconducting tin and lead iodide perovskites with organic cations: phase transitions, high mobilities, and near-infrared photoluminescent properties *Inorg. Chem.* **52** 9019–38

[98] Ogomi Y *et al* 2014 $CH_3NH_3Sn_xPb_{(1-x)}I_3$ perovskite solar cells covering up to 1060 nm *J. Phys. Chem. Lett.* **5** 1004–11

[99] Shockley W and Queisser H J 1961 Detailed balance limit of efficiency of p–n junction solar cells *J. Appl. Phys.* **32** 510–9

[100] Eperon G E and Ginger D S 2017 B-site metal cation exchange in halide perovskites *ACS Energy Lett.* **2** 1190–6

[101] Lin R *et al* 2019 Monolithic all-perovskite tandem solar cells with 24.8% efficiency exploiting comproportionation to suppress Sn(II) oxidation in precursor ink *Nat. Energy* **4** 864–73

[102] Huang H *et al* 2016 Colloidal lead halide perovskite nanocrystals: synthesis, optical properties and applications *NPG Asia Mater.* **8** e328–8

[103] Noh J H *et al* 2013 Chemical management for colorful, efficient, and stable inorganic–organic hybrid nanostructured solar cells *Nano Lett.* **13** 1764–9

[104] Bush K A *et al* 2017 23.6%-efficient monolithic perovskite/silicon tandem solar cells with improved stability *Nat. Energy* **2** 17009

[105] Zhao D *et al* 2017 Low-bandgap mixed tin–lead iodide perovskite absorbers with long carrier lifetimes for all-perovskite tandem solar cells *Nat. Energy* **2** 17018

[106] Hoke E T *et al* 2015 Reversible photo-induced trap formation in mixed-halide hybrid perovskites for photovoltaics *Chem. Sci.* **6** 613

[107] Yu H *et al* 2014 The role of chlorine in the formation process of '$CH_3NH_3PbI_{3-x}Cl_x$' perovskite *Adv. Funct. Mater.* **24** 7102–8

[108] Stranks S D *et al* 2013 Electron–hole diffusion lengths exceeding 1 micrometer in an organometal trihalide perovskite absorber *Science* **342** 341–4

[109] Hieulle J *et al* 2019 Unraveling the impact of halide mixing on perovskite stability *J. Am. Chem. Soc.* **141** 3515–23

[110] Chen B *et al* 2019 Imperfections and their passivation in halide perovskite solar cells *Chem. Soc. Rev.* **48** 3842–67

[111] Pereyra C, Xie H and Lira-Cantu M 2021 Additive engineering for stable halide perovskite solar cells *J. Energy Chem.* **60** 599–634

[112] Abdi-Jalebi M *et al* 2018 Maximizing and stabilizing luminescence from halide perovskites with potassium passivation *Nature* **555** 497–501

[113] Abdi-Jalebi M *et al* 2018 Potassium- and rubidium-passivated alloyed perovskite films: optoelectronic properties and moisture stability *ACS Energy Lett.* **3** 2671–8

[114] Yang C *et al* 2024 Achievements, challenges, and future prospects for industrialization of perovskite solar cells *Light Sci. Appl.* **13** 227

[115] Ahmed Y *et al* 2022 Organic additives in all-inorganic perovskite solar cells and modules: from moisture endurance to enhanced efficiency and operational stability *J. Energy Chem.* **67** 361–90

[116] Wei Q *et al* 2020 Highly stable and efficient perovskite solar cells produced via high-boiling point solvents and additive engineering synergistically *Sci. China Chem.* **63** 818–26
[117] Liu X and Lee E-C 2024 A Lewis-base-based antisolvent method for fabricating triple-cation perovskite solar cells *J. Mater. Chem.* C **12** 4433–41
[118] Yan W *et al* 2022 Machine learning enabled development of unexplored perovskite solar cells with high efficiency *Nano Energy* **99** 107394
[119] Li X *et al* 2024 Machine learning-assisted design of high-performance perovskite photodetectors: a review *Adv. Compos. Hybrid Mater.* **8** 27
[120] Lu D *et al* 2020 Thiophene-based two-dimensional Dion–Jacobson perovskite solar cells with over 15% Efficiency *J. Am. Chem. Soc.* **142** 11114–22
[121] Li Y *et al* 2022 Multifunctional histidine cross-linked interface toward efficient planar perovskite solar cells *ACS Appl. Mater. Interfaces* **14** 47872–81
[122] Chen Q *et al* 2014 Controllable self-induced passivation of hybrid lead iodide perovskites toward high performance solar cells *Nano Lett.* **14** 4158–63
[123] Bi D *et al* 2016 Efficient luminescent solar cells based on tailored mixed-cation perovskites *Sci. Adv.* **2** e1501170
[124] Turren-Cruz S-H, Hagfeldt A and Saliba M 2018 Methylammonium-free, high-performance, and stable perovskite solar cells on a planar architecture *Science* **362** 449–53
[125] Park S M *et al* 2023 Engineering ligand reactivity enables high-temperature operation of stable perovskite solar cells *Science* **381** 209–15
[126] Chen H *et al* 2024 Improved charge extraction in inverted perovskite solar cells with dual-site-binding ligands *Science* **384** 189–93
[127] Zhao P, Kim B J and Jung H S 2018 Passivation in perovskite solar cells: a review *Mater. Today Energy* **7** 267–86
[128] Zhang Z *et al* 2023 Rationalization of passivation strategies toward high-performance perovskite solar cells *Chem. Soc. Rev.* **52** 163–95
[129] Qi J *et al* 2025 Interface field-effect passivation enabled by selectively extruded cations for printable mesoscopic perovskite solar cells with improved performance *Adv. Mater.* **37** 2506114
[130] Feng X *et al* 2025 Dual field passivation strategy for high-performance wide-bandgap perovskite solar cells *ACS Appl. Mater. Interfaces* **17** 25883–93
[131] Ochoa-Martinez E *et al* 2021 Physical passivation of grain boundaries and defects in perovskite solar cells by an isolating thin polymer *ACS Energy Lett.* **6** 2626–34
[132] Dong W *et al* 2022 Surface passivation and energetic modification suppress nonradiative recombination in perovskite solar cells *Nano-Micro Lett.* **14** 108
[133] Menzel D *et al* 2022 Field effect passivation in perovskite solar cells by a LiF interlayer *Adv. Energy Mater.* **12** 2201109
[134] Butler K T *et al* 2018 Machine learning for molecular and materials science *Nature* **559** 547–55
[135] Tao Q *et al* 2021 Machine learning for perovskite materials design and discovery *NPJ Comput. Mater.* **7** 23
[136] Liu H *et al* 2020 Screening stable and metastable ABO_3 perovskites using machine learning and the materials project *Comput. Mater. Sci.* **177** 109614
[137] Li X *et al* 2019 Computational screening of new perovskite materials using transfer learning and deep learning *Appl. Sci.* **9** 5510
[138] Pilania G *et al* 2016 Finding new perovskite halides via machine learning *Front. Mater.* **3** 19

[139] Jain A *et al* 2013 Commentary: the materials project: a materials genome approach to accelerating materials innovation *APL Mater.* **1** 011002

[140] Wang F *et al* 2022 Recent progress of scalable perovskite solar cells and modules *Energy Rev.* **1** 100010

[141] Zandi S and Razaghi M 2019 Finite element simulation of perovskite solar cell: a study on efficiency improvement based on structural and material modification *Sol. Energy* **179** 298–306

[142] Tesfancheal H Y *et al* 2024 Multi-physics device simulations of optimized semi-transparent perovskite solar cells: influence of material types and layer thicknesses on transmittance and electrical performance *Sol. Energy* **284** 113069

[143] Lamichhane A 2023 Energy-gap-refractive index relations in semiconductors—using Wemple–DiDomenico model to unify Moss, Ravindra, and Herve–Vandamme relationships *Solids* **4** 316–26

[144] Lamichhane A and Ravindra N M 2020 Energy gap-refractive index relations in perovskites *Materials* **13** 1917

[145] Mrdjenovich D *et al* 2020 Propnet: a knowledge graph for materials science *Matter* **2** 464–80

[146] Moss T 1985 Relations between the refractive index and energy gap of semiconductors *Phys. Stat. Sol.* B **181** 415

[147] Moss T S 1951 Photoconductivity in the elements *Proc. Phys. Soc. Lond.* A **64** 590

[148] Penn D R 2093 Wave-number-dependent dielectric function of semiconductors *Phys. Rev.* **128** 1962

[149] Gupta V and Ravindra N 1980 Comments on the moss formula *Phys. Stat. Sol.* B **100** 715–9

[150] Ravindra N M 1981 Energy gap-refractive index relation—some observations *Infrared Phys.* **21** 283–5

[151] Kumar V and Singh J K 2010 Model for calculating the refractive index of different materials *Indian J. Pure Appl. Phys.* **48** 571–4

[152] Tripathy S K 2015 Refractive indices of semiconductors from energy gaps *Opt. Mater.* **46** 240–6

[153] Bahadur A and Mishra M 2013 Correlation between refractive index and electronegativity difference for A^NB^{8-N} type binary semiconductors *Acta Phys. Pol.* A **123** 737–40

[154] Hervé P and Vandamme L 1994 General relation between refractive index and energy gap in semiconductors *Infrared Phys. Technol.* **35** 609–15

[155] Manzoor S *et al* 2018 Optical modeling of wide-bandgap perovskite and perovskite/silicon tandem solar cells using complex refractive indices for arbitrary-bandgap perovskite absorbers *Opt. Express* **26** 27441–60

[156] Wemple S H and DiDomenico M Jr 1971 Behavior of the electronic dielectric constant in covalent and ionic materials *Phys. Rev.* B **3** 1338

[157] Whalley L D *et al* 2017 Perspective: theory and simulation of hybrid halide perovskites *J. Chem. Phys.* **220901** 146

[158] Takegahara K 1994 Electronic band structures in cubic perovskite-type oxides: bismuthates and transition metal oxides *J. Electron. Spectrosc. Relat. Phenom.* **66** 303–20

[159] Green M A *et al* 2015 Optical properties of photovoltaic organic–inorganic lead halide perovskites *J. Phys. Chem. Lett.* **6** 4774–85

IOP Publishing

Perovskites
Fundamentals, properties, preparation and applications
N M Ravindra, Priyanka Singh, Leqi Lin and Pankaj Kumar

Chapter 10

Conclusion

A brief overview of the phase diagram of perovskites is presented in this chapter. This is followed by illustrations of process–property–performance correlations in perovskites. Details of perovskite solar cell development throughout the world are described, highlighting some of the major players. Examples of issued and pending patent applications are reported.

10.1 Introduction

Since the discovery of perovskites by Gustav Rose in 1839, it took 170 years to demonstrate the first perovskite solar cell, by Tsutomu Miyasaka and colleagues in 2009. Despite its low efficiency of only 3%, the promise of low-cost, combined with the ability to use hybrid inorganic–organic perovskites, have been the rationale for the enormous growth of these materials, particularly for use in building-integrated photovoltaics, flexible electronics, LEDs and tandem solar cells. From a fundamental materials science perspective, perovskites represent an excellent opportunity to address fundamental challenges such as defects, lead toxicity, scalability and uniformity, stability and durability, and the need to address encapsulation to circumvent these problems.

10.2 Phase diagram

Phase diagrams are crucial for understanding the stability, structural transitions, and crystallization pathways of perovskite materials. In halide perovskites, they describe the relationship between composition, temperature, and pressure, and reveal the conditions under which different structural phases—cubic, tetragonal, orthorhombic, or hexagonal—are stable [1]. Such information is critical for both the optoelectronic performance and long-term stability of perovskite devices. Beyond equilibrium diagrams, non-equilibrium processing, such as rapid crystallization in thin-film deposition, often traps metastable phases. Understanding these phase diagrams is important for designing new perovskite materials with varying

doi:10.1088/978-0-7503-5427-1ch10

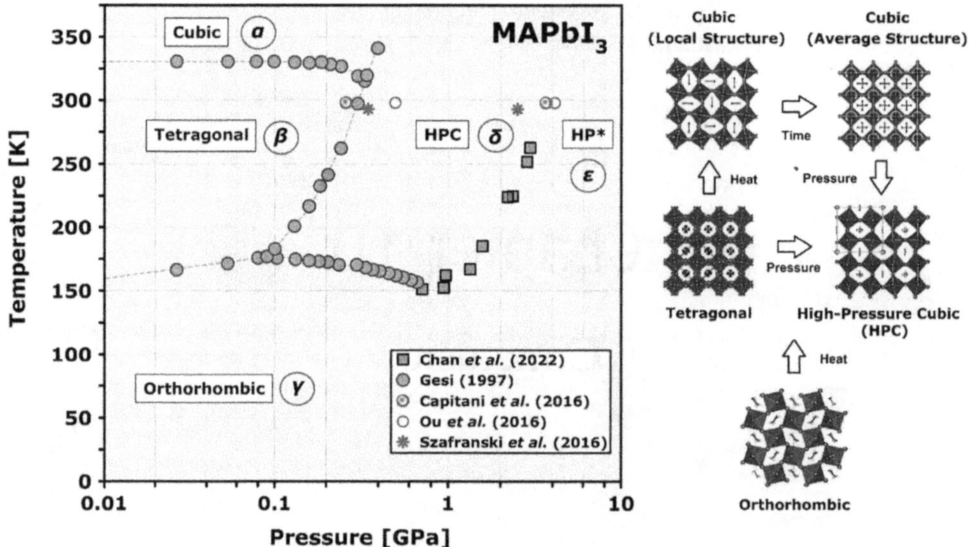

Figure 10.1. Pressure–temperature (*P*–*T*) phase diagram of $MAPbI_3$, constructed from dielectric/infrared spectroscopy ([4, 5]) and x-ray diffraction studies ([6–8]). Greek letters denote the thermodynamically stable phases identified in the literature. Structural projections of the different phases are shown on the right. For the cubic phase, both the average non-equilibrium structure and an instantaneous local structure derived from total-scattering synchrotron diffraction ([9]) are presented for comparison with the average structures of the tetragonal and high-pressure cubic phases ([7]). (Reproduced from [2]. CC BY 4.0.)

compositions and for tailoring processing strategies to achieve desirable, reproducible, high-performance materials. Figure 10.1 represents the pressure–temperature phase diagram of $MAPbI_3$. Three important transitions occur as temperature changes: above ~330 K, it adopts a cubic (α) structure; between ~160 K and 330 K, it is tetragonal (β); and below ~160 K, it transitions into an orthorhombic (γ) phase [2, 3]. These transitions alter lattice symmetry, bandgap, and charge transport properties, therefore influencing the device efficiency. The cubic and tetragonal phases are typically optimal for photovoltaic function, while unwanted phases like the δ-structure of $FAPbI_3$ are non-photoactive and detrimental to device performance. Phase diagrams also guide additive engineering, compositional mixing (e.g. FA/Cs or Br/I mixing as discussed in chapter 9), and processing conditions to suppress unwanted phases such as PbI_2 or δ-$FAPbI_3$, which degrade performance.

Similarly, the Goldschmidt tolerance factor (t), ion size, and chemical composition of A-, B-, and X-site ions determine whether a stable perovskite phase can form at all. Ideal cubic perovskites typically lie in the $0.9 < t < 1.0$ range, while deviations can lead to distorted or non-perovskite structures [10, 11]. In a related study [11], t is calculated for known inorganic ABI_3 compounds using the Shannon radii, following the approach of Cheetham *et al* and others [12–14]. Figure 10.2 presents the distribution of these values, expressed to two decimal places. Blue dots indicate compounds that form perovskites, while red crosses denote non-perovskite

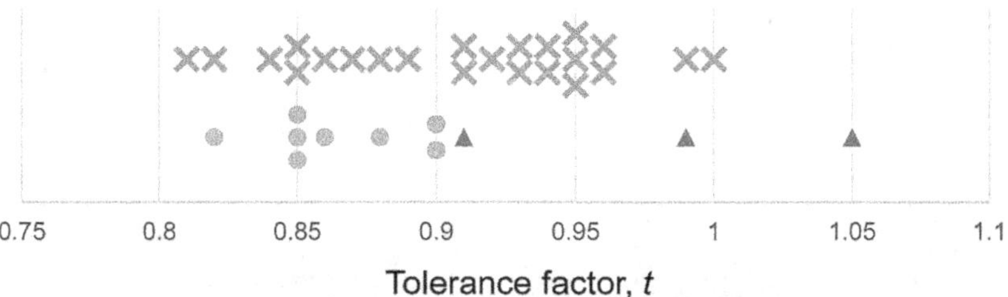

Figure 10.2. Tolerance factors of ABI_3 compounds. Blue dots indicate inorganic ABI_3 compounds that form perovskites under ambient conditions, while red crosses mark those that do not adopt the perovskite structure. Blue triangles represent hybrid $APbI_3$ and $ASnI_3$ compounds forming in the perovskite phase, where molecular ion radii were estimated using the approach of Cheetham *et al* [12, 13]. Notably, no strict tolerance factor boundary exists that clearly separates perovskite-forming from non-perovskite compounds. (Reproduced from [11]. CC BY 3.0.)

structures. Notably, all 32 compounds fall within the range of $0.8 \leqslant t \leqslant 1$, demonstrating that this criterion alone is insufficient for reliably predicting or explaining structural stability [11].

By mapping the stability landscapes, phase diagrams guide processing strategies —from selecting annealing temperatures to choosing additive formulations—ensuring the formation of stable, high-performing perovskite films for solar applications.

10.3 Process–property–performance correlations in perovskites

The development of perovskite materials, especially for photovoltaics and optoelectronic devices, strongly depends on the close relationship between processing conditions, material properties, and overall device performance. Small changes in fabrication routes, such as solvent choice, deposition technique, annealing temperature, or additive engineering, directly influence film morphology, defect density, crystallinity, and stability (discussed in chapter 4). These material properties, in turn, govern the performance metrics such as power conversion efficiency (PCE), carrier lifetime, and operational durability. Understanding these process–property–performance (P–P–P) correlations is therefore essential to optimize scalable fabrication methods while ensuring reproducibility and long-term stability, ultimately bridging the gap between laboratory-scale demonstrations and commercial applications [15–17].

P–P–P correlations are important to understand the influence of fabrication choices on the outcome of perovskite solar cells. In roll-to-roll (R2R) manufacturing, for example, different strategies such as laser scribing or strip registration directly affect the structural properties of the module, including layer uniformity and interconnection accuracy. These properties then determine performance metrics such as geometric fill factor, efficiency, and stability. Thus, optimizing the process not only improves material quality but also ensures reproducibility, scalability, and the commercial viability of perovskite devices (figure 10.3).

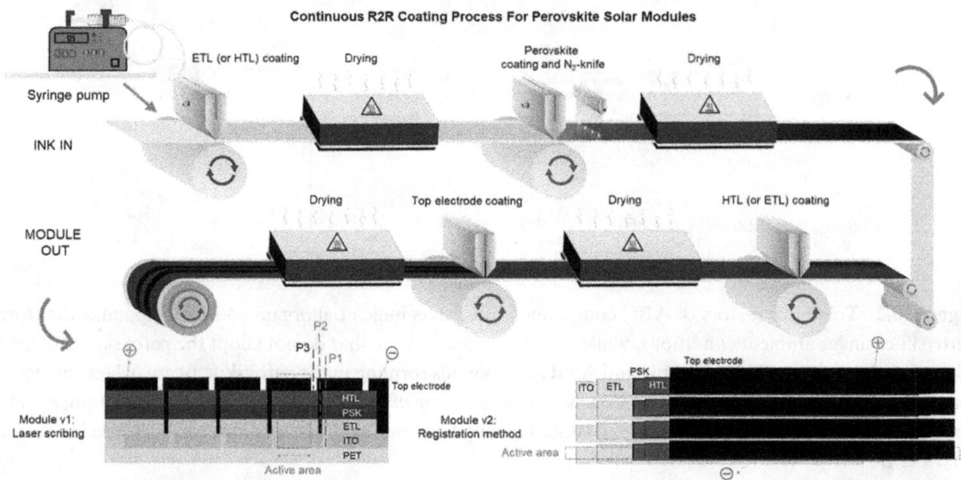

Figure 10.3. Schematic of R2R fabrication for perovskite solar modules using two approaches: laser scribing and registration. In the laser scribing route, an R2R coater equipped with a laser patterns the transparent conductive oxide (P1), the layer below the perovskite (P2), and the top electrode (P3) to form interconnections. In contrast, the registration method relies on strip-coating, where each layer is precisely aligned through a registration system to ensure accurate stacking and interconnections. While registration enables high geometrical fill factors, it requires more meticulous control compared to laser-based patterning. PET—polyethylene terephthalate; ITO—indium tin oxide; N_2—nitrogen; PSK—perovskite; ETL—electron transport layer; HTL—hole transport layer. (Reproduced from [18]. CC BY 4.0.)

10.4 Global perovskite photovoltaic developments

Table 10.4.1 presents a summary of the pilot/pilot-ready (industrial scale, R2R, pre-commercial) efforts in countries across the world.

10.4.1 Pilot/pilot-ready (industrial scale, R2R, pre-commercial)

Table 10.4.1.

Name	Country/region	Focus/product	Status	References
Oxford PV	UK/Germany	Perovskite–Si tandem solar cells	100 MW pilot line > 28% tandem cells	[19]
Saule technologies	Poland	Flexible perovskite films for IoT and BIPV	Small-scale production since 2021	[20]
Microquanta	China	Large-area perovskite modules (20 × 20 cm^2, 17.3%)	Pilot line (100 MW)	[21]
GCL Perovskite	China	Tandem modules targeting GW-scale	Scaling pilot lines	[22, 23]
UtmoLight	China	Flexible and rigid perovskite PV	Pilot line, LONGi partner	[24]

Swift Solar	USA	High-efficiency tandem PV (spin-out from Stanford/ MIT/NREL)	Pilot development	[25]
Tandem PV	USA	Tandem perovskite–Si panels	Rooftop pilot installations	[26]
Solliance Consortium	EU (NL, BE, DE)	R2R perovskite modules (slot-die, screen printing)	Pilot prototyping line	[27]
Heliatek	Germany	Organic PV → hybrid perovskite–organic focus for BIPV	Pilot/demos	[28]
LONGi,	China	Perovskite/silicon TSCs		[29]
P3C Technology & Solutions	India (IIT Delhi incubated)	Glass and flexible perovskite modules (100 kW pilot)	Expanding to 1 MW	[30]
ART-PV (IIT Bombay/ NCPRE)	India	High-efficiency perovskite tandems (~29.8%)	Planning 2 MW pilot	[31]
SolarWindow Technologies	USA	Transparent PV coatings for windows and other surfaces and perovskite /Si TSCs		[32]
IMEC/ EnergyVille	Belgium	Perovskite tandem integration, module stability and scale-up demonstration.		[33]
Fraunhofer ISE (PV-TEC/ module-TEC	Germany	Pilot-line modules, tandem process development and module prototyping.		[34]
Swansea University— SPECIFIC/ R2R team	UK (Wales)	Fully printed/R2R printable perovskite demonstrations, carbon inks and R2R demos.		[35]

10.4.2 Lab prototyping/high-impact research groups

In table 10.4.2, lab prototyping/high-impact research groups are summarized.

Table 10.4.2.

Institution/group	Country	Focus/contribution	References
Ecole Polytechnique Fédérale de Lausanne (EPFL) (Professor Michael Grätzel)	Switzerland	First solid-state perovskite solar cell (2012); ongoing tandem R&D	[36]
Ulsan National Institute of Science and Technology (UNIST) (Professor Nam-Gyu Park)	South Korea	Record efficiencies, stability studies	[37]

(*Continued*)

Table 10.4.2. (*Continued*)

Institution/group	Country	Focus/contribution	References
King Abdullah University of Science and Technology (KAUST) (Professor Stefaan De Wolf, Professor Osman Bakr)	Saudi Arabia	Tandem perovskites (>33% efficiency in lab)	[38]
Interuniversity Microelectronics Centre (IMEC)/EnergyVille	Belgium	Perovskite tandem integration, module stability and scale-up demonstration.	[33]
Fraunhofer Institute for Solar Energy Systems (Fraunhofer ISE) (PV-TEC Module-TEC)	Germany	Pilot-line modules, tandem process development and module prototyping.	[34]
Tsinghua University	China	Large-area/flexible perovskite research and scalable processing (additive and crystallization engineering).	[39]
National Renewable Energy Laboratory (NREL)	USA	NREL maintains official efficiency records; device/module work, stability and scale-up support (perovskite program and patents).	[40, 41]
Commonwealth Scientific and Industrial Research Organization (CSIRO) and University of New South Wales (UNSW)	Australia	R2R printing and scalable deposition. Pilot R2R printing lines, fully printed flexible modules and scaled demonstrators.	[42]
University College London (UCL) and University of Cambridge (Professor Henry Snaith's collaborations)	UK	Pioneer of perovskite PV; Oxford PV spinout	[43]
National Center for Photovoltaic Research & Education (NCPRE)—Indian Institute of Technology (IIT) Bombay	India	National hub for perovskite R&D + tandem scaling	[31]
National Physical Laboratory (NPL) India Council for Scientific & Industrial Research (CSIR)	India	Stability, defect analysis, device engineering, flexible perovskites	[44, 45]
Nanyang Technological University (NTU)/(Singapore Eye Research Institute) SERI/NUS (Singapore)	Singapore	High-efficiency perovskite research, tandem prototyping and pilot collaborations in the region.	[46]
Sekisui Chemical	Japan	Mass production of perovskite solar films.	[47]

10.5 Examples of patent applications/patents issued

The following table represents examples of patent applications/patents issued. The complete list of patents on perovskites continues to be extremely dynamic and represents one of the most active areas of research in materials science and engineering, across the world.

Patent number/Publication date/Description/ Applicant(s)	Abstract
Patent No.: US 8.419 967 B2 Pub. date: 16 April 2013 Perovskite oxide, oxide other publications composition, oxide body, piezoelectric device Fujifilm Corporation, Tokyo (JP)	A perovskite oxide represented by a general expression, (Aa, Bb)(Cc, Dd, Xx)O_3. (where, A: an A-site element, A = Bi, $0 < a$, B: one or more types of A-site elements, $0 \leqq b < 1.0$, C: an B-site element, C = Fe, $0 < c < 1.0$, D: one or more types of B-site elements, $0 \leqq d < 1.0$, $0 < b + d$, X: one or more types of B-site elements, the average valence of which is greater than the average valence of C and D in chemical formula, $0 < x < 1.0$, (average valence of A-site in chemical formula) + (average valence of B-site in chemical formula) > 6.0, O: oxygen, and standard molar ratio among A-site elements, B-site elements, and oxygen is 1:1:3, but it may deviate from the standard within a range in which a perovskite structure is possible.).
Patent No.: US 2020/0240000 A1 Pub. date: 30 July 2020 Mxene layers as substrates for growth of highly oriented perovskite thin films Drexel University, Philadelphia, PA (US)	The present disclosure is directed to using MXene compositions as templates for the deposition of oriented perovskite films, and compositions derived from such methods. Certain specific embodiments include methods preparing an oriented perovskite, perovskite-type, or perovskite-like film, the methods comprising: (a) depositing at least one perovskite, perovskite-type, or perovskite-like composition or precursor composition using chemical vapor deposition (CVD), physical vapor deposition (PVD), or atomic layer deposition (ALD) onto a film or layer of a MXene composition supported on a substrate to form a layered composition or precursor composition; and either (b) (1) heat treating or annealing the layered precursor composition to form a layered perovskite-type structure comprising at least one

(*Continued*)

(*Continued*)

Patent number/Publication date/Description/ Applicant(s)	Abstract
	oriented perovskite, perovskite-type, or perovskite-like composition; or (2) annealing the layered composition; or (3) both (1) and (2).
Patent No.: US 2009/0295876A1 Pub. date: 3 December 2009 Ferroelectric oxide structure, method for producing the structure, and liquid-discharge apparatus Hiroyuki Kobayashi Yukio Sakashita	A ferroelectric oxide structure includes a substrate and a ferroelectric thin-film deposited on the substrate. The ferroelectric thin-film has a thickness of greater than or equal to 200 nm and a tetragonal crystal system. The ferroelectric thin-film has (100) single-orientation crystal orientation.
Patent No.: US 10 796 857 B2 Pub. date: 6 October 2020 Inorganic/organic hybrid perovskite compound film, and method for manufacturing same Jang Won Seo, Seoul (KR); Jun Noh, Daejeon (KR); Nam Joong Jeon, Gwang	The present invention relates to an inorganic/organic hybrid perovskite compound film. An inorganic/organic hybrid perovskite compound film according to the present invention is polycrystalline, and has a discontinuous (100) plane scattering intensity on a grazing incidence wide angle x-ray scattering (GIWAXS) spectrum obtained using an x-ray wavelength of 1.0688 Å.
Patent No.: US 9 187 842 B2 Pub. date: 17 November 2015 Oriented perovskite oxide thin film Tomohiko Nakajima, Ibaraki (JP); Tetsuo Tsuchiya, Ibaraki (JP): Takaaki Manabe, Ibaraki (JP)	A thin film which comprises an organic metal salt or an alkoxide salt or an amorphous thin film is formed on a substrate, wherein each of the thin films enables the formation of a Dion–Jacobson perovskite-type metal oxide represented by the composition formula A(B. M.O.) (wherein n is a natural number of 2 or greater; A represents one or more monovalent cations selected from Na, K, Rb and Cs; B comprises one or more components selected from a trivalent rare earth ion, Bi, a divalent alkaline earth metal ion and a monovalent alkali metal ion; and M comprises one or more of Nb and Ta; wherein a solid solution may be formed with Ti and Zr) on a non-oriented substrate. The resulting product is maintained at the temperature between room temperature and 600 °C.; and crystallization is achieved while irradiating the amorphous thin film or the thin film comprising the organic metal salt or the alkoxide salt on the substrate with

ultraviolet light such as ultraviolet laser. In this manner, it becomes possible to produce an oriented Dion–Jacobson perovskite type oxide thin film characterized in that thin film can be oriented on the substrate in a (001) direction.

Pub. No.: US 2019/0173025 A1
Pub. date: 6 June 2019
Solar cell, photoabsorber layer, and forming method of photoabsorber layer
Ryuusuke Uchida. Osaka (JP); Michio Suzuka, Kyoto (JP)

A solar cell comprises a first electrode, a second electrode opposite to the first electrode, and a photoabsorber layer located between the first electrode and the second electrode and including a first layer and a second layer. The first layer contains a first compound which has a perovskite structure represented by the composition formula $A' M' X' z$, where A' is a monovalent cation, M' is a divalent cation, and X' is a halogen anion. The second layer contains a second compound which has a perovskite structure represented by the composition formula $A^2MPX^2 z$, where A^2 is a monovalent cation, M^2 is a divalent cation, and X^2 is a halogen anion, and has a different composition from the first compound. At least one of the first compound in the first layer and the second compound in the second layer has a single orientation.

Patent No.: US 7 795 170 B2
Pub. date: 14 September 2010
Catalyst
Hirohisa Tanaka, Shiga (JP); Isao Tan, Shiga (JP); Mari Uenishi, Shiga (JP): Masashi Taniguchi, Shiga (JP); Mareo Kimura, Shizuoka (JP); Satoshi Matsueda, Shizuoka (JP); Akimasa Hirai, Shizuok

Provided is a catalyst with a noble metal efficiently supported on the surfacemost thereof. A composite oxide-containing layer is formed on a catalyst carrier so as to contain a perovskite-type composite oxide represented by the following general formula (1) and an other composite oxide, and a noble metal layer is furtherformed on the catalyst carrier so as to be supported on the surfacemost of the catalyst carrier by immersing the catalyst carrier formed with the composite oxide-containing layer in an aqueous noble metal salt solution to impregnate the catalyst carrier with the aqueous noble metal salt solution: ABO_3. (1) (wherein A represents at least one element selected from rare earth elements and alkaline earth metals; B represents at least one element selected from transition elements

(*Continued*)

(*Continued*)

Patent number/Publication date/Description/ Applicant(s)	Abstract
	(excluding rare earth elements); X represents an atomic ratio of less than 1; *y* represents an atomic ratio of 1.0; and O represents an oxygen excess or an oxygen deficiency.)
Patent No.: US 8 471.442 B2 Pub. date: 25 June 2013 Piezoelectric ceramic, method for producing same, and piezoelectric device Keiichi Hatano, Takasaki (JP); Yutaka Doshida, Takasaki (JP)	Disclosed is a piezoelectric ceramic which is characterized by containing Ki–Na, LiNb–TaSb, IO (wherein *x*, *y*, *Z* and *w* each represents a molar ratio and satisfies 0 sXs 1, 0.sys 1, 0s Zs 1, Osws 1) as the main phase and KNbOSiO, as a sub-phase, while containing, as an additive, a Cu compound in an amount of 0.02–5.0 mol in terms of CuO relative to 100 mol of the main phase.
Patent No.: US 9 656 878 B2 Pub. date: 23 May 2017 Perovskite related compound Masatomo Yashima, Kanagawa (JP); Kotaro Fujii, Kanagawa (JP); Kazuki Omoto, Tokyo (JP); Yuichi Esaki, Tokyo (JP); Chihiro Saito, Kanagawa (JP)	Perovskite related compound of the present invention have layered structures in which perovskite units and A-rare earth structure units are alternately arranged. The reduced cell parameters a–c, and C–*Y*, and the reduced cell volume *V*, are within the following ranges: a 6.05 + 0.6 A. b = 8.26 + 0.8 A. c = 9.10 + 0.9 A, C, 103.4 + 10, B, 90 + 10, *Y* = 90 + 10, and *V*, 442.37 + 67 A. At least one of the reduced cell parameters a–c, can be *m*/*n* times as large as the aforementioned values, where *m* and *n* are independent natural numbers, the square roots of 2 or 3 or integral multiples thereof. Values of a, b, and c, can be replaced with one another.
Patent No.: US 9 082.551 B2 Pub. date: 14 July 2015 High dielectric nanosheet laminate, high dielectric element and method for producing the same Minoru Osada, Ibaraki (JP); Yasuo Ebina, Ibara	A high dielectric nanosheet laminate is produced by laminating nanosheets, each of which has a thickness of 10 nm or less and is formed of an oxide that has a perovskite structure wherein at least four NbO_6 octahedrons, TaO_6 octahedrons or TiO_6 octahedrons are included in a unit lattice. Consequently, the high dielectric nanosheet laminate can achieve a high dielectric constant and a satisfactory insulation property, which are preferable for high dielectric nanosheet multilayer capacitors or

the like, at the same time even if formed very thin.

Patent No.: US 8 100 513 B2
Pub. date: 24 January 2012
Ferroelectric film, process for producing the same, ferroelectric device, and liquid discharge device
Takami Arakawa, Kanagawa-ken (JP)

A ferroelectric film having a columnar structure constituted by a plurality of columnar grains, and containing as a main component a perovskite oxide which has a composition expressed by a compositional formula Als (Zr,Ti,), M. O, where A represents one or more A-site elements including lead (Pb) as a main component, M represents one or more of vanadium (V), niobium (Nb), tantalum (Ta), and antimony (Sb) as one or more B-site elements, zirconium (Zr) and titanium (Ti) are also B-site elements.

Patent No.: US 9 876 235 B2
Pub. date: 23 January 2018
Fuel cell including cathode (58) having main phase including perovskite oxide and sub phase
Ayano Kobayashi, Nagoya (JP); Makoto Ohmori, Nagoya (JP)

A fuel cell includes an anode, a cathode and a solid electrolyte layer that is disposed between the anode and the cathode. The cathode includes a main phase and a sub phase. The main phase is composed mostly of perovskite oxide which is expressed by the general formula ABO, and includes at least Sr at the A site. The sub phase is composed mostly of strontium sulfate. An occupied area ratio of the sub phase in a cross section of the cathode is no more than 10.2 %.

Patent No.: US 2009/0023580 A1
Pub. date: 22 January 2009
Catalyst composition
Hirohisa Tanaka, Shiga (JP); 4 March 2005 (JP). 2005–061008 Kiyotomi Mitsumori, Shiga (JP); 16 June 2005 (JP) 2005–177047 Isao Tan, Shiga (JP); Ichiro Takahashi, Shiga (JP); Mari Uenishi, Shiga (JP); Nobuhiko Kajita, Shiga (JP); Masashi Taniguchi, Shiga (JP); Kazuya Naito, Shiga (JP); Mareo Kimura, Shizuoka (JP); Keiichi Narita, Shizuoka (JP); Akimasa Hirai, Shizuoka (JP); Hiromasa Suzuki, Shizuoka (JP); Satoshi Matsueda, Shizuoka (JP); Hiroki Nagashima, Shizuoka (JP); Yoshinori Ishii, Shizuoka (JP); Norihiko Aono, Shizuoka (JP)

An object of the present invention is to provide a catalyst composition containing a perovskite-type composite oxide which exhibits a satisfactory catalytic performance over a long time even in a high temperature atmosphere and has a stable quality in which Rh and/or Pt dissolves to form a solid solution at a high rate. To achieve the object described above, in the present invention the catalyst composition is prepared to comprise an Rh containing perovskite-type composite oxide represented by the following general formula (I) and/or a Pt-containing perovskite-type composite oxide represented by the following general formula (II) and a thermostable oxide optionally containing a noble metal.

(Continued)

(*Continued*)

Patent number/Publication date/Description/ Applicant(s)	Abstract
Patent No.: US 2021/0366662 A1 Pub. date: 25 November 2021 Solar cell module Tamotsu Horiuchi, Shizuoka (JP); Nozomu Tamoto, Shizuoka (JP); Takahiro Ide, Shizuoka (JP); Yuuji Tanaka, Shizuoka (JP); Naomichi Kanei, Shizuoka	A solar cell module (100) including: a substrate (1); and a plurality of photoelectric conversion elements disposed on the substrate (1), each of the plurality of photoelectric conversion elements including a first electrode (2a, 2b), an electron transport layer (3, 4), a perovskite layer (5), a hole transport layer (6), and a second electrode (7a, 7b), wherein, within at least two of the photoelectric conversion elements adjacent to each other, the hole transport layers (6) are continuous with each other, and the first electrodes (2a, 2b), the electron transport layers (3, 4), and the perovskite layers (5) are separated by the hole transport layer (6) within the at least two of the photoelectric conversion elements adjacent to each other.
Patent No.: US 12 014 885 B2 Pub. date: 18 June 2024 Molecular doping enabled scalable blading of efficient hole transport layer-free perovskite solar cells Jinsong Huang, Chapel Hill, NC (US); Wuqiang Wu, Chapel Hill, NC (US)	A method of forming a photoactive device includes steps of: forming a photoactive layer, the photoactive layer comprising a perovskite material and a dopant; wherein the photoactive device comprises a positive electrode and a negative electrode; wherein said photoactive layer is directly or indirectly in electronic communication with the positive electrode and directly or indirectly in electronic communication with the negative electrode; and wherein the photoactive device is free of a hole transport layer between the photoactive layer and the positive electrode.
Patent No.: US 11 626 527 B2 Pub. date: 11 April 2023 Solar cell Tomoyasu Yokoyama, Osaka (JP); Kenji Kawano, Osaka (JP); Yumi Miyamoto, Tottori (JP)	Provided is a solar cell including a first electrode, a second electrode, a light-absorbing layer located between the first electrode and the second electrode, and an intermediate layer located between the light-absorbing layer and at least one electrode selected from the group consisting of the first electrode and the second electrode. The light-absorbing layer contains a perovskite compound represented by a chemical formula ASnX, (where A is a monovalent cation and X is a halogen anion). The

intermediate layer is in contact with the light-absorbing layer. The at least one electrode selected from the group consisting of the first electrode and the second electrode has light-transmissive property. The intermediate layer contains at least one selected from the group consisting of (4-(1′,5′-dihydro-1′—methyl-2′H—[5, 6] fullereno-C60-Ih[1,9-c]pyrrol-2′-yl) benzoic acid) and fullerene C60.

Patent No.: US 11 837 971 B2
Pub. date: 5 December 2023
Systems for driving the generation of products using quantum vacuum fluctuations
Garret Moddel, Boulder, CO (US)

Described herein are systems incorporating a Casimir cavity, such as an optical Casimir cavity or a plasmon Casimir cavity. The Casimir cavity modifies the zero-point energy density therein as compared to outside of the Casimir cavity. The Casimir cavities are paired in the disclosed systems with product generating devices and the difference in zero-point energy densities is used to directly drive the generation of products, such as chemical reaction products or emitted light.

Patent No.: US 12 274 109 B2
Pub. date: 8 April 2025
Methods of making semiconductor perovskite layers and compositions thereof
Jing-Shun Huang, Pasadena, CA (US); John Iannelli, San Marino, CA (US); Jiunn Benjamin Heng, Los Altos Hills, CA (US); Jeong Chul Lee, Monrovia, CA (US); Liam Sohngen, Pasadena, CA (US); Brian D Hunt, La Crescenta, CA (US); Eric W Wong, Los Angeles, CA (US); Chenyu Chou, Pasadena, CA (US)

The present disclosure may provide semiconductor perovskite layers and method of making thereof. In some cases, the perovskite layer may comprise a composition of MA, FA,2Cs, 3PbX3. MA may be methylammonium, FA may be formamidinium, *n*1, *n*2, and *n*3 may independently be greater than 0 and less than 1, and $n1 + n2 + n3$ may equal 1.

Patent No.: US 2022/0064 524 A1
Pub. date: 3 March 2022
Stabilized perovskite quantum dot material
Lutfan Sinatra, Thuwal (SA); Marat Lutfullin, Southampton (GB); Osman Mohammed BAKR, Jeddah (SA)

Described herein are materials comprising (1) a monomer or a polymer; (2) perovskite quantum dots interspersed in the monomer or the polymer, each of the perovskite quantum dots independently having the formula:

Csa(MA)b(FA)cRbdPbpSnrBisClxBryIz,

wherein:

MA is CH_3NH_3;

FA is $HC(NH_2)_2$;

(Continued)

(*Continued*)

Patent number/Publication date/Description/ Applicant(s)	Abstract
	a, b, c, and d are each independently a number from 0 to 1, provided that the sum of a, b, c, and d is 1; p, r, and s are each independently a number from 0 to 1, provided that the sum of p, r, and s is 1; and x, y, and z are each independently a number from 0 to 3, provided that the sum of x, y, and z is 3; and (3) an additive interspersed in the monomer or the polymer, the additive comprising: a halide-based additive; a light scattering agent having the formula: M_2O_3, wherein M is, at each occurrence, independently, a metal, provided that at most one instance of M is a group 13 element; or both. Also described are devices comprising such materials, as well as methods of forming such materials.
Patent No.: US 11 591 516 B2 Pub. date: 28 February 2023 Luminescent group 1 a copper halides and uses thereof Bayram Saparov, Norman, OK (US); Tielyr D Creason, Norman, OK (US); Rachel M Roccanova, Frisco, TX (US); Aymen Yangui, Norman, OK (US)	High photoluminescence, high stability, inorganic perovskite compounds comprising an alkali metal selected from potassium (K), rubidium (Rb), and cesium (Cs), copper (Cu); and at least one halogen selected from chlorine (C1), bromine (Br), and iodine (I). The perovskites may be free of lead (Pb). The inorganic perovskite compound may be used in an optoelectronic device. The optoelectronic device optionally contains a phosphor such as a blue-emitting phosphor. The inorganic perovskite compound may be used as an anti-counterfeiting nanotaggant applied on or within an object that susceptible to counterfeiting to enable confirmation of an authentic object.
Patent No.: US 2018/0166 741 A1 Pub. date: 14 June 2018 Composite membrane comprising solid electrolyte, method of making said composite membrane, and electrochemical cell comprising said composite membrane Hui Xu, Acton, MA (US); Jarrod D Milshtein, Cambridge, MA (US); Katherine Harrison,	A composite membrane that is suitable for use in an electrochemical cell, an electrochemical cell including the composite membrane, and a method of making the composite membrane. In one embodiment, the composite membrane includes a porous support and a solid electrolyte. The porous support is a unitary structure made of a polymer that is non-conductive to ions. The porous support is

Arlington, MA (US); Mario Moreira, Hudson, MA (US)

shaped to include a plurality of straight-through pores. The solid electrolyte has alkali ion conductivity and preferably completely fills at least some of the pores of the porous support. A variety of techniques may be used to load the solid electrolyte into the pores. According to one technique, the solid electrolyte is melted and then poured into the pores of the porous support. Upon cooling, the electrolyte re-solidifies, forming a monolithic structure within the pores of the porous support.

Patent No.: US 9 394.214 B2
Pub. date: 19 July 2016
Oxygen storage and production of C hydrocarbons
Juan D Henao, Houston, TX (US); Paul F Keusenkothen, Houston, TX (US)

Disclosed are reactors and reaction processes for contacting hydrocarbon reactant in the presence of oxygen stored and released within a thermal mass region of the reactor, and catalytically converting at least a portion of alkane, e.g., methane, in the hydrocarbon reactant to produce a reaction mixture comprising a Cs composition. Oxygen storage and release for carrying out the catalytic conversion is achieved by including an oxygen storage material in a thermal mass region of the reactor. Flow-through reactors can be used to carry out oxygen storage and the hydrocarbon conversion reactions. Reverse-flow reactors are examples of flow-through reactors, which can be used to carry out oxygen storage and the hydrocarbon conversion reactions.

Patent No.: US 2010/0227 758 A1
Pub. date: 9 September 2010
Alumina-based perovskite catalysts and catalyst supports
Xiaolin D Yang, Edison, NJ (US); Doan Lieu, Cedar Crove, NJ (US); Qingyuan Hu, East Brunswick, NJ (US)

An alumina-based perovskite is formed by mixing a lanthanide source with a transitional alumina to form a dual phase composition comprising *in situ* formed LnAIO dispersed in alumina. A second metal can be also included to form LaMO perovskite on alumina. The lanthanide content of the composition may range from about 6–35 wt.%, and the second metal from about 0.5–20 wt.%, to yield a high surface area composition which is useful as a catalyst or catalyst support such as for precious metals.

Patent No.: US 10 950 794 B2
Pub. date: 16 March 2021
Methods for forming a perovskite solar cell

A perovskite thin film and method of forming a perovskite thin film are provided. The perovskite thin film includes a substrate, a

(*Continued*)

(*Continued*)

Patent number/Publication date/Description/ Applicant(s)	Abstract
Thad Druffel, Louisville, KY (US); Brandon Lavery, Hamburg, NY (US); Krishnamraju Ankireddy, Louisville, KY (US); Amir Hossein Ghahremani, Louisville, KY (US); Blake Martin, Louisville, KY (US); Gautam Gupta, Prospect, KY (US)	hole blocking/electron transport layer, and a sintered perovskite layer. The method of forming the perovskite solar cell includes depositing a perovskite layer onto a substrate and processing (for example, by sintering) the perovskite layer with intense pulsed light to initiate a radiative thermal response that is enabled by an alkyl halide additive.
Patent No.: US 2017/0133 163 A1 Pub. date: 11 May 2017 Perovskite-containing solar cells comprising fulleropyrrolidine interlayers Thomas P Russell, Amherst, MA (US); Todd Emrick, South Deerfield, MA (US); Yao Liu, Amherst, MA (US); Zachariah A Page, Goleta, CA (US)	Perovskite-containing solar cells are described herein. An inverted perovskite solar includes an anode substrate, a photoactive layer including a perovskite, a hole transport layer disposed between the anode substrate and the photoactive layer, an electron transport layer, a metal cathode layer, and an interlayer disposed between the electron transport layer and the metal cathode layer. A tandem solar cell includes a first sub-cell, a second sub-cell, and an interconnecting layer disposed between the first sub-cell and the second sub-cell. The first subcell includes a perovskite layer having a thickness of 50–200 nanometers. The second sub-cell includes a photoactive layer and an interlayer disposed on the photoactive layer. The interlayers and the interconnecting layer each include a fullerpyrrolidine having a structure as defined herein.
Patent No.: US 2022/0231 233 A1 Pub. date: 21 July 2022 Perovskite solar cells with near-infrared sensitive layers Jinsong Huang, Chapel Hill, NC (US); Yuze Lin, Chapel Hill, NC (US); Shangshang Chen, Chapel Hill, NC (US)	The present disclosure is directed to perovskite-based solar cell device structures and compositions comprising one or more near infrared sensitive semiconducting materials. The near infrared sensitive semiconducting materials can extend the photoresponse spectra of the devices to the near infrared region, thereby improving the power conversion efficiency of the solar cell.
Patent No.: US 11 515 491 B2 Pub. date: 29 November 2022	Stable perovskite films having substantially no phase transition within a predetermined

Perovskite film, method for producing the same, light-emitting device and solar cell
Chuanjiang Qin, Fukuoka (JP); Toshinori Matsushima, Fukuoka (JP); Chihaya Adachi, Fukuoka (JP)

temperature range are disclosed. In the films, formation of carrier traps is suppressed. Thermally stable perovskite solar cells and light-emitting devices using the films are also disclosed.

Patent No.: US 11 452 970 B2
Pub. date: 27 September 2022
Calcium cobalt zirconium perovskites as oxygen-selective sorbents for gas separation
Qinghe Zheng, Durham, NC (US); Marty Lail, Raleigh, NC (US)

The present disclosure provides methods to use calcium cobalt zirconium perovskites as oxygen-selective sorbents for the separation of oxygen from a gas mixture such as air. Systems and high temperature oxygen detectors are also provided. In a preferred embodiment, the perovskite is configured as a membrane.

Patent No.: US 2025/0113 722 A1
Pub. date: 3 April 2025
Surface treatments for perovskite films to improve efficiency/stability of resulting solar cells
Jinsong Huang, Chapel Hill, NC (US); Ye Liu, Chapel Hill1, NC (US); Yuze Lin, Chapel Hill, NC (US)

Described herein are surface treatment methods for removing one or more surface defect layers from polycrystalline films, polycrystalline films that are free of one or more defect surface layers, and use of the films in solar cells. In certain embodiments, the method is conducted by means of an adhesive tape or mechanical polishing. As described herein, solar cells containing the surface treated perovskite films show enhanced efficiency and stability.

Patent No.: US 8 138 673 B1
Pub. date: 20 March 2012
Radiation shielding
Carol Ann Wedding, Toledo, OH (US)

Apparatus and method using a gas discharge device for shielding an object and/or person from electromagnetic (EM) radiation including radar, microwaves, x-rays, and/or gamma rays. The device comprises multiple gas discharge cells, each cell being within a gas-filled hollow shell. The gas is selected to absorb radiation particularly when the gas is in a discharge state. The shell may be composed of a radiation absorption material.

Patent No.: US 2021/0113 996 A1
Pub. date: 22 April 2021
Promoted mixed oxides for 'low-temperature ' methane partial oxidation
Fanxing LI, Raleigh, NC (US); Arya Shafiefarhood, Raleigh, NC (US)

Redox catalysts are provided for 'low-temperature' methane partial oxidation in absence of gaseous oxidants. Methods of converting the methane to syngas using the catalysts are also provided. In some aspects, the conversion takes place at temperatures of about 400 °C to about 950 °C. The methods can be used to convert methane to syngas containing carbon monoxide and hydrogen

(*Continued*)

(*Continued*)

Patent number/Publication date/Description/ Applicant(s)	Abstract
	gas. In some aspects, the methods are carried out in a fixed bed reactor with reverse flow.
Patent No.: US 9 013 102 B1 Pub. date: 21 April 2015 Radiation detector with tiled substrates Daniel K Wedding, Toledo, OH (US); Arens Carol Ann Wedding, Toledo, OH (US); King Oliver M Strbik, III, Holland, OH (US)	Tiled substrates containing hollow gas filled plasma-shells for radiation detecting or sensing. The gas filled plasma shells are placed on or within the surface of a substrate which may be a printed circuit board. Multiple substrates containing plasma-shells are tiled together edge to edge to form a self supporting structure such as a dome or hemisphere for radiation detection.
Patent No.: US 11 848 396 B2 Pub. date: 19 December 2023 Surfactant additive for solution coating large-area high-efficiency perovskite solar cells and other devices Jinsong Huang, Lincoln, NE (US); Yehao Deng, Chapel Hill, NC (US)	In an aspect, a method of forming a photoactive device comprises: providing a perovskite–surfactant solution, said perovskite–surfactant solution comprising a perovskite ink and a surfactant; and coating said perovskite–surfactant solution onto a receiving surface of a substrate thereby forming a layer of said photoactive device; wherein said layer comprises a perovskite material; and wherein an active area of said photoactive device is at least 1 cm^2.
Patent No.: US 2020/0048 106 A1 Pub. date: 13 February 2020 Redox-active oxide materials for thermal energy storage Sean M Babiniec, Arvada, CO (US); Andrea Ambrosini, Albuquerque, NM (US); Eric N Coker, Albuquerque, NM (US); James E Miller, Albuquerque, NM (US)	Thermochemical storage materials having the general formula A, A.B.B′03–8, where A = La, Sr, K, Ca, Ba, Y and B = Mn, Fe, Co, Ti, Ni, Cu, Zr, Al, Y, Cr, V, Nb, Mo, are disclosed. These materials have improved thermal storage energy density and reaction kinetics compared to previous materials. Concentrating solar power thermochemical systems and methods capable of storing heat energy by using these thermochemical storage materials are also disclosed.
Patent No.: US 9 777 953 B2 Pub. date: 3 October 2017 Apparatus for thermally cycling an object including a polarizable material Ahmet Erbil, Atlanta, GA (US); David F Walbert, Atlanta, GA (US)	An apparatus includes a thermal chamber, a first reservoir containing a first liquid/vapor two-phase system, a second reservoir containing a second liquid/vapor two-phase system and conduits connecting the first reservoir and second reservoir to the thermal chamber. The first and second liquid vapor two-phase systems include a liquid phase and

a separate vapor phase. The apparatus also includes a conduit connecting the vapor phases of the first and second reservoirs. The apparatus can be used to thermally cycle an object placed in the thermal chamber or the vapor region of the first reservoir. The object can include one or more layers of an electrically or magnetically polarizable material.

Patent No.: US 2019/0276 696 A1 Pub. date: 12 September 2019 Formulations for printed electronic devices, preparation methods and uses thereof Junyou Pan, Guangzhou, Guangdong (CN); Xi Yang, Guangzhou, Guangdong (CN); Xiaolin Yan, Guangzhou, Guangdong (CN)	A formulation, which is used for forming a functional material thin film in a printed electronic component, comprises a functional material and a solvent, and the solvent can be vaporized from the formulation to form the functional material thin film. The solvent is formed by mixing at least two organic solvents including a first solvent and a second solvent. The first solvent and the second solvent are soluble with each other, the boiling point of at least one of the first solvent and the second solvent is higher than or equal to 160 °C, the boiling point of the first solvent is higher than that of the second solvent, the viscosity of the first solvent is higher than that of the second solvent, and a difference between the viscosity of the first solvent and the viscosity of the second solvent is at least 2 cPs. The solvent system containing at least two solvents can effectively dissolve the functional material without the need of adding an additive, and can also effectively prevent the occurrence of a 'coffee ring effect', and accordingly the thin film having a uniform thickness.
Patent No.: US 2008/0246116A1 Pub. date: 9 October 2008 Symmetrical programmable crossbar structure Blaise Laurent Mouttet, Arlington, VA (US)	A crossbar structure includes a first layer or layers including first p-type regions and first n-type regions, a second layer including second p-type regions and second n-type regions, and a resistance programmable material formed between the first layer(s) and the second layer(s), wherein the first layer(s) and the second layer(s) include first and second intersecting wiring portions forming a crossbar array.

(*Continued*)

(*Continued*)

Patent number/Publication date/Description/ Applicant(s)	Abstract
Patent No.: US 10 872 734 B2 Pub. date: 22 December 2020 Solvent-free, solid phase synthesis of hybrid lead halide perovskites with superior purity Javier Vela-Becerra, Ames, IA (US); Bryan A Rosales, Ames, IA (US)	A method of synthesizing a mixed-halide perovskite is disclosed herein. The method includes the steps of mixing a first single-halide perovskite and a second single-halide perovskite to form a solid phase mixture and heating the solid phase mixture at a temperature below a first decomposition temperature of the first single-halide perovskite and below a second decomposition temperature of the second single-halide perovskite for a time sufficient to form the mixed-halide perovskite. During the mixing, the first and second single-halide perovskite are both in the solid phase. A mixed-halide perovskite made according to the method is also disclosed herein. The mixed-halide perovskite is free of amorphous and/or semicrystalline phases. The mixed-halide perovskite can be utilized in a photovoltaic cell in a solar panel.
Patent No.: US 2024/0083 755 A1 Pub. date: 14 March 2024 CO2 hydrogenation catalysts for the commercial production of SYNGAS Dennis Schuetzle, Grass Valley, CA (US); Robert Schuetzle, Sacramento, CA (US); Anja Rumplecker Galloway, San Rafael, CA (US); Orion Hanbury, Sacramento, CA (US)	The present invention is generally directed to the production of low-carbon syngas from captured CO_2 and renewable H_2. The H_2 is generated from water using an electrolyzer powered by renewable electricity, or from any other method of low-carbon H, production. The improved catalysts use low cost metals, they can be produced economically in commercial quantities, and they are chemically and physically stable up to 2100 °F. CO_2 conversion is between 80% and 100% with CO selectivity of greater than 99%. The catalysts don't sinter or form coke when converting H_2:CO, mixtures to syngas in the operating ranges of 1300–1800 °F, pressures of 75–450 psi, and space velocities of 2000–100 000 h 1. The catalysts are stable, exhibiting between 0% and 1% CO conversion decline per 1000 h. The syngas can be used for the synthesis of low-carbon fuels and chemicals, or for the production of purified H_2. The H_2 can

be used at the production site for the synthesis of low-carbon chemical products or compressed for transportation use.

Patent No.: US 2020/0313 093 A1
Pub. date: 1 October 2020
Method for manufacturing perovskite solar cells and multijunction photovoltaics
Nitin P Padture, Providence, RI (US); Yuanyuan Zhou, Providence, RI (US)

A laminated structure is prepared by providing a first substrate having a n-type oxide layer on a first surface thereof and a second substrate having a p-type oxide layer on a first surface thereof. The first surface of the first substrate, the first surface of the second substrate, or both has a liquid halide layer thereon. The first substrate is pressed into contact with the second substrate such that the first surface of the first substrate contacts the first surface of the second substrate. The halide layer is then solidified to form the laminated structure.

Patent No.: US 2011/0183 834 A1
Pub. date: 28 July 2011
Modified perovskite type composite oxide, method for preparing the same, and composite dielectric material
Shinji Tanabe, Tokyo (JP)

Provided is a modified perovskite type composite oxide in which the dielectric characteristics are equal to or better than those prior to modification, there is no substantial elution of coating components from the modifying coating components, and elution of the A-site metals is suppressed effectively, while the cracking traits are good. A modified perovskite type composite oxide in which the particle surface of a perovskite type composite oxide is firstly coated with at least one selected from a group consisting of TiO, Al_2O, ZrO_2. and Nd_2O, wherein the first coating is formed by hydrolyzing at least one selected from a group consisting of a hydrolyzable TiO precursor, a hydrolyzable Al–O precursor, a hydrolyzable ZrO precursor, and a hydrolyzable Nd_2O precursor, and then calcining it at 700 °C–1200 °C.

Patent No.: US 2023/0247 845 A1
Pub. date: 3 August 2023
Lead-absorbing materials for the sequestration of lead in perovskite solar cells

Described herein are solar cells, comprising: an active layer comprising a perovskite composition, wherein the perovskite composition comprises lead; and, a lead-absorbing material. In certain embodiments,

(*Continued*)

(*Continued*)

Patent number/Publication date/Description/ Applicant(s)	Abstract
Jinsong Huang, Chapel Hill, NC (US); Shangshang Chen, Chapel Hill, NC (US)	the lead-absorbing material is an ion exchange material. The lead-absorbing material helps prevent lead leakage in damaged solar cells and solar modules under severe weather conditions.
Patent No.: US 2018/0321 393 A1 Pub. date: 8 November 2018 Perovskite-type halides and methods thereof Yuntao Wu, Knoxville, TN (US); Mariya Zhuravleva, Knoxville, TN (US); Luis Stand, Knoxville, TN (US); Charles L Melcher, Oak Ridge, TN (US)	Metal halide optical materials (e.g., scintillator materials or persistent phosphors) are described. More particularly, the optical materials include codoped perovskite-type halides, wherein the codoping ion is present at a molar ratio of 5000 parts per million (ppm) or less with respect to all cations. For example, the optical material can be a codoped trihalide having the formula ABX, where A is one or more alkali metal, B is one or more alkali earth metal, and X is one or more halide that is doped with up to about 10 atomic percent of a dopant ion and codoped with up to about 5000 ppm of one or more isovalent or aliovalent codopant ion, such as a tetravalent ion (e.g., Zr++), a trivalent ion (e.g., Sc3+, Y3+, GdB +, or La+ ion) or a divalent ion (e.g., Mg2+). The codoped material can have modified afterglow compared to a noncodoped material.
Patent No.: US 11 530 303 B2 Pub. date: 20 December 2022 Perovskite–polymer composites and methods Yajie Dong, Winter Springs, FL (US); Shin-Tson Wu, Oviedo, FL (US); Caicai Zhang, Orlando, FL (US); Ziqian He, Oviedo, FL (US)	Perovskite–polymer composites including perovskite nanocrystals dispersed in a polymer matrix, wherein the perovskite nanocrystals have an average size of from about nm to about 20 nm. Methods for producing a perovskite–polymer composites that may include contacting a solid material comprising a polymer matrix with a solution comprising a perovskite precursor; allowing the solution to penetrate the solid material to yield a swollen solid material comprising the perovskite precursor dispersed within the polymer matrix; optionally contacting the swollen solid material with an antisolvent; and annealing the swollen solid material to

crystallize the perovskite precursor perovskite nanocrystals dispersed in the polymer matrix.

Patent No.: US 10 177 705 B2
Pub. date: 8 January 2019
Composite solar cell, solar cell module, and concentrating solar cell
Hisashi Uzu, Settsu (JP); Mitsuru Ichikawa, Settsu (JP); Masashi Hino, Settsu (JP); Tomomi Meguro, Settsu (JP); Kenji Yamamoto, Settsu (JP)

A composite solar cell comprises a spectroscopic element, a first photoelectric conversion element, and a second photoelectric conversion element. The first photoelectric conversion element is positioned in a first direction of the spectroscopic element and the second photoelectric conversion element is positioned in a second direction of the spectroscopic element. The first photoelectric conversion element is a perovskite-type photoelectric conversion element containing, in a light absorbing layer, a perovskite crystal structure material represented by a general formula R 'NHZM' X3. A band gap of a light absorbing layer of the second photoelectric conversion element is narrower than the band gap of the light absorbing layer of the first photoelectric conversion element. The spectroscopic element preferentially outputs the short wavelength light of the incident light in the first direction and preferentially outputs the long wavelength light of the incident light in the second direction.

Patent No.: US 10 344 210 B2
Pub. date: 9 July 2019
Luminescent substance particle having perovskite-type structure
Tatsuya Hishiki, Nagoya (JP)

A luminescent substance particle including BaSnOz having a perovskite-type structure, wherein the luminescent substance particle contains one of 0.07% by mass or less of Fe (iron), 0.005% by mass or less of Cr (chromium) and 0.02% by mass or less of Ni (nickel). A wavelength conversion film including the luminescent substance particle for converting a light in an ultraviolet region to a light in an infrared region. A wavelength conversion device including a substrate and the wavelength conversion film formed on the substrate.

(*Continued*)

(*Continued*)

Patent number/Publication date/Description/ Applicant(s)	Abstract
Patent No.: US 10 115 494 B2 Pub. date: 30 October 2018 Composite body, honeycomb structural body, and method for manufacturing composite body Yunie Izumi, Nisshin (JP); Yoshimasa Kobayashi	A composite body including a substrate and a forming portion which is composed of a composite phase containing a perovskite oxide and a metal oxide different from the perovskite oxide and which is formed on the substrate. The composite body may be a composite body manufactured by a manufacturing method including a forming step of firing in an oxidizing atmosphere, a laminated body in which an inorganic raw material powder containing a compound powder and a metal powder is disposed on a substrate so as to form a forming portion composed of a composite phase containing a perovskite oxide and a metal oxide different from the perovskite oxide on the substrate.
Patent No.: US 2025/0104 934 A1 Pub. date: 27 March 2025 Ligand modified perovskite optoelectronic devices Chik Ho Wallace Choy, Hong Kong (CN); Hong Zhang, Hong Kong (CN)	A method of ligand-induced regional modification of a perovskite film of perovskite optoelectronic device can include generating a ligand atmosphere, exposing a perovskite optoelectronic device in the ligand atmosphere, and removing the perovskite optoelectronic device from the ligand atmosphere. Methods for improving the performance and stability of perovskite optoelectronic devices are performed by using a ligand-induced modification of complete devices at room temperature. This post-device treatment, completely separated from the fabrication process of common perovskite optoelectronic devices, provides a general strategy to improve the stability of different completed perovskite optoelectronic devices (i.e., perovskite solar cells, perovskite light-emitting diodes, and photodetectors) without introducing any undesirable impurities during device fabrication.
Patent No.: US 9 537 082 B2 Pub. date: 3 January 2017	In a piezoelectric ceramic which has an alkali-containing niobate-based perovskite structure in which constituent elements are

Piezoelectric ceramic, piezoelectric ceramic component, and piezoelectric device using such piezoelectric ceramic component
Keiichi Hatano, Takasaki (JP); Hiroyuki Shimizu, Takasaki (JP)

Li, Na, K, Nb, and O, a LiNbO crystal phase is intentionally deposited on a piezoelectric ceramic having an alkali-containing niobate-based perovskite structure, whereby sintering becomes possible at about 1000 °C rather than at the normally required sintering temperature of 1050 °C or greater.

Patent No.: US 2013/0200 457 A1
Pub. date: 8 August 2013
Strongly correlated oxide field effect element
Yasushi Ogimoto, Higashiyamato-City (JP)

Provided is a strongly correlated oxide field effect element demonstrating a phase transition and a switching function induced by electrical means. The strongly correlated oxide field effect element is a strongly correlated oxide field effect element 100 including a channel layer 2 constituted by a strongly correlated oxide film, a gate electrode 14, a gate insulating layer 31, a source electrode 42, and a drain electrode 43. The channel layer 2 includes an insulator-metal transition layer 22 of a strongly correlated oxide and a metallic state layer 21 of a strongly correlated oxide that are stacked on each other. The thickness t of the channel layer 2, the thickness t1 of the insulator-metal transition layer 22, and the thickness t2 of the metallic state layer 21 satisfy the following relationship with critical thicknesses t1c and t2c for respective metallic phases of the layers: $t - t1 + t22t1ct2c$, where $t1 < t2C$.

Patent No.: US 2014/0055 009 A1
Pub. date: 27 February 2014
Piezoelectric ceramic and piezoelectric element
Keiichi Hatano, Takasaki-shi (JP); Sumiaki Kishimoto, Takasaki-shi (JP); Yutaka Doshida, Takasaki-shi (JP)

A piezoelectric ceramic is constituted by a polycrystal whose main phase is an alkali-containing niobate perovskite structure, where both elemental nickel and elemental manganese are present at the grain boundary of the polycrystal. The piezoelectric ceramic is such that drop in piezoelectric characteristics due to application of high electric field is suppressed.

Patent No.: US 2021/0296 618 A1
Pub. date: 23 September 2021
Perovskite light emitting device with multiple emissive layers

A light emitting device is provided. The device comprises a first electrode, a second electrode and at least two emissive layers. A first emissive layer of the at least two emissive

(*Continued*)

(*Continued*)

Patent number/Publication date/Description/ Applicant(s)	Abstract
Peter Levermore, Sedgefield (GB)	layers is disposed over the first electrode. A second emissive layer of the at least two emissive layers is disposed over the first emissive layer. The first emissive layer is in contact with the second emissive layer. The second electrode is disposed over the second emissive layer. At least one emissive layer of the at least two emissive layers comprises a perovskite light emitting material. The device comprises at least one further emissive layer of the at least two emissive layers, wherein the at least one further emissive layer comprises a perovskite light emitting material, an organic light emitting material or a quantum dot light emitting material.
Patent No.: US 2020/0017 364 A1 Pub. date: 16 January 2020 Halide ABX3 perovskite particles and their application in controlling photo-flux Yanan Li, Montreal (CA); Dawei Zhang, Lachine (CA); Shiyong Zhao, Longueuil (CA); Shuyong Xiao, St-Laurent (CA)	The present invention provides a light valve containing ABX, perovskite particles; more specifically is related to a light valve containing halide ABXz perovskite particles that can control light transmittance. The preferable halide ABXz perovskite particles in this invention consist of A being at least one of Cs+, CH3NH37, and Rb+, B being at least one of Pb2+, Ge2+, and Sn2+, and X being at least one of C1−, Br′, and I′. This kind of halide ABXz perovskite particles were suspended in a liquid suspension to make a light valve with a light transmittance control, which discloses a completely new application for ABX, perovskite materials.
Patent No.: US 11 676 771 B2 Pub. date: 13 June 2023 Perovskite solar cell and method for manufacturing same Hyun Suk Jung, Seoul (KR); Taekyu Ahn, Seongnam-si (KR); Zhu Jun, Suwon-si (KR); Bonghyun Jo, Seoul (KR); Han Gill Sang, Gunpo-si (KR); Dong Hoe Kim, Seoul (KR); Jidong Kim, Anseong-si (KR)	A method for manufacturing a perovskite solar cell, includes disposing an electron transport layer on a transparent conductive substrate, disposing an additive-doped perovskite light absorption layer on the electron transport layer, disposing a hole transport layer on the additive-doped perovskite light absorption layer, and disposing an electrode on the hole transport layer. The disposing of the additive-doped perovskite light absorption layer includes adding an additive having

hydrophobicity to a perovskite precursor solution, and applying the additive-added perovskite precursor solution onto the electron transport layer to form the additive-doped perovskite light absorption layer.

Patent No.: US 2021/0104 686 A1
Pub. date: 8 April 2021
Materials and methods to improve publication classification the stability of metal halide perovskites
Yanfa Yan, Toledo, OH (US); Chongwen Li, Toledo, OH (US); Baicheng Weng, Toledo, OH (US)

Materials and methods for improving the stability of perovskites are described.

Patent No.: US 10 401 023 B2
Pub. date: 3 September 2019
Perovskite catalysts enhanced combustion on porous media
Nina Orlovskaya, Orlando, FL (US); Ruey-Hung Chen, Orlando, FL (US); Manuel Robayo, Oviedo, FL (US)

The effects of different perovskite catalysts, catalytic active materials with a crystal structure of ABO2, on matrix stabilized combustion in a porous ceramic media are explored. Highly porous silicon carbide ceramics are used as a porous media for a catalytically enhanced matrix stabilized combustion of a lean mixture of methane and air. A stainless steel combustion chamber was designed incorporating a window for direct observation of the flame within the porous media. Perovskite catalytic enhancement of SiC porous matrix with La0. 75Sr0. 25Fe0. 6Cr0. 35Ru0. 0503; La0. 75Sr0. 25Fe0. 6Cr0. 403; La0. 75Sr0. 25Fe0. 95Ru0. 0503; La0. 75Sr0. 25Cr0. 95Ru0. 0503; and LaFe0. 95Ru0. 0503, for example, were used to enhance combustion. The flammability limits of the combustion of methane and air were (56) explored using both inert and catalytically enhanced surfaces of the porous ceramic media. By coating the SiC porous media with perovskite catalysts it was possible to lower the minimum stable equivalence ratio.

Patent No.: US 2009/0208 396 A1
Pub. date: 20 August 2009

An alumina-based perovskite is formed by mixing a lanthanide source with a transitional alumina to form a dual-phase

(*Continued*)

(*Continued*)

Patent number/Publication date/Description/ Applicant(s)	Abstract
Alumina-based perovskite catalysts and catalyst supports Xiaolin David Yang, Edison, NJ (US); Doan Lieu, Cedar Grove, NJ (US)	composition comprising *in situ* formed $LnAlO_3$ dispersed in alumina. The lanthanide content of the composition ranges from about 12–24 wt. % to yield a high surface area composition which is useful as a catalyst or catalyst support such as for precious metals.
Patent No.: US 9 057 022 B2 Pub. date: 16 June 2015 Luminescent nanosheets Tadashi Ozawa, Tsukuba (JP); Takayoshi Sasaki, Tsukuba (JP)	A luminescent nanosheet has perovskite octahedral crystal units combined together in a planar configuration. The octahedral crystal units each have a multistacked crystal sheet structure. The octahedral crystal units are multistacked over at least three high in a direction vertical to a sheet plane, and an element providing a luminescence center is solid-solubilized between the multistacked octahedral crystal units.
Patent No.: US 2024/0090 316 A1 Pub. date: 14 March 2024 Mixtures and applications thereof in optoelectronic field Junyou Pan, Taizhou (CN); Jiahui Tan, Taizhou (CN)	Disclosed are mixtures including an organic compound H and a perovskite emitter E. Also provided are formulations containing the mixtures and at least one solvent. Further provided are organic light-emitting devices containing the mixtures.
Patent No.: US 2021/0036 249 A1 Pub. date: 4 February 2021 Systems and methods for bulk-semiconductor sensitized solid-state upconversion Lea Nienhaus, Tallahassee, FL (US); Sarah Wieghold, Tallahassee, FL (US)	Systems and methods for upconversion based on bulk semiconductor sensitizers are provided. In some aspects, issues with previous upconversion approaches are overcome using bulk-semiconductor thin films as sensitizers for the triplet state to achieve efficient upconversion based on triplet triplet annihilation. Varying the film thickness shifts the threshold of efficient upconversion to subsolar incident powers, enabling practical applications for solar energy harvesting. Systems and methods are provided for upconversion of light in a solid state electronic device, the methods including exposing a bulk semiconductor to a first light source comprising light of a first wavelength,

wherein the bulk semiconductor is associated with an organic material capable of upconversion via triplet–triplet annihilation from triplet states in the organic material, and observing light emitted from the organic material at a second wavelength, wherein the second wavelength is shorter than the first wavelength. A one-step synthesis of solid-state upconversion devices is also provided.

Patent No.: US 2023/0264 970 A1
Pub. date: 24 August 2023
Compositions comprising perovskite and non-perovskite
Masaru Kuno, South Bend, IN (US); Michael Brennan, South Bend, IN (US); Anthony Ruth, South Bend, IN (US); Ilia Pavlovetc, South Bend, IN (US); Jeffrey A Christians, Golden, CO (US); Taylor Hennessey Moot, Golden, CO (US); Joseph Matthew Luther, Golden, CO (US)

Compositions comprise a perovskite and a non-perovskite. Perovskites comprise AAA″BX3, and non-perovskites may comprise A″, B and *X*, where A is a first cation, A′ is a second cation, A″ is a third cation, B is a fourth cation. *X* is an anion. In some instances, A, A′, and A″ are each independently (NH_2), CH+, CH3NH, +, Cs+, Rb+, or (NH)(CNH)+, with the proviso that A, A′, and A″ are each different. The perovskite may have a first crystal structure in which the anion is corner-sharing, the non-perovskite may have a second crystal structure comprising at least one of an orthorhombic structure, a hexagonal structure, or a perovskite-like structure, and $1-x-$y may be greater than about 0.15.

10.6 Concluding remarks

In recent years, research highlights in materials science and engineering encompass significant global efforts to develop technologies relating to perovskites—from materials fundamentals to concept to design to product to prototyping to scaling to manufacturing. For example, the National Renewable Energy Laboratory maintains an extensive portfolio of patents on perovskites [48] that focus on eight technology areas relating to the commercialization of perovskite solar cells. Research institutions focusing on perovskite research include EPFL (Switzerland), Huazhong University of Science and Technology (China), and Korea Research Institute of Chemical Technology (South Korea). Some of the other major industry players are Fujifilm Holdings (Japan), LG Corp (South Korea), Microquanta Semiconductor (China), Oxford PV (UK), Sekisui Chemical (Japan), and Trina Solar (China). Once the key challenges relating to their durability, reliability, scalability, stability, toxicity and uniformity are fully addressed, perovskite-based devices such as LEDs, solar cells and perovskite-based photonics will transition to full scale manufacturing in countries across the world. For example, perovskite tandem solar cells present an excellent opportunity for the rapid growth of perovskites. Solar cell manufacturers such as Hanwha QCells, Longi and Trina Solar are making enormous investments in this endeavor. They are joined by Microquanta Semiconductor, Oxford PV and Tandem PV. The primary drivers and incentives continue to be along the lines of cost-effectiveness, enhanced efficiency and power density.

References

[1] Mohanty A *et al* 2019 Phase diagram and dielectric properties of $MA_{1-x}FA_xPbI_3$ *ACS Energy Lett.* **4** 2045–51

[2] Marin-Villa P *et al* 2025 Methylammonium lead iodide across physical space: phase boundaries and structural collapse *J. Phys. Chem. Lett.* **16** 184–90

[3] Whitfield P S *et al* 2016 Structures, phase transitions and tricritical behavior of the hybrid perovskite methyl ammonium lead iodide *Sci. Rep.* **6** 35685

[4] Chan Y T *et al* 2022 High-pressure investigations in CH_3NH_3PbX (X = I, Br, and Cl): suppression of ion migration and stabilization of low-temperature structure *Phys. Rev.* B **106** 214106

[5] Gesi K 1997 Effect of hydrostatic pressure on the structural phase transitions in $CH_3NH_3PbX_3$ (X = Cl, Br, I) *Ferroelectrics* **203** 249–68

[6] Capitani F *et al* 2016 High-pressure behavior of methylammonium lead iodide ($MAPbI_3$) hybrid perovskite *J. Appl. Phys.* **119** 185901

[7] Szafrański M and Katrusiak A 2016 Mechanism of pressure-induced phase transitions, amorphization, and absorption-edge shift in photovoltaic methylammonium lead iodide *J. Phys. Chem. Lett.* **7** 3458–66

[8] Ou T *et al* 2016 Visible light response, electrical transport, and amorphization in compressed organolead iodine perovskites *Nanoscale* **8** 11426–31

[9] Beecher A N *et al* 2016 Direct observation of dynamic symmetry breaking above room temperature in methylammonium lead iodide perovskite *ACS Energy Lett.* **1** 880–7

[10] Mazumdar S, Zhao Y and Zhang X 2021 Stability of perovskite solar cells: degradation mechanisms and remedies *Front. Electron.* **2** 712785

[11] Travis W *et al* 2016 On the application of the tolerance factor to inorganic and hybrid halide perovskites: a revised system *Chem. Sci.* **7** 4548–56

[12] Kieslich G, Sun S and Cheetham A K 2014 Solid-state principles applied to organic–inorganic perovskites: new tricks for an old dog *Chem. Sci.* **5** 4712–5

[13] Kieslich G, Sun S and Cheetham A K 2015 An extended tolerance factor approach for organic–inorganic perovskites *Chem. Sci.* **6** 3430–3

[14] Li C *et al* 2008 Formability of ABX_3 (X = F, Cl, Br, I) halide perovskites *Struct. Sci.* **64** 702–7

[15] Jeon N J *et al* 2014 Solvent engineering for high-performance inorganic–organic hybrid perovskite solar cells *Nat. Mater.* **13** 897–903

[16] Saliba M *et al* 2016 Incorporation of rubidium cations into perovskite solar cells improves photovoltaic performance *Science* **354** 206–9

[17] Abdi-Jalebi M *et al* 2018 Potassium- and rubidium-passivated alloyed perovskite films: optoelectronic properties and moisture stability *ACS Energy Lett.* **3** 2671–8

[18] Parvazian E and Watson T 2024 The roll-to-roll revolution to tackle the industrial leap for perovskite solar cells *Nat. Commun.* **15** 3983

[19] Jowett P 2024 Oxford PV starts commercial distribution of perovskite solar modules *PV Magazine* https://www.pv-magazine.com/2024/09/05/oxford-pv-starts-commercial-distribution-of-perovskite-solar-modules/

[20] Spaes J 2021 Saule Technologies opens perovskite solar cell factory in Poland *PV Magazine* https://www.pv-magazine.com/2021/06/17/saules-technologies-opens-perovskite-solar-cell-factory-in-poland/

[21] Shaw V and Hall M 2022 Chinese PV industry brief: Microquanta builds 12 MW ground-mounted project with perovskite solar modules *PV Magazine* https://www.pv-magazine.com/2022/02/18/chinese-pv-industry-brief-microquanta-builds-12-mw-ground-mounted-project-with-perovskite-solar-modules/

[22] Ng E 2024 China's GCL investing US$98 million in next-generation solar technology to revive sector *South China Morning Post Publishers* https://www.scmp.com/business/china-business/article/3273455/chinas-gcl-investing-us98-million-next-generation-solar-technology-revive-sector

[23] Weaver J F 2024 GCL says perovskite solar module passes silicon degradation tests *PV Magazine* https://www.pv-magazine.com/2024/06/20/gcl-says-perovskite-solar-module-passes-silicon-degradation-tests/

[24] Shaw V 2025 Chinese PV industry brief: Utmolight starts perosvkite module production *PV Magazine* https://www.pv-magazine.com/2025/02/07/chinese-pv-industry-brief-utmolight-begins-perosvkite-solar-module-production-at-gw-scale-facility/

[25] Perovskite-info 2025 Swift Solar https://perovskite-info.com/swift-solar

[26] Fischer A 2025 Tandem PV moves toward perovskite manufacturing with $50 million funding *PV Magazine* https://www.pv-magazine.com/2025/03/06/tandem-pv-moves-toward-perovskite-manufacturing-with-50-million-funding/

[27] Solliance 2020 Scaled perovskite solar modules pass three critical stability tests *IMEC* https://www.imec-int.com/en/articles/scaled-perovskite-solar-modules-pass-three-critical-stability-tests

[28] Overton G 2018 Heliatek completes largest organic photovoltaics BIPV installation to date *Laser Focus World* https://www.laserfocusworld.com/detectors-imaging/article/16571605/heliatek-completes-largest-organic-photovoltaics-bipv-installation-to-date

[29] Wikipedia 2024 L ONGi *Wikipedia* https://en.wikipedia.org/w/index.php?title=LONGi&oldid=1249092836

[30] Thompson V 2025 Indian startup testing 15%-efficient perovskite solar modules *PV Magazine* https://pv-magazine.com/2025/03/14/indian-startup-testing-15-efficient-perovskite-solar-modules/

[31] Peleg R 2025 IIT Bombay startup's perovskite solar cell seen by government as 'game-changer' to Indian economy *Perovskite-Info* https://perovskite-info.com/iit-bombay-startup-s-perovskite-solar-cell-seen-government-game-changer-indian

[32] SolarWindow Technologies Inc. 2020 SolarWindow first-ever: electricity-generating flexible glass using high-speed manufacturing process *GlobalNewsWire* https://globenewswire.com/news-release/2020/11/23/2131850/0/en/SolarWindow-First-Ever-Electricity-Generating-Flexible-Glass-Using-High-Speed-Manufacturing-Process.html

[33] IMEC and partners show outdoor stability of highly anticipated perovskite solar modules *IMEC* https://imec-int.com/en/press/imec-and-partners-show-outdoor-stability-highly-anticipated-perovskite-solar-modules

[34] PV-TEC—Photovoltaic Technology Evaluation Center *Fraunhofer ISE* https://www.ise.fraunhofer.de/en/rd-infrastructure/tecs/pv-tec.html

[35] Beynon D *et al* 2023 All-printed roll-to-roll perovskite photovoltaics enabled by solution-processed carbon electrode *Adv. Mater.* **35** 2208561

[36] Kim H-S *et al* 2012 Lead iodide perovskite sensitized all-solid-state submicron thin film mesoscopic solar cell with efficiency exceeding 9% *Sci. Rep.* **2** 591

[37] Kim Y C *et al* 2017 Printable organometallic perovskite enables large-area, low-dose x-ray imaging *Nature* **550** 87–91

[38] Bellini E 2023 KAUST claims 33.2% efficiency for perovskite/silicon tandem solar cell *Pv-Magazine* https://www.pv-magazine.com/2023/04/13/kaust-claims-33-2-efficiency-for-perovskite-silicon-tandem-solar-cell/

[39] Li M *et al* 2024 High-efficiency perovskite solar cells with improved interfacial charge extraction by bridging molecules *Adv. Mater.* **36** 2406532

[40] NREL 2025 Perovskite solar cells *Photovoltaic Research* https://nrel.gov/pv/perovskite-solar-cells

[41] NREL 2025 Best research-cell efficiency chart *US Department of Energy, Office of Energy Efficiency and Renewable Energy* https://www2.nrel.gov/pv/cell-efficiency

[42] CSIRO 2025 Printable solar cells for lightweight energy https://csiro.au/en/research/technology-space/energy/Electricity-transition/Photovoltaics/Printable-solar-cells

[43] Xia R and Hu Y 2025 Light people: Prof. Henry Snaith's (FRS) perovskite optoelectronics journey *Light Sci. Apps.* **14** 7

[44] Kumar P and Chauhan A K 2020 Highly efficient flexible perovskite solar cells and their photo-stability *J. Phys. D: Appl. Phys.* **53** 035101

[45] Kumar P 2024 Role of *in-situ* substrate heating on the growth of perovskite films in printed perovskite solar cells *Appl. Phys.* A **130** 566

[46] Perovskite solar cells developed by NTU Singapore scientists record highest power conversion *EurekAlert* https://eurekalert.org/news-releases/702348 (Accessed: 7 September 2025)

[47] Japan's $1.5 billion solar bet: can perovskite cells challenge China's dominance? *E+E Reader* https://environmentenergyleader.com/stories/japans-15b-solar-bet-can-perovskite-cells-challenge-chinas-dominance,64907 (Accessed: 7 September 2025)

[48] Perovskite patent portfolio, national renewable energy laboratory *NREL* https://nrel.gov/pv/perovskite-patent-portfolio (Accessed: 7 September 2025)

www.ingramcontent.com/pod-product-compliance
Lightning Source LLC
Chambersburg PA
CBHW080521220326
41599CB00032B/6156

* 9 7 8 0 7 5 0 3 5 4 2 8 8 *